FISHERIES

Contributing Authors

(in order of appearance)

Dr. Frederick G. Johnson, Senior Lecturer
School of Fisheries
University of Washington, WH-10
Seattle, WA 98195

Dr. Marc L. Miller, Professor
Institute for Marine Studies, HR-05
University of Washington
Seattle, WA 98195

Dr. Donald R. Gunderson, Associate Professor
Fisheries Research Institute, WH-10
University of Washington
Seattle, WA 98195

Katherine W. Meyers, M.S., Fishery Biologist
Fisheries Research Institute
University of Washington, WH-10
Seattle, WA 98195

Dr. Bruce S. Miller, Director, Fisheries Science and
Management, and Professor, School of Fisheries
University of Washington, WH-10
Seattle, WA 98195

Dr. Robert R. Stickney, Director and Professor
School of Fisheries
University of Washington, WH-10
Seattle, WA 98195

Dr. Frieda B. Taub, Professor
School of Fisheries
University of Washington, WH-10
Seattle, WA 98195

Dr. Robert C. Francis, Director, Fisheries Research
Institute, and Professor, School of Fisheries
University of Washington, WH-10
Seattle, WA 98195

William High, Fishery Biologist
National Marine Fisheries Service
7600 Sand Point Way N. E., Building 4
Seattle, WA 98115–0070

Dr. Thomas R. Loughlin, Fishery Biologist
National Marine Mammal Lab
National Marine Fisheries Service
7600 Sand Point Way N. E.
Seattle, WA 98115

Dr. R. V. Miller, Fishery Biologist
National Marine Mammal Lab
National Marine Fisheries Service
7600 Sand Point Way N. E.
Seattle, WA 98115

Dr. Chris Kohler, Professor
Fisheries Research Laboratory
Southern Illinois University
Carbondale, IL 62901–6511

Dr. Gilbert B. Pauley, Assistant Leader
Washington Cooperative Fish and Wildlife Research Unit
U.S. Fish and Wildlife Service,
and Associate Professor, School of Fisheries
University of Washington, WH-10
Seattle, WA 98195

Charlie White, President
Charlie White Productions, Ltd.
P.O. Box 2003
Sidney, B.C. V8L 3S3
Canada

John Thomas
Luhr Jensen Co.
P.O. Box 297
Hood River, OR 97031

FISHERIES

HARVESTING LIFE FROM WATER

edited by

Frederick G. Johnson
Robert R. Stickney

University of Washington
School of Fisheries

KENDALL/HUNT PUBLISHING COMPANY
2460 Kerper Boulevard P.O. Box 539 Dubuque, Iowa 52004-0539

Dedication

This book is dedicated to students and to Ernie Salo, University of Washington Professor and Fishery Biologist.

Contents

Preface

Anyone who appreciates nature should like this book.
Anyone who likes food should appreciate this book.
Anyone who loves to fish should love this book.

This book was designed for use as an introductory textbook for students at an advanced secondary level or beginning college level who are interested in fisheries and natural resource sciences. It was also designed to satisfy the curiosities and interests of those who like to go fishing.

Acknowledgments

This book would not have been possible without the generous assistance of Marcus Duke and Ken Adkins. Marcus Duke, Editor for the School of Fisheries at the University of Washington, gathered all the many sources of effort together to produce a coherent whole, kept track of myriad details and deadlines, and smoothed the rougher edges of the authors' styles wherever necessary. Ken Adkins, Instructional Services Coordinator at the School of Fisheries, created the graphs, diagrams and computer graphics. Unless noted otherwise, all of the figures in this book are credited to Ken Adkins. Photographs are credited to the chapter authors, unless noted otherwise. (Front cover photo courtesy of Allan Hartt; back cover courtesy of Dick Kocan.) Additional production assistance was provided by Maxine Davis, Dorothy Kemp, Barry McFarland, Marianne McClure, Abby Simpson, and Cheryll Sorensen.

The authors also wish to thank the following reviewers for their helpful suggestions: Bruce Campbell, Marianne McClure, Steve Mathews, Ted Pietsch, Thomas Quinn, John Skalski, Lynwood Smith, and Isabel Tinoco. Finally, we would like to thank Mariel Damaskin, Associate Editor at Kendall/Hunt Publishing Company, for her enthusiasm and encouragement.

1

Fishery Science and Management

1

Fisheries
Harvesting Life from Water

Frederick G. Johnson

Introduction

Fisheries involve the taking of living organisms from water for human use, primarily as food. Other uses for these resources include the manufacture of jewelry, furs, purses, buttons, drugs and the pursuit of pleasure. Fisheries exist for species other than finfish and shellfish, including such things as seaweeds, sponges, corals, jellyfish, sea urchins, frogs and turtles. Fisheries supply industries and provide occupations and recreation.

In all, people are major predators upon fishes, shellfishes and other aquatic organisms, and have been for many thousands of years. There is little reason to think of fisheries as "unnatural" sources of loss to living aquatic resources. Predation, whether by humans or not, is a rule in aquatic ecosystems rather than an exception. The sustaining principle in fisheries is that organisms are generally able to reproduce enough to repopulate following heavy losses, whether those losses are due to predation, environmental extremes, or other causes.

Humans can be extremely effective predators, however, and we now have the capability to harvest stocks beyond their abilities to recover. We also realize that, and we can adjust our predatory impacts over the short term in order to maximize them over the long term. This idea, sacrificing the short-term gain to assure greater success over the long term, brings forth a host of conflicts among participants in a fishery. Most of these conflicts are unique to human predators.

The scale and degree of capitalization of fisheries varies considerably—from heavily capitalized, "high-tech" offshore operations where much of the product is processed at sea and sold throughout the world, to "low-tech" nearshore operations, often operating at subsistence levels and without the benefit of refrigeration, commercial long-distance transportation and product processing. Fisheries of the latter type are common in developing countries that have fishery resources but lack a sophisticated economic infrastructure. Such fisheries are called **artisanal** fisheries. Finally, we have the practice of fishing for pleasure as an end in itself to deal with. This spectrum of possibilities offers a great challenge to fishery scientists owing to the wide range of separate, often conflicting uses of fishery resources.

People tend to use contrivances rather than their own body parts to capture their prey (Figure 1.1). We are somewhat like spiders in that way, but we hang our webs in the waters of the world.

Fishery Science

A fishery is the combination of an aquatic resource with an organized harvest system. For both parts to remain healthy and sustainable, we must pay careful attention to the status of the resource and the level of harvest. That kind of attention is provided by fishery science.

Fishery science is a multidisciplinary applied science that draws information from a number of so-called "pure" sciences in order to apply that knowledge to the solution of fishery problems. The most common fishery problem is that there aren't enough fish for everybody.

Figure 1.1. The human fishing machine. (Courtesy of Eric Warner.)

The disciplines that contribute to fishery science include biological, physical, social, mathematical and technological sciences. The following outline describes some of the more significant components of fishery science.

Biological Sciences

Ichthyology	—	the study of finfish
Invertebrate Zoology	—	the study of animals that lack backbones
Botany	—	the study of plants
Physiology	—	the study of how living things function internally
Ecology	—	the study of how organisms interact with each other and their environment
Genetics	—	the study of inherited characteristics
Pathology	—	the study of diseases

Physical Sciences

Oceanography	—	the study of oceans
Limnology	—	the study of freshwater habitats

Social Sciences

Sociology   —   how people interact with each other
Economics   —   how money moves among people
Public Policy   —   how laws and regulations affect and modify
                           human behavior

Technological Sciences

Technical Developments —   improved means for stock location and assessment
                       —   improved processing at sea—new products and
                           product analogues
                       —   remote sensing
                       —   improvements in fishery capture gear and navigation

Mathematics

Population Dynamics   —   how populations change over time
Biostatistics   —   population sampling and analysis
Simulation   —   computerized techniques for modeling
                           population changes

Interdisciplinary Applications

Aquaculture   —   ''farming'' aquatic organisms
Fishery Management   —   controlling the human side of fisheries

Obviously, then, fishery science includes much more than fishery biology, and as we will learn later, fishery management includes much more than fishery science.

Prehistory of Fisheries

The practice of fishing was developed long before the practice of keeping historical records. We know that people around the world have engaged in fisheries for food, trade, ornaments and religious customs for many thousands of years. Even the relatively sophisticated practice of aquaculture of freshwater fishes has been practiced in China for over 4000 years. Numerous examples of ancient art and the long-held customs of native peoples attest to the importance of fisheries to human cultures over the ages.

Tribes of Native Americans in the Pacific Northwest, for example, treat the first returning salmon each year with reverence, holding a communal ceremony for the fish before it is returned to the sea. In this ceremony it is believed that the spirit of the first fish will carry a message to the other fish that their upriver migration will be welcomed and that each returning fish will be treated with respect. Many native peoples from both eastern and western cultures believe that fishes

Figure 1.2. A large freshwater fish, the arapaima, shown after capture by an Amazonian native. These are the largest exclusively freshwater fish and can be as much as 9 feet long.

a b

Figure 1.3. Native American designs depicting (a) a salmon and (b) a crab. (Courtesy of Ken Adkins.)

and shellfishes have significance far beyond their ability to sustain human life. When some of these people speak of fish, they sound more like they are speaking of members of their own families than about things to prey upon.

Modern people throughout the world are now realizing the importance and fragility of our fishery resources. One of the more encouraging indications of this is an increasing tendency for parties and nations that so frequently find fault with each other to reach agreement over fishery issues of mutual concern.

History of Fisheries

Our history of fisheries will be brief and selective, covering events since 500 B.C. It begins with a book on a relatively sophisticated practice, the aquaculture of common carp. Written by Fan Lee, the book was published in China in 473 B.C. It would take more than 2000 years for the western world to begin to catch up with Asia's lead in aquaculture. From 500 B.C. to 500 A.D. the culture of common carp in freshwater ponds flourished in China. After that, however, the Tang Dynasty took power and the culture of common carp for human consumption was temporarily banished because the name of the fish sounded just like the name of the emperor. This proved beneficial in the long run, since the technology for culturing the common carp alone was then adjusted to related species. This led to **polyculture** of carps, where different species were grown together in a pond, each species feeding on a different group of plants or animals. The polyculture of carps, in practice by 1000 A.D., still continues on a large scale in Asia and other parts of the world, often in combination with agricultural practices. At first, young fishes to stock the ponds were caught in the wild. Now the ponds are stocked with fishes produced in hatcheries by fish farmers. Aquaculture now constitutes the fastest-growing portion of world fisheries output.

The western world's early history of fisheries centered around the English common law system and concerned wild fish stocks rather than fishes grown in aquaculture. Developed over many years and elaborated in the Magna Charta (1215 A.D.) and in Parliaments, the common law concept held that wild fishes belonged to the Commons (or the Crown, in England's case at that time). Regulations on fishing were then put into force and the highest priority for access to fisheries went to royal uses and entertainment. It is noteworthy that many of the earliest examples of regulation of fishing effort dealt with recreational fisheries.

Privilege and allocation of fish stocks in the American Colonies were not pressing issues due to the abundance of fishes that were initially present. The Commons in America came to mean the property of the common people. In 1639, however, a policy of public allocation emerged in the Massachusetts Bay Colony after declines were noted in the abundance of fish stocks that were heavily harvested.

It was another 150 years before the rich fishery resources on the Pacific coast began to receive regulatory attention. Long before the Europeans had settled the American Colonies in the east, the salmon and halibut resources in the west had been fished by native Indians for subsistence and trade. Their fishing practices were guided by tribal customs and beliefs, but apparently without regulatory allocation. The fact that the Native American system worked well was evident to early white explorers who wrote enviably about the abundance of fishes in the west.

By the time of the Civil War, recreational fishermen in the United States were insisting that the states exercise their duty to protect the public trust by placing legislative restrictions on fishing. This led to regulations restricting activities of recreational fishermen in Massachusetts in 1865, and shortly thereafter in New Hampshire and California. Canneries for commercially caught salmon had been set up on the Pacific coast by then, and so had trout hatcheries to enhance recreational fishing on the east coast and in England.

Before the turn of the 20th century, the industrialization of fisheries hastened with the advent of steam-powered capture vessels. Purse seiners fished for menhaden off the U.S. Atlantic coast, trawlers fished for mixed stocks in the North Sea, and longliners fished for halibut in the North Pacific. Later developments that also had considerable industrial impact include the use of diesel-powered capture vessels, the hydraulic power block for net retrieval and refrigeration systems on vessels.

With the progressing industrialization of marine fisheries in the early 1900s, evidence began to mount that fish stocks were declining under increasing fishing pressure, particularly Pacific halibut and those stocks fished by trawl in the North Sea. Concern over these fisheries led to the formation of several international fishery commissions. The International Council for the Exploration of the Sea (ICES) was established in 1902, and was followed by the International Pacific Halibut Commission, the International Whaling Commission and the International Pacific Salmon Fisheries Commission. Fisheries laboratories and fisheries commissions proliferated to bring scientific information and international cooperation to bear on fishing problems. The Halibut Commission was quite effective in that regard while the Whaling Commission was not.

By the 20th century, regulation of fishing effort to serve the special interests of recreational fishermen had been in place for some time, more-or-less on a local scale. With the rapid industrialization of commercial marine fisheries, however, effective fishery management became necessary on a global scale. World War II gave fishery scientists a good look at the importance of fishing effort on stock size since a number of fish stocks rebounded after they were fished lightly during the war years.

After 1950, the industrialization of fisheries included the use of factory methods for processing on shipboard, where large stern trawlers with freezing capabilities extended fisheries far offshore to major fish stocks throughout the world. The need for effective management of oceanic fisheries was by then obvious, after inshore fisheries had received earlier management attention. Collapses among some important fisheries, including Atlantic herring, California sardines, Peruvian anchoveta, and Alaska king crab drew even more attention to the fragility of fishery stocks. The United Nations Law of the Sea Conferences began in 1958 and continued for many years to grapple with resource issues. Ecuador was the first to emerge from the fray by declaring sovereignty over fishery resources within 200 miles of her coastline. Most coastal nations soon followed, including the United States in 1976 and Canada in 1977.

The United States declared its 200-mile "Fisheries Conservation Zone" with the Fishery Conservation and Management Act (FCMA) of 1976. Also called the Magnuson Act after Washington State Senator Warren Magnuson, the new law created a number of significant changes in how management of commercial fishing in the rich coastal shelf waters of the United States would take place. The FCMA created eight regional management councils to oversee fisheries in the 200-mile zone and allowed foreign nations to harvest only those stocks present in surplus beyond the amounts that U.S. harvesters would take. The act also directed fishery managers to take economic, social and political factors into account, in addition to scientific criteria, in establishing fishery policy. After 1983, the Fisheries Conservation Zone came to be known as the "Exclusive Economic Zone," or EEZ.

At about the same time that the Magnuson Act was taking shape, another event that would dramatically change the way fishery resources were allocated took place in the Pacific Northwest. Rather than an act of legislation, however, it involved a 1974 landmark judicial decision, *U.S. vs. Washington,* or the Boldt decision. The Boldt decision was an historic and precedent-setting affirmation of Native American treaty rights that resulted in a 50–50 allocation of salmon and steelhead between Treaty Indian tribes and other citizens in Washington State.

Looking Ahead

In the following chapters, we explore a wide range of topics that span our interests in harvesting life from water. We begin by taking a look at human motivations and mechanisms for establishing fisheries. Then we examine fundamental biological processes that determine how fish stocks came to be and continue to exist in nature. After that we bring ourselves back into the story by considering the ways we analyze fish stocks from the standpoint of harvesting them sustainedly. We then take a close look at the various marine and freshwater fisheries, including aspects of their management and means of harvest. Finally, we consider some of the emerging issues in fisheries, including aquaculture, marine mammals and their protection, and conflicts over the use of our natural resources.

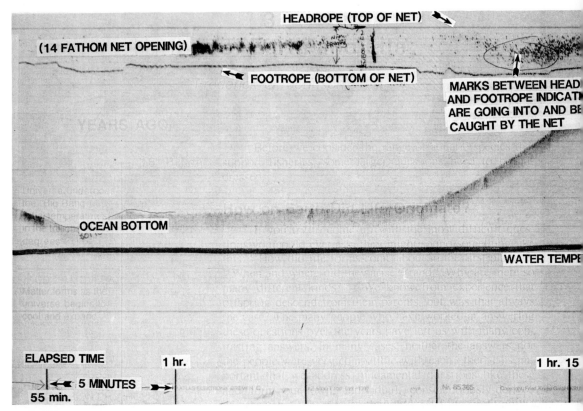

Figure 1.4. A depth sounder tracing from a modern commercial fishing vessel that trawls for pollock. Transducers attached to the trawl net, together with equipment on the vessel itself, provide information on water temperature, net shape and net fullness, in addition to fish location and depth. This is an example of the application of sophisticated technology to modern fishing methods. (Courtesy of Ken Adkins.)

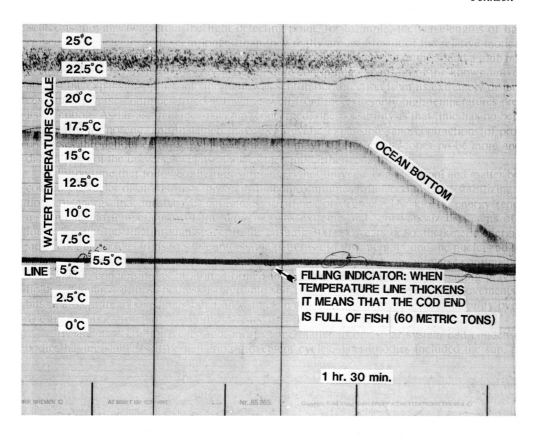

WATER TEMPERATURE SCALE

25°C
22.5°C
20°C
17.5°C
15°C
12.5°C
10°C
7.5°C
5.5°C
LINE 5°C
2.5°C
0°C

OCEAN BOTTOM

FILLING INDICATOR: WHEN
TEMPERATURE LINE THICKENS
IT MEANS THAT THE COD END
IS FULL OF FISH (60 METRIC TONS)

1 hr. 30 min.

2

Fish and People

Marc L. Miller and Frederick G. Johnson

About 2.5 million years ago, humans stopped living as scavengers, relying on the natural foods of the savannas and what other predators left behind, and began to design more systematic ways to gather and then hunt food. From that time until the present—and with the transitions in major procurement patterns to slash-and-burn horticulture, plow-and-irrigation farming, pastoralism and industrialism—we have adapted to changing environmental constraints and opportunities with interesting mixtures of technologies and institutions.

For virtually all of our existence, people have utilized fishes in some fashion. This chapter discusses the roles fishes play in nutrition and commerce, and introduces the manner in which societies have adjusted their social structures around fishery resources.

Value of Fish

Fish are important to us for the protein they provide in our diet, for the satisfaction we find in culinary and dining experiences, for raw materials used in the manufacture of many non-food products and, of course, for the attainment of business and policy objectives.

Food and Nutrition

By 1987 there were over five billion people in the world. Most people eat every day, and supplying the world's people with enough high-quality food is no small concern. Food that is high in animal protein is also rich in essential amino acids like lysine and methionine. Diets that are low in animal protein can severely limit human well-being. This is often the case in developing countries. Global dietary studies show that as personal income increases, so does the average number of calories consumed. The proportion of overall calories contributed by animal protein, animal fat and sugar is also relatively high for people with higher incomes. In contrast, with lower levels of personal income, fewer calories are consumed and the proportion of those calories from animal protein is very low. Most of the calories in the diets of people in low income regions come from carbohydrates, primarily from grains, which may not supply all of the essential amino acids. Amino acids are the building blocks of proteins. The amino acid composition of fish flesh is very close to our dietary requirements.

Fishes and shellfishes provide the type of food that people in both rich and poor nations need—relatively low in fat and high in protein. The protein content of fish flesh is 19% (of wet weight), which is relatively high, while the caloric content is relatively low. Among the fats found in seafood are some special ones that are thought to be particularly beneficial to human health.

An epidemiological study comparing the health of Eskimos, who have diets rich in seafood, with Danes, whose diets are high in chicken eggs, milk products and red meat showed that the Eskimos had significantly fewer instances of coronary heart disease, psoriasis and bronchial asthma than the Danes. Since that study, the kinds of fats and oils present in fishes and shellfishes have received considerable research attention regarding their potential health benefits. Much of that attention has focused on **cholesterol.**

Cholesterol is a fatty substance, a **lipid**, and despite its unhealthful reputation, cholesterol is a necessary component of cell membranes. Cell membranes control which substances can enter and leave cells, largely because they have a lipid bilayer that acts as a **hydrophobic** ("water fearing") barrier for the cell (Figure 2.1). Cholesterol occupies spaces in the hydrophobic regions of cell membranes. Portions of the cell membrane that interact freely with water are called **hydrophilic** ("water loving"). We will take a closer look at cell membranes when we learn more about their functions in Chapters 6 and 7. The problems associated with cholesterol, however, have little to do with its presence in the cell membranes, where it is needed, but much to do with its presence in the human circulatory system.

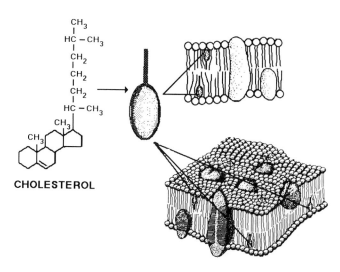

CHOLESTEROL

Figure 2.1. Cholesterol and lipid regions of cells. This diagram shows hydrophobic, lipid-rich regions in the interior of the cell membrane and hydrophilic regions at the inner and outer surfaces of the membrane. The large globular structures are proteins.

Cholesterol in the blood may accumulate at tiny breaks in the lining of blood vessels, forming a mass called plaque that can grow to block off a portion of the inner diameter or lumen of the blood vessel. Blood clotting elements called platelets, white blood cells and calcium may also add to the plaque as it forms. In the coronary arteries, those that supply the heart with blood, the growth of plaque can restrict the flow of blood enough to cause coronary heart disease and heart attacks. In one of the most publicized health food discoveries of our era, seafood consumption seems to minimize this risk.

Some of the lipids present in fish oils can help lower the blood cholesterol levels of people who eat fish regularly and may reduce the risk of heart attacks. The specific lipids thought to be responsible for this benefit are the omega-3 fatty acids. These fish oils are available on the market in encapsulated form, but the greatest benefits probably accrue from simply eating seafood. Lipids found in fish oils allow the cell membranes of fishes that live in cold water to remain flexible at low temperatures. Animals that have red meat are **homeothermic**, or "warm blooded," and have higher body temperatures and higher concentrations of saturated fat that would solidify at colder temperatures. Saturated fats have hydrogen atoms bound to the carbon atoms in the fat molecules at all possible bonding sites, hence these fats are considered to be "saturated" with hydrogen and they have relatively high melting points. "Unsaturated" fats have lower melting points, have some double bonds between the carbon atoms that might otherwise have been occupied by hydrogen bonds and, therefore, have less opportunity to be "saturated" with hydrogen atoms (Figure 2.2).

The complete story on how seafood in the diet might confer health benefits remains to be told. Besides heart disease-related benefits, fish and shellfish consumption is also associated with reduced synthesis of **prostaglandins**. Prostaglandins are "mini-hormones" associated with inflammatory responses in arthritis, bronchial asthma and other problems. Even without the benefits attributed to fish oils, however, the fact remains that fishes and shellfishes are very good foods—high in protein, low in "bad" fat, high in "good" fat and rich in many vitamins, zinc and phosphorus.

$$CH_3-CH_2-CH_2-CH_2-CH_2-CH_2-CH_2-CH_2-CH_2-COOH$$

Capric acid

$$CH_3-CH_2-CH_2-CH_2-CH_2-CH=CH-CH_2-CH=CH-CH_2-CH_2-CH_2-CH_2-CH_2-CH_2-CH_2-COOH$$

Linoleic acid

Figure 2.2. Chemical formulas for capric acid and linoleic acid. Linoleic acid is an omega-6 polyunsaturated fatty acid (two or more double bonds are indicated by C=C) found in vegetable oil. Capric acid is a saturated fatty acid typical of animal fat.

In the United States, fat provides 30% to 40% of the calories in the diet. The average American now eats about 16 pounds of fishes and shellfishes per year, a figure that has been rising in recent years corresponding with increases in health and diet consciousness. In Japan, the per capita consumption of seafood is much higher by a factor of five or more. The proportion of calories provided by fat in the Japanese diet is only 3%. Americans have a high incidence of coronary heart disease, obesity and other health problems not shared equally by the Japanese. Other factors besides diet, such as genetic differences, certainly contribute to differences in health revealed by epidemiological studies, but diet appears to be of primary importance to improving health.

About 70% of the world's fishery harvest is eaten directly by people. Most of the rest is reduced to fish oil or fish meal and fed to other animals that people will eat later, like poultry, pigs and cows. The entire global fisheries harvest has amounted to about 70 to 90 million metric tons per year since 1973. These figures represent live weights. In 1973, the world's largest fishery, the Peruvian anchoveta fishery, collapsed, and gains in global harvests since then have been small. Compared with the American per capita consumption of fishery products of 16 pounds per year, the world average per capita consumption is 27 lbs/year, or about 12 kilograms. These figures were compiled by the Food and Agriculture Organization of the United Nations (FAO) and reflect data from 1986. The 27 lbs/yr world average represents 23% of the animal protein consumed, the largest single source of high quality protein. Beef is second. We now take a look at the forms of fishery products that reach the consumer.

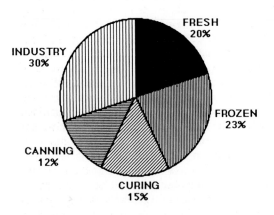

Figure 2.3. Worldwide forms of fishery products; industrial forms include fish meal and fish oil. (Source: FAO 1987.)

Fishery Products

The primary mode of processing fishery products is now freezing, which has increased in importance over the years and is used for 23% of fishery production. Fresh products account for 20% of global production followed by curing and canning at 15% and 12% (Figure 2.3). Curing implies preservation practices like drying, smoking and salting, which, like canning, avoid the need for refrigeration. The remaining 30% of fishery production is used industrially, mainly in the form of fish meal, and is not directly eaten by people.

Among the majority of fishery products (those used directly as food), there is a wide range in the portion of the product that is consumed. Larger fishes range from 25% to 75% edible portion of total weight. In other words, fishes with big heads and fins like rockfish might yield fillets that weigh only about one fourth the weight of the live fishes, but meatier fishes like salmon yield up to three quarters edible weight. Of course, some small fishes like anchovies and sardines are eaten whole, either dried or canned. Among shellfishes, shrimp have about 50% edible portions while oysters may have only about 7%. Seaweed products are usually 100% edible.

There are over 200 taxonomic families of animals and plants that yield fishery products. These organisms include algae, finfish, crustaceans, molluscs, mammals and others. In most cases, it is the muscle tissue of an animal that is eaten, cooked or uncooked, but there are as many variations for eating fishes and shellfishes as one can imagine. Often we consume fishery products without realizing it, unless we read labels very carefully. Substances extracted from seaweeds occur in a wide range of food and nonfood products.

Among sea urchins, sturgeon, herring and other fishes, the eggs (**caviar**) or gonads (**roe**) can be important products. This brings us to the three "s" words in fishery products: **sushi, sashimi** and **surimi.** Sushi refers to a rice preparation that is sticky and slightly sweet/sour over which various fishery products, usually raw, are placed. Sea urchin roe, raw squid, fish roe, boiled octopus, raw shrimp and raw pieces of fish of many kinds are served with sushi rice in sushi bars. Often the sushi preparations are wrapped with a type of seaweed called **nori**. Sushi bars are now popular in the United States and many countries around the world after having been important in Japan for many years.

Sashimi refers to fish or other seafood that is thinly sliced and eaten raw. Eating sashimi is a long-held dietary practice in Asia and is gaining popularity in the United States. The prime fish for sashimi is tuna, but many types of fishes and shellfishes are used. The seafood is usually dipped in soy sauce and wasabi, a green Japanese horseradish paste. The consumption of uncooked fishes raises the issue of parasites that might infect humans and lead to health problems. Carefully prepared seafood sashimi is generally safe to eat. Eating freshwater fishes raw, however, can be dangerous because they may contain certain parasites, such as liver flukes.

While sushi and sashimi are becoming much more familiar to western nations as tastes in food are broadening, the big news that is changing the global marketing of fishery products in recent years is surimi.

Surimi is a processed fish product pioneered by the Japanese. The surimi process utilizes relatively inexpensive marine fishes with white flesh like Alaska pollock, *Theragra chalcogramma,* which are present in enormous numbers in the North Pacific Ocean. These fishes used to be considered "trash" fish by American fishermen until the potential for surimi products was more fully realized. The making of surimi is essentially a high-technology industrial process that results in a minced and gelled fish protein product that can be modified by the addition of flavorings and textures into countless consumer products. Many of these products are analogues of other seafood, such as crab. Surimi can be processed at sea, stored by freezing for long periods and yields products of high and predictable quality for different market tastes around the world. The fishes are delivered immediately after capture at sea to factory processing ships by harvest vessels, or increasingly, are harvested by large trawl nets deployed by the factory ships themselves. Automatic fillet machines adjust mechanically to the dimensions of each fish and send the flesh through a series of mechanical steps that results in a product that meets strict quality control standards. The surimi revolution in fishery products has only begun. Many new products and new sources of fishes will be involved with surimi technology before this revolution has run its course. For example, the potential for using surimi technology on menhaden is being studied. Menhaden are caught off the U.S. Atlantic coast in large amounts, over one million metric tons per year. They represent the largest U.S. fishery by tonnage but have so far been used only for reduction to fish meal and fish oil. As we will learn in later chapters, consuming seafood directly is a far more efficient use of food than feeding it to other animals and then eating those animals later.

Development of new food processing equipment, in addition to the new processes themselves, continues to broaden markets for fishery products. Automatic fillet machines have opened up a number of new possibilities. Together with the ability of factory ships to process catches at sea shortly after capture, the new equipment has made "fresh-frozen" or "flash-frozen" fillets possible for fish like hake and others having flesh that undergoes rapid decomposition after capture. For shellfish, the automatic shelling machines now used to clean small shrimp have been a major breakthrough. Technology for edible seaweed production is also spreading from its centers in China and Japan. The Japanese have developed a culture system for nori that allows them to use floating machines that harvest the new growth from submerged nets in a manner much like mowing a lawn.

Not all fishery products become food. Jewelry, cosmetics, clothing, a host of industrial materials and pets are provided by fisheries. Techniques for culturing pearls in both freshwater and seawater have yielded elaborate industries for bivalve aquaculturists, primarily in Japan, where surgical implantation practices are used to speed up the natural pearl-forming process. Bivalve, abalone, and other mollusc shells are also used for jewelry, buttons and various decorative purposes. Other fishery products, like precious corals, are used for jewelry as well.

Many fishery products are used for personal hygiene. Natural sponges used to be familiar household products before they were replaced by synthetic sponges. Some of us remember when "hard-hat" sponge divers descended the depths, at considerable danger to themselves, to gather bathtub accessories. Thickening agents extracted from algae are used in cosmetics and toothpaste. Pearlescent additives for cosmetics and shampoo are made from ground up fish scales.

Eel, salmon and other fish skins have recently become popular leather-like fishery byproducts, used for the manufacture of things like wallets, belts, handbags and bathing suits. Furs, of course, are valuable clothing items that have led to the decline of some marine mammals over the years, including fur seals and sea otters. Some fishery products, like mollusc shells, have even been used as money by Native Americans.

Oils extracted from whales were used extensively as fuel for lamps in earlier years. Other materials from whales were used to make corsets. Presently, a wide range of industrial materials are either extracted as a primary product from fishery harvests or are fishery byproducts. Foremost among these substances are the thickening agents like carrageenin, algins and agar, which come from seaweeds. Besides their use in drugs, paints, foods, beverages (including beer) and cosmetics, they have been used for microbiological research for many years and have broad industrial importance. Waste shells from crab and other crustacean shellfish are even finding industrial applications. Uses of these shell materials may come to include clarification treatment for sewage effluents. In economic terms, the sum of the industrial uses of fishery products is considerable and growing.

The cuttle bones in pet stores that are used in bird cages are fishery products, but so are many of the pets themselves. The market for aquarium fishes has become a very big business throughout the world. In many cases, wild fishes are harvested by people in developing countries to sell to aquarists in developed countries, and this has caused many problems for the natural populations of some species. In other cases, these fishes are being supplied by aquaculturists. To give an idea of the economic importance of "ornamental" fishes, the total value of world imports in 1986 alone was $68,000,000 (U.S.). The major importing countries were the U.S. ($27,000,000), followed by West Germany ($11,000,000) and Japan ($9,000,000). The world market for ornamental fishes has been increasing steadily in recent years and can help developing countries with their balance of trade.

Perhaps the most significant non-food fishery product, even in terms of economic impact in developed countries like the United States and Canada, is the fishing "experience." That is the major product sought by those who fish recreationally. The value of sport fishing is difficult to assess in terms of dollars because sport-caught fishes are not bought and sold like commercially caught fishes, even though both are usually taken home and eaten. To the fish, of course, it makes little difference, but to the management of fisheries the distinction makes a great deal of difference. In general, the economic contribution and "value" of sport-caught fishes is considerably greater than that of commercial fishes when all of the many factors of the recreational fishing experience are taken into account. Avid anglers are known to spend thousands of dollars for the opportunity to catch a trophy fish. We look further into these issues in the next section and in later chapters on sport fishing.

Fishery Economics and Trade

Only two nations, Japan and the Soviet Union, together harvest about one fifth of the world fisheries total production each year. Japan is the world leader and the Soviet Union is second. China is next, followed by Chile, Peru and the United States (Figure 2.4).

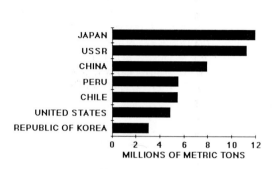

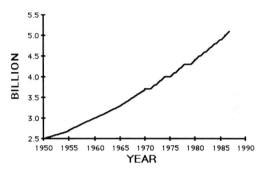

Figure 2.4. Leading nations in fishing harvests. (Source: FAO 1987.)

Figure 2.5. World population. (Source: FAO 1987.)

Despite having very rich fishing grounds and a relatively low per capita consumption rate, the United States nevertheless imports far more than it exports in terms of value of fishery products. United States citizens favor expensive products like shrimp and lobster. The United States has run a trade deficit in fishery commodities since 1895. The Japanese also consume expensive fishery commodities and with the recent rise of their yen against the U.S. dollar they are better prepared to compete for those commodities on the world market.

Many factors determine the relative levels of consumption of fishery products among nations. Those factors include proximity to productive places to capture fishes or to grow fishes by aquaculture; the availability of refrigeration, transportation and processing; population; personal tastes; religious influences; and economic factors like ability to pay. Economic factors have become increasingly significant in shaping the international distribution of fishery commodities. Shortly after World War II, about 20% of the world fisheries harvest was traded internationally. International trade now incorporates nearly 40% of the total world harvest. The two leading importing nations during the 1980s were Japan and the United States. The two leading exporters were Canada and the United States (Table 2.1). Figure 2.5 shows world population growth since 1950 and suggests that the rate of population growth in recent years has not leveled off and, in fact, may be increasing. Some of the more populous countries, like China, seem to have brought the rate of their population growth under control. Others, most notably in Africa, have not.

The total annual yield from the world's fisheries over the same period is shown in Figure 2.6. The good news is that the increase in world fisheries production was about 6% per year between 1950 and 1970. The bad news is that the rate of increase since then has been modest, despite considerable advances in aquaculture and wild harvest technology. Hope that fisheries might provide widespread relief from hunger and malnutrition in developing countries has also moderated since then. In terms of world fish production per capita (Figure 2.7), we can see that there has been no net increase since 1973.

Table 2.1. International Trade in 1986—Top Ten Importer and Exporter Countries (Value in U.S. Dollars).

Country	Imports	Country	Exports
Japan	6,593,515,000	Canada	1,744,189,000
United States	4,748,692,000	United States	1,480,990,000
France	1,510,431,000	Denmark	1,381,460,000
Italy	1,264,513,000	Korean Republic	1,188,391,000
United Kingdom	1,264,042,000	Norway	1,171,170,000
Germany FR	1,113,211,000	Thailand	1,011,896,000
Spain	721,977,000	Japan	867,851,000
Hong Kong	624,726,000	Iceland	857,994,000
Denmark	595,950,000	Netherlands	766,379,000
Canada	433,113,000	China	645,813,000

Source: FAO 1987.

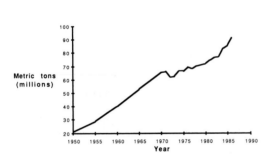

Figure 2.6. World fish production. (Source: FAO 1987.)

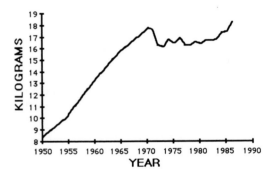

Figure 2.7. World fish production per capita. (Source: FAO 1987.)

Other major trends in global fisheries include the following:

- The use of freezing for fishery commodities has increased rapidly since 1950, with frozen products now being the major form for international trade. Frozen products are less available for use in developing countries where refrigeration is scarce. Primary forms in those regions are fresh and cured products.
- Technological advances now allow fisheries to exist anywhere in the world, with major expansions now being unlikely. Few marketable species are underutilized. Remaining underutilized resources present in large amounts, like krill, lampfish and some stocks of squid, are either too expensive to harvest, too low in market value or both. Technological advances continue in detection, harvest, processing and marketing, and will provide better utilization of stocks that are already fished moderately to heavily.
- Ninety percent of fishery production is from marine environments, nearly all of that from within the 200-mile limits of coastal nations; 10% is from freshwater. Joint ventures be-

tween developed and developing countries can facilitate full utilization of stocks within the exclusive economic zones of the developing countries.

- Smaller increases in world fishery production will lead to higher prices as demand and human population continue to increase.
- With greater portions of fishery production going into international trade, prices within the producing countries will increase further (Figure 2.8).
- More advanced processing technology will increase product quality, predictability and marketing, decrease spoilage and waste, and increase prices for the processed products.
- With higher values of fishery products and more ways to process them, less fishery production will be used for animal feeds.
- As the value of fishery commodities increases, international competition for fishery resources will intensify, and developing countries will have reduced opportunities to buy them (Figure 2.9).
- Aquaculture will continue to expand in importance. Aquaculture currently accounts for about 10% of overall production and constitutes an even greater proportion of products that are eaten directly. Aquaculture products are usually premium value products and tend to be consumed in developed countries, even if they are produced in developing countries.

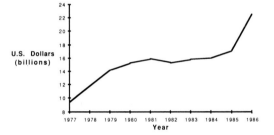

Figure 2.8. Value of total exported world catch. (Source: FAO 1987.)

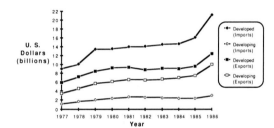

Figure 2.9. Value of total world catch: developed and developing countries. (Source: FAO 1987.)

In the United States, commercial landings over the last 40 years have been relatively unchanged, but imports during that same period have increased. Among U.S. commercial fishery landings, shrimp are the most valuable, followed by salmon, crab and tuna.

The data considered above do not include recreational harvests, which are small when compared with commercial harvests in landings but large, in some countries, when compared with commercial harvest values. In the United States, for example, where one third of the adult population fishes recreationally, the direct contribution of recreational fisheries to the national economy amounts to about $25 billion per year. That figure rivals or exceeds the combined value of all fishery products that are traded internationally each year.

Commercial value of a fishery commodity refers simply to its market value. Establishing recreational value is more difficult because it equals the value of the fish plus many more qualitative or aesthetic factors involved with the "recreational experience." Typically, recreational value is translated into the amount that someone is willing to spend to catch that sport-caught fish, and that money enters a nation's economy directly. Adding the indirect economic benefits of that money tends to double the direct amounts, or in other words, the total direct and indirect impact of recreational fisheries in the United States adds up to about $50 billion per year and generates the equivalent of about 600,000 full time jobs. The most popular recreational fishery in the United

States involves freshwater bass (smallmouth and largemouth bass, *Micropterus dolomieui* and *M. salmoides,* respectively).

For stocks that are fished both commercially and recreationally, it can be unwise to exclude one use or the other since both serve necessary and valuable purposes. We need to establish a good balance between allocations to the different user groups and the health of the resources that serve them. Striking that balance can be quite a challenge. The mandate for the world's fishery scientists and managers is clear even if the means to accomplish it are not—we need to make the best of what we have now and, more importantly, we need to assure that our natural resources will be healthy enough to meet our needs in the future.

Human Structure of Fisheries

In the simplest of ecological definitions, a **fishery** is a union of fishes and humans. Over the ages, societies of many different types have tested the quality, abundance and accessibility of diverse fishery stocks. Results of this experimentation show that humans usually behave as short-horizon predators (motivated by several, sometimes competing, goals) toward fishes deemed "valuable," although some societies also have experience with long-horizon conservation and stewardship roles. To understand fisheries, then, it is essential to know not only what fishes do, but what we do. This has prompted cultural anthropologists, sociologists, economists, political scientists and legal scholars, among others, to study variations in the human commitment to fisheries.

Fishing Societies

It is generally thought that there are some 20,000 species of fishes on Earth, a number at least equal to that of all other vertebrate animals combined. There is much more confusion about the number of human societies, in large part due to difficulties in ascertaining the influence of common origin and diffusion, and in imposing temporal boundaries on societies. Accordingly, answers to questions concerning the number and size of societies which depend upon fishing require some qualification.

Murdock and White's Standard Cross-Cultural Sample of 186 societies is a useful source for appreciating the role of fishing in non-industrial societies. In preparing a representative sample of the world's known and well-described cultures—each "pinpointed" to the smallest subgroup of the society in question at a specific point in time—Murdock and White examined the ethnographic reports of explorers, travelers, missionaries and anthropologists for more than 1,250 societies. They then selected their standard sample of 186 societies in a way that gave balanced attention to the six major regions of the world: Sub-Saharan Africa, Circum-Mediterranean, East Eurasia, Insular Pacific, North America, and South and Central America. The pinpointed dates for the standard sample ranged from 1750 B.C. to A.D. 1965, and the heaviest concentration fell between 1850 and 1950, the formative period of professional anthropology.

Murdock and White's work reveals that fishing is the dominant mode of production—in some instances supplemented by hunting, gathering and advanced agricultural activities—in 9% of the non-industrial sample. In 12% of the sample, fishing generates more than one quarter of the food supply, but is subdominant to another subsistence activity (e.g., horticulture, shifting cultivation, domestication of animals). A subsequent study by Murdock and Morrow analyzing the same societies shows that fishing—taken to encompass the capture of fishes, large aquatic mammals, shellfishes and small aquatic fauna—is absent as a food source in 14% of the sample and contributes less than 10% to the food supply in another 43% of the societies (Table 2.2).

Table 2.2. Contribution of Fishing in the Standard Cross-cultural Sample of Non-industrial Societies.

Fishing Contribution to the Food Supply	Number of Societies	Percentage
No fishing conducted	27	15
Less than 10%	79	43 58%
Greater than 10%, but less than that of other techniques	55	30
Greater than that of any other technique, but less than half of total	10	5
Greater than that of all other techniques combined	13 184*	7

Derived from Murdock and Morrow 1970.

*Data not available for two societies.

These studies confirm that fishing plays some kind of subsistence role in the great majority of non-industrial culture types. (Reviewing the ethnographic literature, Ruddle and Akimichi report that part-time fishermen in developing countries are, in many instances, farmer-fishermen who are "quintessential peasants" in that fishing is characterized by elaborate risk-sharing institutions, competitive factionalism, complex client-patron relationships and dependence on the larger society.) While fishing can still be found in many subsistence configurations, global modernization—a complex product of industrialization, urbanization, population growth, technological change and, of course, the rise of bureaucratic authority—has profoundly influenced relationships of people to fishes.

One of the most elementary questions in sociology is that which asks how it is that cultures hang together; a variant of this query asks how it is that cultures fall apart. Emile Durkheim (1858–1917), a sociologist, began to answer these questions by theorizing that social solidarity—the glue which makes society possible—is of two kinds: **Mechanical solidarity**, found in the most undifferentiated of cultures where primary group interaction and repressive law predominate, has as its essence widely held moral values forming a "totality of beliefs and sentiments" across similar people. **Organic solidarity**, which is to be found in more differentiated or stratified cultures where secondary group interaction and restitutive law are more the rules than the exception, has as its essence cooperation of functionally interdependent and dissimilar people. Durkheim used the term "mechanical" to emphasize that individuals in homogeneous traditional societies are, in a functional sense, relatively interchangeable, or highly substitutable. By contrast, individuals in "organic" industrial societies perform quite different, but interrelated functions, as the various organs in a living organism collaborate. Observing modernization, Durkheim concluded that mechanical solidarity was steadily being displaced by organic solidarity, in no small part due to increasing specialization in the division of labor.

In modern applications of Durkheim's distinction, the functional unity of traditional or non-industrial societies is seen to be underwritten by mechanical solidarity, while that of industrial societies is supported by organic solidarity. Interestingly, communities highly dependent on fishing have been less vulnerable to modernization than those dependent on other productive enterprises. Studies of contemporary fishing communities—whether they be in industrial or developing nations—stress the persistence of those "traditional" features which fishing shares with hunting and gathering: uncertainty, danger, fraternity and voluntary organization, low levels of bureaucratization, territorial recruitment rooted in kinship and a low degree of productive complexity. Using Durkheim's terminology, the distinct and unifying local character of fishing communities resists the mechanical-organic transformation. Disturbingly—and perhaps because fishing communities are so often geographically remote and fishermen are so frequently separated from shoreside affairs—fishermen throughout the world, as reported by Smith, are often treated by those who practice landbased occupations as if they were socially marginal, somehow less essential to society.

Roles of Fishing

Fishing may be said to have three sociological roles. First, fishing can be regarded as a **subsistence activity**. In communities characterized by mechanical solidarity, subsistence fishing themes are widely reflected in kinship, religious and political institutions. Native Americans in the Pacific Northwest, for example, developed their most elaborate subsistence ritual to celebrate the annual return of the first salmon; similarly, native islanders in the Torres Straits area between New Guinea and Queensland, Australia conducted highly stylized rites in connection with the dugong and turtle, the two most important food animals of the region. Second, fishing can be treated as a **work activity**. In contemporary commercial fisheries, fishing becomes an occupational option and "work" can invoke images of either mechanical or organic solidarity. Third, fishing can be viewed as a **recreational activity**. Throughout the world, this aspect of fishing is increasingly seen to compete with fishing as subsistence or work.

A few comments expand on the nature and interrelationships of the subsistence, work and recreational roles of fishing. First, each of these roles shows variation. Subsistence fisheries, for example, frequently differ in the division of labor between men and women, as well as in the time and effort devoted to fishing. Focusing on work organization, Van Maanen et al. have found contemporary commercial fishing to have both "traditional" and "modern" forms (Table 2.3). Looking at the reasons people become commercial fishermen, Miller and Van Maanen have contrasted the social identities of American "traditional" fishermen born into fishing families (e.g., Swamp Yankees, Cape Codders) with those of "non-traditional" fishermen of several types—Educated Fishermen, Hippies and Outlaws, and Part-timers and Seasonals. Recreational fishing, of course, takes many forms, which encompass the play of children, the casual party-boating adults and the serious angling of sportsmen.

Second, these roles typically tend to rise and fall in an evolutionary sequence. Fishing is initially important as a food source, then it becomes one occupation among many, then an aspect of leisure. Because this pattern unfolds at different rates and with blurred transitions in different societies, subsistence, commercial and recreational activities are juxtaposed and intertwined in the contemporary world. An expanding human population coupled with the emergence of powerful fishing technologies (e.g. enormous factory trawlers, sophisticated distant water seiners, synthetic nets, computers) has created fisheries which are sociologically complex. This multiple-use reality is one of the reasons for the management of fisheries (see Chapters 14, 18 and 19).

In passing, it must be emphasized that there is no easy way to tell which kind of fishing is most important, or most deserving of protection or promotion. Unfortunately, the statistical information which would help in this regard is rarely available and often of low quality. The catch and number of participants in the subsistence and very localized small-scale commercial fisheries which abound in the world are, for example, extremely difficult to measure. Considering only developing countries, World Bank analysts Sfeir-Younis and Donaldson have estimated that roughly 12 million fishermen work full-time, that twice that number work part-time, and that perhaps a number exceeding the total number of harvesters are employed in the processing, marketing and distribution of fishes and in boat building and net making. In the commercial fisheries, it is equally hard to determine the catch and fishing commitment of full-time, not to mention part-time and seasonal,

Table 2.3. Contemporary Forms of Commercial Fishing.

Social Organization		
Backgrounds of fishermen	Homogeneous	Heterogeneous
Ties among fishermen	Multiple	Single
Boundaries to entry	Social	Economic
Number of participants	Stable	Variable
Social uncertainty	Low	High
Relations with competitors	Collegial & individualistic	Antagonistic and categorical
Relations to port	Permanent with ties to community	Temporary with no local ties
Mobility	Low	High
Relation to fishing	Expressive (fishing as a job)	Instrumental (fishing as job)
Orientation to work	Long-term optimizing (survival)	Short-term maximizing (seasonal)
Tolerance for diversity	Low	High
Nature of disputes	Intra-occupational	Trans-occupational
Economic Organization		
Relation of boats to buyers	Personalized (long-term, informal)	Contractual (short-term, formal)
Information exchange	Restrictive and private	Open and public
Economic uncertainty	Low (long-term)	High (long-term)
Capital investment range	Small	Large
Profit margins	Low	High
Rate of innovation	Low	High
Specialization	Low	High
Regulatory mechanisms	Informal and few	Formal and many
Stance toward authority	Combative	Compliant

Source: Van Maanen et al. 1982.

fishermen. (Exacerbating the problem, the competitive nature of commercial fishing all too frequently leads entire industries and nations to misrepresent landing figures). Recreational fishing is also difficult to monitor. The United Nations Food and Agriculture Organization, for example, reports catch statistics (for 840 "species items" ranging from marine and freshwater fishes to marine mammals to sponges and algae) in commercial, industrial and subsistence—but not recreational—categories. The net result of these and other data inadequacies is that the condition of the human side of fisheries is too often poorly understood.

Social Organization and Change

Over the last half century, coastal nation states have dramatically intensified their interest in living marine resources by establishing exclusive economic zones stretching 200 nautical miles to sea. While this has simplified management problems concerning the fishing of foreigners in the waters of coastal states (foreign fleets are frequently entitled to catch only what the domestic fleet cannot, or are denied access to fishing grounds altogether), it sometimes has added to local problems (no longer able to fish in distant waters, fishermen now vie with their compatriots for resources inside the 200-mile limit, leading to an escalation of capitalization). Moreover, subsistence, small-scale and hitherto isolated fishing communities have steadily come to figure as potential players in fishery development plans and as beneficiaries of national fishing policies. Further complicating the picture, a trillion dollar a year world travel and tourism industry is beginning to sense the potential of undeveloped recreational fisheries, particularly in marine waters. When fisheries become congested, fishing becomes political.

Miller et al.'s concept of a **natural resource management system** composed of four interacting elements—natural resources, profit-seeking industries, management bureaucracies and diverse publics—is helpful in understanding complex, crowded and controversial fisheries. **Natural resources** in fisheries encompass a rich variety of fishes, marine mammals, seabirds, plants and other biological populations. **Profit-seeking industries** in fisheries include those which focus on the commercial harvest, processing and marketing of resources; those which focus on the recreational or ornamental harvest of resources; and those which provide related services (e.g., maintenance, insurance) and equipment (e.g., vessels, gear). **Management bureaucracies** in fisheries consist of regulatory authorities as local, provincial, regional, national and international levels of government. Typically, these bureaucracies utilize scientific expertise in fisheries management. **Diverse publics** in fisheries involve organizations associated with social movements (e.g., Greenpeace and environmental advocacy groups), as well as relatively unorganized public constituencies (e.g., consumers and the general public).

This framework fosters a holistic appreciation of how fisheries change over time, forcing questions about how elements of the fishing industry forge or break alliances, how industry and elements of government cooperate or clash, how policies are designed with a strong or weak adherence to fishery science, how environmental and consumer groups make their interests known. Answers, of course, confirm that fishery management systems have life histories of their own that are the synergistic products of not just the behavior of fishes and the environment, but the social, political and economic behavior of people as well.

Another concept useful in understanding social change in fisheries is that of **logistic growth**, an S-shaped curve in which a process (e.g., economic development, growth of an epidemic, growth of a stock size) ultimately exhausts the available productive resources or the potential market. (One mathematical representation of a logistic growth function is $t = 1/(1+e^{-t})$. Additional logistic growth functions along with their associated curves are presented in Chapters 8 and 13). Looking at the human side of fisheries, we see that logistic growth describes situations such as those in which the number of fishermen (or vessels) in a fishery is initially very small, then expands in an exponential way for a period, then tapers off, having reached a "carrying capacity" or "saturation" level. Changes in the number (the quantity) of fishermen may result from changes in the fish stock or changes in the market demand for fishes. Of course, logistic curves do not reveal how the kinds (the quality) of fisherman in a fishery change, as, for example, when small-scale market economy fisherman displace subsistence fishermen and are, in turn, displaced by fishermen with grander technologies.

Discussion

This chapter has provided an introduction to the many ways in which fishes have become important to people and to the different ways societies have been organized to harvest fishes. To conclude, several speculative remarks are offered about the future of fishes as sources of food and a basis for lifestyles.

For almost all of human history, people have perceived the supply of fishes to be inexhaustible. Over the last hundred years, the limiting factors of fish production have become better known. Thus, at the beginning of the century, the problem of economic overfishing in common property fisheries began to be widely discussed. Today, we extract some 80 million metric tons of fishes, shellfishes and algae from the oceans each year. This number is close to the 100 million metric ton figure generally considered to be the sustainable yield ceiling.

For fishes to make any great jump in feeding the world, developments will have to take place on two fronts. On the first, we will have to look for new sources of fishes. In this regard, marine "underutilized" species such as squids, lampfishes and krill come to mind, as do many kinds of fish (both of marine and inland origin) having aquacultural potential. Progress in those areas will depend on technological innovation in the harvesting and processing sectors. On the second front, we will have to persuade people to eat new kinds of fishes and fishery products. Progress in this area will depend on advances in fishery food science and processing and marketing advances in fitting fishery products to the culturally based preferences of consumers.

Insofar as fishing is evaluated as a way of life, it is abundantly clear that subsistence, commercial and recreational constituencies will continue to encounter one another in the policy arena, as well as on water. Whatever the results of the sparring of harvesters, a larger question to be faced by the public as a whole concerns the proper nature and performance of fishery management regimes which—in the sense they are paid for with taxpayers' dollars—effectively subsidize fishing communities.

The interwoven issues of how to make fishes widely available as food and how to fairly allocate opportunities to catch fishes across user groups will, of course, be settled differently around the world. Fishes will not be debating the outcomes, but people most definitely will.

Additional Reading

Brown, L.R. 1985. Maintaining world fisheries. Pages 73–96 in State of the World 1988: A Worldwatch Institute Report on Progress Toward a Sustainable Society. Norton and Co., New York.

Durkheim, E. 1893. Division of Labor in Society. (Translated by G. Simpson, 1933). MacMillan, New York.

Miller, M.L., R.P. Gale and P.J. Brown, eds. 1987. Social Science in Natural Resource Management Systems. Westview Press, Boulder, Colorado.

Miller, M.L. and J. Van Maanen. 1982. Getting into fishing: Observations on the social identities of New England fishermen. Urban Life 11(1):27–54.

Murdock, G.P. and D.O. Morrow. 1970. Subsistence economy and supportive practices. Ethnology 9:302–330.

Murdock, G.P. and D.R. White. 1969. Standard cross-cultural sample. Ethnology 8:329–369.

Ruddle, K. and T. Akimichi. Introduction. In K. Ruddle and T. Akimichi (eds.), Maritime Institutions in the Western Pacific. National Museum of Ethnology, Osaka.

Sfeir-Younis, A. and G. Donaldson. 1982. Fishery: Sector Policy Paper. The World Bank, Washington, D.C.

Smith, E. 1977. Comments on the heuristic utility of maritime anthropology. Pages 2–8 in The Maritime Anthropologist. Dep. Sociology and Anthropology, East Carolina University.

United National Food and Agriculture Organization (FAO). 1987. Yearbook of Fishery Statistics—Catches and Landings. Fishery Commodities. Rome.

Van Maanen, J., M.L. Miller and J.C. Johnson. 1982. An occupation in transition: traditional and modern forms of commercial fishing. Work and Occupations 9(2):193–216.

3
History of Earth and Life

Frederick G. Johnson

YEARS AGO

— 16 Billion

Universe undergoes
the "Big Bang,"
initial temperatures
in the trillions of
degrees

Matter forms as the
universe begins to
cool and expand

— 15 Billion

Expansion continues,
cooling progresses and
more matter forms

— 14 Billion

Before we consider the range of aquatic species that
support fisheries, some larger questions need to be addressed.

How on Earth Did Life Originate?

If you try to come up with the most difficult questions you could ever ask, they might sound something like
this: "How did life forms come about in the first place?",
"When and where did they arise?" and "Why are there so
many different kinds?" We know from experience that
offspring descend from their parents, but was that always
the case? The many people who have worked at answering
these questions over the years have left us with many contrasting answers. In many cases, neither the answers nor
the people were very compatible with each other. It is important that we address fundamental questions like these
and form our own, individual, conscientiously-derived
opinions about them, even though our efforts might not be
rewarded with absolute answers. When we do this, we
enter the realm of theory, the art of argument and the raw
material for scientific and non-scientific inquiry.

The scientific approach to problem solving is to
develop hypotheses that challenge or support a theory and
then put these to the test by experimentation. That is not
the only approach that is practiced, though, for issues as
broad and important as these. We can begin our understanding of natural systems with a leap of the imagination,
backward in time, to the origin of our solar system.

We need to turn our clock back by about 15 billion
years, to a time when astronomers and physicists say that
the universe underwent an enormous explosion—or "big
bang." They infer that an event of that kind took place because the universe seems to be expanding, based upon the
"red-shift" of light reaching the earth from distant solar
systems. Red light has longer wavelengths than other
colors of the visible spectrum. A shift of original wavelengths to longer ones is evidence of a "Doppler effect"
for relative motion between bodies. If the light-emitting

source is moving away from the light-detecting point, for example, the wavelengths of light perceived at the detector will seem longer than they really are as a result of the relative motion. (The same thing happens with sound waves, and you can hear the difference in the sound a train makes as it moves toward you and then away.) Besides the continuing expansion of the known universe (or at least the expansion at the time the light was sent in our direction from its extremely distant sources), the universe continues to cool down as well from the extremely high temperatures present at its inception. As far as time and space are concerned, one possibility is that time has no beginning and space has no end, and that there have been cyclic expansions and contractions of proximate universes throughout eternity. Whatever the case may be, as our telescopes probe more and more distant parts of the universe, we also look farther and farther back in time.

In the time that followed the big bang, a very hot universe began to cool and matter began to condense. Nuclei of atoms first appeared in about 3 minutes, after cooling and stabilizing from a flood of energy that must have amounted to trillions of degrees. The kinetic energy (energy of movement and temperature) of the universe was extreme and sent vaporized streams of newly formed matter hurtling through space. In the ensuing 300,000 years or so, the kinetic energies subsided enough for electrons to settle into stable patterns of movement around the nuclei and the first atoms were formed of hydrogen, helium and later, other elements. As the universe continued to cool and expand, condensations of atoms formed into bodies bound by their augmenting gravitational fields, and many of these bodies also fell into self-perpetuating cycles of mutual movements, or orbits, that kept the bodies in association with each other and gave them relative stability over time. The first galaxies formed 2 or 3 billion years after the big bang.

By about 4.5 billion years ago, most of the matter in our solar system had condensed into bodies having stable, self-perpetuating movement cycles. These bodies included the sun, planets,

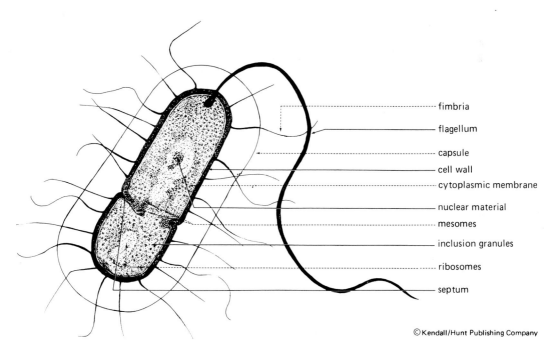

fimbria
flagellum
capsule
cell wall
cytoplasmic membrane
nuclear material
mesomes
inclusion granules
ribosomes
septum

© Kendall/Hunt Publishing Company

Figure 3.1. A diagram of a bacterium, a prokaryotic cell. (Courtesy of Kendall/Hunt.)

YEARS AGO

14 Billion

Universe continues to
expand and cool

13 Billion

Galaxies form

12 Billion

comets and asteroids. Our earth at that time was hot and lifeless. Within these bodies, particularly the sun, atoms and molecules continued to move between states of stability and instability, mixing and reassociating in new combinations. Our clock will turn only another billion years or so until a new form of stability occurs on earth—the self-perpetuating cycles of living things.

The most ancient fossils discovered on earth are dated at 3.2 to 3.8 billion years old. There are several ways of establishing these ages, but radioisotope studies are most commonly used. Other methods for aging include studies on magnetic characteristics and the layering, or stratification of materials deposited long ago by sedimentation. There is confidence within the scientific community that the various aging techniques, taken together, yield fairly accurate estimates.

Now we have a problem. How on earth did those first organisms get here? They were simple in structure and microscopic in size. They are called **prokaryotes**, which means "before the cell nucleus," and organisms of that kind are still present on earth in the form of bacteria and photosynthetic blue-green algae. A number of theories are invoked to explain their presence on earth over 3 billion years ago.

Some of the 18th and 19th century ideas are fun. One is called **spontaneous generation**, which stems from the observation that if you leave food out too long there are all kinds of living things that begin to grow on it, seemingly out of nowhere. Mice "spontaneously" appear in grain stores, molds on bread and meat, and all sorts of tiny **animalcula** grow in liquids that are left out for a while. (Early microscopists began studying microorganisms and cells about 300 years ago.) Explanations for human reproduction were also imaginative. Some 18th century microscopists believed they saw a tiny person, the **homunculus**, curled up inside each sperm cell and ready to grow inside the womb. Louis Pasteur laid the theory of spontaneous generation neatly to rest with elegant and simple experiments which showed that the "germs" came from the air and could not be seen until they had multiplied considerably, and that these microbes could be killed by heating ("pasteurization"). His hay-infusion flasks treated in this way and then sealed off at the top are still clear and germ-free, even though he performed his experiments over 100 years ago. However, if the earth's first organisms did not arise spontaneously, how did they come about?

We need to return to our time clock to consider what the earth was like when living things first appeared. When the earth formed it was very hot and gradually became cooler. The early atmosphere was **reducing** in chemical terms, not **oxidizing** as it is now. (When a chemical substance is reduced it gains electrons; when it is oxidized, it loses electrons.) The early atmosphere held primarily gaseous hydrogen (H_2), methane (CH_4), ammonia (NH_3) and water vapor (H_2O). These gases were continuously being reintroduced to the atmosphere by volcanic activity, which was much more violent than it is now. The atomic elements indicated in these gases are the same as those that make up most of the molecules in today's organisms. As the earth cooled, the kinetic energy of its atmospheric molecules declined, and that allowed much of the water vapor to condense and precipitate to the earth's surface, forming the primordial sea. The primordial sea held the same gases that the atmosphere held, but they were dissolved and mingled, forming countless chemical combinations with the addition of radiant energy from the sun and from recurrent electrical storms.

At some point in the earth's first billion years, land masses rose from the primordial sea. By 200 million years ago, a single continent had formed, called **Pangaea**, which split into two continents, **Laurasia** and **Gondwanaland**, about 135 million years ago. Today's continents are still moving through a process called **plate tectonics** (discussed in Chapter 10). After the prokaryotes flourished, the atmosphere began to change from reducing to oxidizing in character as the amount of oxygen gas in the atmosphere increased gradually to the 21% that now occurs. The remainder of the present atmosphere is primarily nitrogen gas (N_2) with smaller amounts of carbon dioxide (CO_2) and water.

Biological activity of the prokaryotes caused the shift in the atmosphere to the oxidizing state; therefore, the conditions under which the first organisms appeared would not reoccur on a global scale. The oxygen was released by the photosynthetic activity of the early prokaryotes that gleaned their needed energy from sunlight. They were the earth's first **autotrophs**, or "self-feeders." The other prokaryotes were **heterotrophs**, or "other-feeders" and had to consume either each other, the autotrophs, or dissolved organic molecules to meet their energy needs. We will examine three major theories that address the origin of these organisms. A fourth, the "extraterrestrial" explanation, will not be considered further because entry of objects into the earth's atmosphere ordinarily produces more than enough heat to sterilize them.

Chemical Evolution

The theory that a gradual chemical evolution preceded biological evolution is probably the "classic"

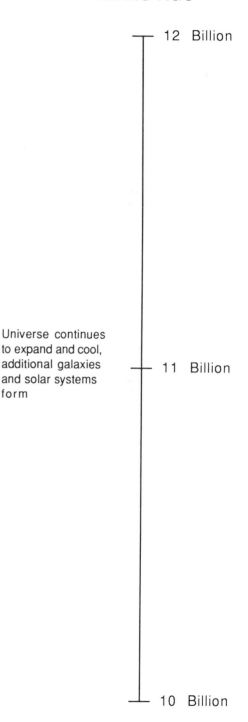

YEARS AGO

— 12 Billion

Universe continues to expand and cool, additional galaxies and solar systems form

— 11 Billion

— 10 Billion

YEARS AGO

— 10 Billion

— 9 Billion

Universe continues
to expand and cool,
additional galaxies
and solar systems
form

— 8 Billion

scientific theory to address the origin of life. It was developed during the 1920s by A.I. Oparin, a Russian scientist. The basic premise of the theory is that chemical conditions of the early earth's atmosphere and primordial sea, together with ultraviolet irradiation, electrical energy from lightning and lots and lots of time, were sufficient to cause enough arrangements and rearrangements of inorganic matter to produce the first self-replicating organisms. Oparin held that the possibilities of chemical recombination over the period preceding biogenesis were so enormous that ordinary chemical reactions and aggregative processes were sure to produce a winner sooner or later, which, when formed, would go ahead and maintain its own integrity (having **homeostasis**) and produce offspring that later might diverge, or **speciate**, by a biological process described by Charles Darwin. Oparin's prebiotic entities were droplets containing a range of more complicated "organic" molecules, some held in suspension and some in solution and surrounded by a barrier or membrane of hydrophobic molecules that allowed some degree of selectivity over what entered and left the droplet. Oparin called his droplets **coacervates** and assumed that some of the coacervates developed metabolic capabilities, the ability to conduct internally-controlled chemical reactions. Some of these reactions could break molecules down into smaller parts to yield energy (**catabolism**) and other reactions used up energy to synthesize larger molecules (**anabolism**). Eventually, there would be a selection for the most successful coacervates at the expense of less successful ones. Some coacervates might "swallow" others and inherit new benefits and later on the ability to replicate (produce biochemical copies of) the molecules that controlled the chemical reactions, and the reproduction of the coacervates themselves would allow them to take over the primordial territory for their own kind.

Oparin's hypotheses were tested, to some extent, by the experiments of Stanley Miller in the 1950s and other researchers as well. These scientists duplicated the chemical conditions of the early earth atmosphere in containers, adding water and energy in the form of ultraviolet radiation or electrical sparks. These experiments did result in the synthesis of "organic" molecules, including the building blocks of proteins and nucleic acids. In today's organisms, nucleic acids make up the **genes**, or genetic material found in the chromosomes, that replicate before an organism reproduces itself. Each copy of the gene carries a code for the synthesis of specific proteins, most of which become **enzymes** and control the progress of chemical reactions, or **metabolism**.

Despite the success of these experiments at lending credence to Oparin's hypotheses, none of the experiments ever produced a real, live, jumping and slimy critter. Supporters of the theory said it took more time. Opponents held that there was too much organization among the components of a living thing to accrue merely from chance recombinations, regardless of the time allowed, even with all the primordial soup and lightning strikes one might drum up. Both of the next two theories address this conceptual difficulty.

Life from Clay Crystals

Problems relating to organization, as well as the abilities to grow, reproduce, disperse and then grow and reproduce again, are bridged by the clay crystal theory for the origin of life on earth. It is an elegant, largely untested theory put forth by A.G. Cairns-Smith and others. This theory, put forth in 1985, suggests the first organisms were clay crystals.

Cairns-Smith also confronted the organization problem, citing as an analogy the possibility of forming an arch by dropping rocks from the sky. No matter how many times you drop a batch of boulders, you will never have them fall into a perfect arch like the arch that a rock footbridge makes over a creek. The boulders will always fall into a disorganized heap into the creekbed. Cairns-Smith asserted that Oparin's assumption—that surely, sooner or later, some number of chance events affecting primordial coacervates would result in such organization—is stretched too far. It is much easier to imagine the initial presence of an organized inorganic structure, onto which the "organic" structure is simply added. If we assume, for example, that there is a big pile of fallen wood debris lying over the creek, and then we drop our boulders on top of the woodpile, allowing them to rest and settle together there, we may well be left with a rock arch years later after the wood has rotted away.

Now translate your imaginary woodpile into something far smaller, a group of mineral crystals growing slowly as water, bearing dissolved substances, percolates slowly through spaces in the matrix where the crystals are growing. Dissolved minerals carried in solution by the water add to a crystal and cause it to grow in a completely organized fashion. If there is some disturbance to the matrix or a piece of the growing crystal breaks off and, carried by the water, lodges in a crevice somewhere else in the system, it will continue to grow, provided that it is supplied with the right dissolved minerals. The new growth also retains the organizational characteristics of the old.

YEARS AGO

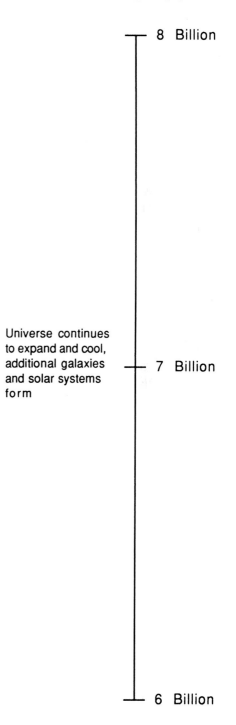

8 Billion

Universe continues
to expand and cool,
additional galaxies   7 Billion
and solar systems
form

6 Billion

YEARS AGO

— 6 Billion

— 5 Billion

Our solar system forms

Earth forms

Earth begins to cool

— 4 Billion

That gives us organization, growth, reproduction and dispersal. So far, we have nothing we did not already know regarding crystal formation.

Now we incorporate organic molecules into this inorganic framework. We begin with the building-block molecules thought to have been present in the primordial soup and produced in Stanley Miller's experiments. We conceptually percolate these substances through crystal-lined interstices along with the dissolved minerals and see what happens. We then find that some of the "organic" molecules bind to the surface of the growing crystals and "grow" as the crystal grows.

One of the nucleic acids, **RNA (ribonucleic acid)** is known to bind to mineral surfaces. RNA is the messenger from **DNA (deoxyribonucleic acid)**, which directs the synthesis of protein enzymes in modern cells. DNA makes up the genes of modern cells. The first genes were probably RNA, but neither RNA nor DNA have been fully synthesized by early-earth simulation experiments.

Getting back to the crystal theory, the idea is that an early association was formed between the organic and mineral components such that both components grew together. Later the mineral component became less necessary to the growth of the organic component and eventually disappeared. In other words, the wood under the bridge was no longer present and the boulders could stand on their own. (Note in the analogy that we reversed the roles of the mineral and the organic components) At any rate, according to this theory, the organic component by then had established the necessary means to perpetuate itself and exist on its own, and further changes to the organic component amounted to biological, or Darwinian evolution.

There are a number of other factors that add appeal to the clay crystal theory. One is that mineral crystal surfaces are in many cases used to catalyze biochemical reactions by the chemical industry because their presence can greatly enhance the rate of chemical reactions, even those involving organic substances. Another appealing factor is that once a stable organic system becomes established, it can be protected from the destabilizing influences of ultraviolet radiation by shielding from the mineral phase. It is necessary, for our assumptions on biogenesis to hold, that disruptive influences that yield successful new combinations do not destroy those new combinations.

Other scientists suggest that life may have evolved more than once, where organisms good at metabolizing may have combined with others good at replicating, to yield a combined form that was good at both, as today's organisms are. The idea that unlike organisms combined in the past to yield successful combinations has considerable merit, and we know of many contemporary examples including lichens, corals, giant clams and perhaps even modern cells (**eukaryotes**).

There are many more interesting ideas we could discuss here, but after all, this is a book about fish. We will take a look at one more theory on the origin of life before we get to fish.

Creation

Many people believe that life did not come about by itself, but instead was guided into existence by a deity, creator, or god. In times past, people believed that many gods were involved in creation. Often, these beliefs stem from religious writings, including the Bible and others, that also hold answers to the questions we have just considered. Most scientists, however, consider creationism to be mythology. Unlike scientific beliefs that rest largely upon the formulation and testing of hypotheses, religious beliefs are based upon faith in the creator and upon the validity of the religious writings. These differences in opinion on how to base one's beliefs constitute a very contemporary issue. According to recent polls conducted in the United States, Americans are fairly evenly divided in their beliefs about fundamental biological issues.

Creationists put forth a number of arguments against the scientific theories. First of all, many of the intricacies of biological existence, including the ability of humans to think, of birds to fly and sing, and of bees to instinctively conduct their lives seem too ordered to creationists to have come about by an independent process. The ways in which all the facets of nature come together so neatly, in such a "beautiful" form seem, to the theologically minded, to be impossible without supreme intervention. Many perplexing problems relating to the nature of the universe as well as living phenomena can be explained and understood by assuming divine intervention.

It is unlikely that more than one of these theories is correct. When it comes to conflicting theories, usually either one is correct and the others incorrect, or all are wrong.

Whatever the basis for the origin, or origins of living beings might be, one thing is certain—in biochemical terms they are all built essentially the same. The energy-processing and information-storing biochemical machinery are quite similar among organisms as widely diverse as bacteria, green plants, fish and people. That makes it much easier to learn how living things "tick."

YEARS AGO

—— 4 Billion

Reducing atmosphere present on earth

Early seas form as earth cools

First organisms evolve

Land masses form

Prokaryotes leave fossil remains

—— 3 Billion

Active volcanism continues

Prokaryotes colonize more habitats

Atmosphere changes from reducing to oxidizing

—— 2 Billion

YEARS AGO

(expanded scale)

— 2 Billion

Oxygen in
atmosphere
increases

Prokaryotes
diversity

Movement of land
masses continues

— 1 Billion

The bottom line to these theories is this: think about them and form your own opinion, but never lose your sense of wonder for whatever must have happened.

Jaws

Now that we have secured our earth-bound perpetuating living units, by whatever means—what next? First of all, we should consider what life means, then consider what it means to define a group of living things as a **species** and, finally, we will try to get some understanding of the diversity of species in general, and of fishes in particular.

As it turns out, what it means to be alive is not at all clear-cut. Definitions of life typically include a description of the abilities living things have, like the ability to grow, reproduce, metabolize and respond to stimuli. We have already seen that mineral crystals can grow and reproduce, but they are not considered to be alive. Then we have viruses, which have no metabolism of their own but can infect host cells and cause those cells to reproduce more copies of the virus. Are viruses alive? They probably are not, but what it comes down to is that it does not matter, they can kill you and they need to be reckoned with. Countless spores, cysts, seeds and other organic entities may not be alive in a strict sense, but nevertheless bear self-perpetuating competencies. This brings us to a key concept that is often hidden in our efforts to understand and catalogue biological phenomena—that living things, however defined, will continue to exist if they perpetuate themselves and will become extinct if they do not. Whether they meet our definition of life or not is inconsequential. If an organism looks like a plant, but behaves like an animal it makes no difference—organisms do not succeed or fail on the basis of our descriptions of them but only on how well they secure their places in the future. Oparin, in his classic *Origin of Life* stated this concept succinctly:

The biologist, unlike the layman, knows no lines of demarcation separating plant life from animal life, nor for that matter living from non-living material, because such differentiations are purely conceptual and do not correspond to reality.

When we move on to consider the many kinds of fishes and other aquatic life, we should remember that efforts to classify them are our efforts, not their efforts. It is also true that no organism is "higher," nor more "lowly" than another, just as neither is "good" nor "bad." They simply exist.

The definition of a **species** is somewhat easier and for the most part corresponds to functional reality. A species consists of a group of organisms capable of interbreeding and bearing fertile offspring. This is not to say that some different, yet closely related species cannot interbreed. If they do, however, the resulting **hybrid** offspring would be sterile. (In rare cases, certain fish hybrids can reproduce.) Also, within the members of a species there can be considerable variation in appearance and other characteristics. Recognition of that variation led Charles Darwin to formulate his famous theory of **speciation**, or the process by which new species arise through divergence of the parent stocks.

Darwin published his evolutionary masterpiece, *The Origin of Species,* in 1859. Since then his book and theory on speciation by **natural selection** have been hailed as the greatest unifying theory in biology—one that lends a cohesive understanding to the enormous amount of previously unexplained and diverse biological observations. Darwin began with the realization that living things are so good at reproducing under optimal conditions that any given species could overrun the earth if its reproductive successes were not kept closely in check by mortalities. He understood that some members of a species would live longer than others, based on their individual characteristics, and would thereby have a greater likelihood of passing on the characteristics that enhanced their survivability to their offspring. He saw that nature, by natural selection (sometimes called "survival of the fittest") could act as a very effective filter for the many different traits expressed in a population of organisms. He also saw how effective humans could be by their own **selective breeding**, or "artificial selection" at creating new strains of animals and plants for various purposes.

Darwin's theory of biological evolution through natural selection provided a functional framework for the taxonomists, who had long been cataloguing and classifying the many groups of animals and plants. The theory predicts that organisms in more closely related groups, or **taxa**, are actually also more closely related in terms of their evolutionary ancestry. In *The Origin of Species,* Darwin stated:

> The affinities of all the beings of the same class have sometimes been represented by a great tree. I believe this simile largely speaks the truth. The green and budding twigs may represent existing species; and those produced during former years may represent the long succession of extinct species. At each period of growth all the growing twigs have tried to branch out on all sides, and to overtop and kill the surrounding twigs and

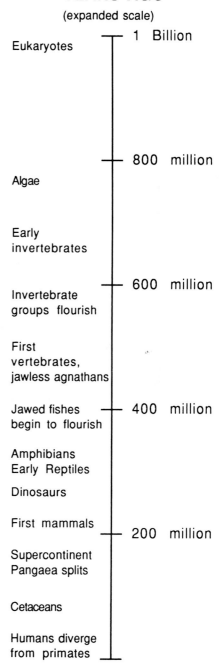

YEARS AGO

(expanded scale)

Eukaryotes — 1 Billion

Algae — 800 million

Early invertebrates

Invertebrate groups flourish — 600 million

First vertebrates, jawless agnathans

Jawed fishes begin to flourish — 400 million

Amphibians
Early Reptiles

Dinosaurs

First mammals — 200 million

Supercontinent Pangaea splits

Cetaceans

Humans diverge from primates

Present

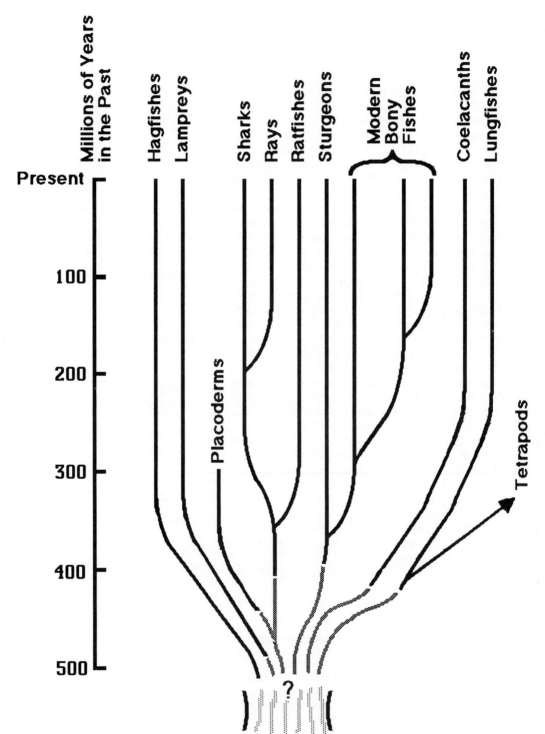

Figure 3.2. A "tree" showing the evolution of major fish groups. The tetrapod branch led to the amphibians, reptiles and mammals.

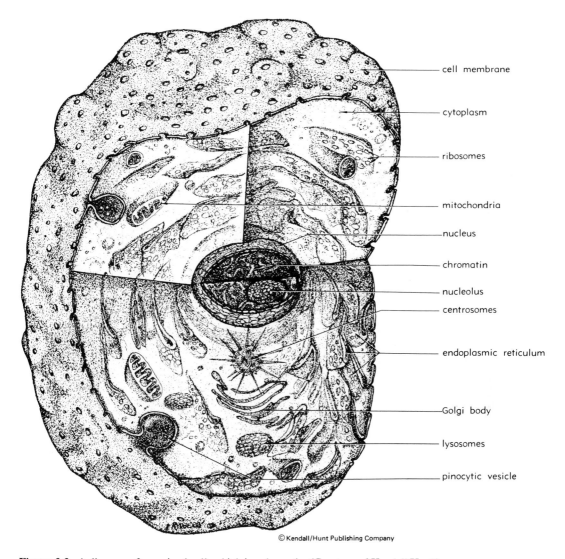

cell membrane

cytoplasm

ribosomes

mitochondria

nucleus

chromatin

nucleolus

centrosomes

endoplasmic reticulum

Golgi body

lysosomes

pinocytic vesicle

© Kendall/Hunt Publishing Company

Figure 3.3. A diagram of an animal cell, which is eukaryotic. (Courtesy of Kendall/Hunt.)

Figure 3.4. A placoderm, an example of the first group of jawed fishes. (Courtesy of Barry McFarland.)

branches, in the same manner as species and groups of species have at all times overmastered other species in the great battle for life. The limbs divided into great branches, and these into lesser and lesser branches, were themselves once, when the tree was young, budding twigs, and this connection of the former and present buds by ramifying branches may well represent the classification of all extinct and living species in groups subordinate to groups. Of the many twigs which flourished when the tree was a mere bush, only two or three, now grown into great branches, yet survive and bear the other branches; so with the species which lived during long-past geological periods, very few have left living and modified descendants. From the first growth of the tree, many a limb and branch has decayed and dropped off; and these fallen branches of various sizes may represent those whole orders, families, and genera which have now no living representatives, and which are known to us only in a fossil state.

The topic of fossils brings us back to our time clock, which we left running through all this, hoping to get sooner to the topic of fishes. Darwin thought that biological evolution took place very slowly and gradually, with the addition of only small increments of change at each step over geological time. Other scientists now believe that biological evolution and speciation may not transpire evenly and gradually, but instead may occur in spurts. Stephen Jay Gould, for example, described a process that may be more accurate, called **punctuated equilibrium**, in which evolution among groups of organisms proceeds with Darwinian sluggishness as our clock ticks by until, presto, a new feature arises and allows the rate of evolution to increase dramatically. Suppose, for example, that a new cell type arises through a combination of some existing ones, or that in a predatory world, some predators suddenly develop JAWS!

Our time clock now shows our earth with an oxidizing atmosphere, plenty of prokaryotes and organisms with a new type of cells—the **eukaryotes**. Eukaryote means "true nucleus", and the new cells had a nucleus and chromosomes, along with other formed elements called **organelles** ("little organs") within the cells. **Chloroplasts,** for photosynthesis, and **mitochondria,** for metabolic functions, are examples of organelles. Both chloroplasts and mitochondria have their own DNA and are thought to have been separate organisms at one time. The eukaryotes may represent the combination and beneficial coexistence of two or more prokaryotes. All new major groups of organisms would have cells of the eukaryotic type (prokaryotic and eukaryotic cells are shown in Figure 3.1 and 3.2).

When the clock reaches nearly 600 million years ago (not billion), a relatively sudden burst of fossil-leaving organisms appears, including examples from most major taxonomic groups. It is the Cambrian period, with plants and **invertebrate** ("without backbone") animals well-represented. There are no **vertebrates,** or animals with backbones, and no jaws present during the Cambrian. There must have been plenty of predation going on, however, since many of the invertebrates had hard parts or shells that protected their bodies, and that is why they left such magnificent fossilized remains.

When we reach 500 million years ago, the Ordovician period, the first vertebrates appear, belonging to a group still present on earth called the agnathous ("without jaws") fishes. By the Devonian period, about 400 million years ago, we find remains of the first fishes with jaws, the **placoderms** (Figure 3.4). We can be fairly certain that the addition of jaws conferred a considerable advantage and, after the placoderms, a relatively rapid **adaptive radiation** occurred, giving the increasing number of vertebrates an increasing chunk of the world's real estate. The cartilaginous fishes (**Chondrichthyes**), the bony fishes (**Osteichthyes**), and other jawed vertebrates followed. Jaws were very handy back then and still are. You probably used yours today.

Before we end this chapter we should let our time clock catch up with us. Around 135 million years ago, the continent Pangaea split into Laurasia and Gondwanaland. About 2 million years ago humans diverged from the primates, and here we are. Let us not forget that we are made of stardust.

4
Forms of Aquatic Life I

Frederick G. Johnson

In this chapter and the one following, we examine the different kinds of aquatic organisms that support fisheries. We will consider how these organisms are classified into taxonomic groups by learning about their **systematics**—their classification into a system that implies natural relationships like those described by Darwin. We will also learn about the diversity of life forms and about some of their major features and habits.

Classification Systems

We have already encountered two ways to classify organisms. One was by cell type—the prokaryotes having cells with a relatively simple structure and the eukaryotes having more complicated structures including cell nuclei and organelles. The other was by their source of energy. The autotrophs obtain their energy directly from inorganic sources, and most capture their energy from sunlight by the process of photosynthesis. Organisms that cannot capture their own energy from inorganic sources are called heterotrophs. While animals are heterotrophs, they can be classified more specifically by feeding type. Some of the terms used in this context include **carnivores** (meat eaters), **herbivores** (plant eaters), **omnivores** (eaters of both plant and animal materials), **piscivores** (eaters of fishes), **planktivores** (eaters of plankton) and **detritivores** (eaters of decomposing bits of animal or plant remains). Other terms that classify animals by feeding type refer to how they feed rather than what they feed upon. Terms of this kind include **filter feeders**, which obtain nutrition by filtering food particles out of the water, and **deposit feeders**, which feed on organic materials deposited on the sea or lake bottom. Some of the terms used above are not mutually exclusive. For instance, most filter feeders are also planktivores.

Organisms can also be classified by their physical position in the aquatic environment. Those that live on the bottom are called **benthos**, or **benthic organisms**. There are two categories for organisms that live within the water column, **plankton** and **nekton**. Plankton are at the mercy of the currents, so water movements control their distribution. Many planktonic organisms can move on their own power, but often only enough to maintain their vertical position in the water column. Plankton are often microscopic, but some are quite large, like some of the jellyfishes. Plankton are conveniently divided into two groups: **phytoplankton** are plants, and **zooplankton** are animals. Nekton have the ability to move through the water from place to place on their own. Whales, porpoises and most fishes are members of the nekton community.

The categories covered above are functional groupings only; they do not imply genetic relationship. The primary classification system for organisms, **taxonomy**, implies that organisms grouped together in the same **taxon** are related genetically. Systematists try to make sure that organisms assigned to a taxon are genetically related (meaning they share a common evolutionary history). It is assumed that such organisms evolved from a common ancestor—the smaller or more specific the taxonomic group, the more recent the common ancestor. Systematists, then, try to assure that taxonomy (classification) reflects **phylogeny** (evolutionary history).

As time goes on, we learn more and more about genetic and evolutionary relationships and, as a consequence, systematists are continuously re-evaluating the assignment of organisms to various taxonomic groups. Often these scientists disagree with each other on how to proceed.

The system used to name species was developed by Carolus Linnaeus in the 18th century and is called **binomial nomenclature** (each species being defined by two words, genus and species). Early classification systems were based primarily on morphological similarities (similarities in form and appearance). Linnaeus classified and catalogued many species of plants and animals and published his findings in *Species Plantarum* (1753) and *Systema Naturae* (1758).

Linnaeus' system was not the first attempt at the classification of organisms. Even in prehistoric times, there were probably attempts at finding order within the diversity of living things. Aristotle developed his *Scala Naturae* around 350 B.C., a system referred to as a "ladder of nature" where simple-bodied forms appeared low on the ladder and humans were at the top. Linnaeus classified organisms in a similar fashion, based upon his "great scale of being." The notion that evolution develops organisms toward some kind of "higher" state has persisted, even though Darwin's theory clearly refuted the idea that there was any unidirectional progression or driving force behind evolution.

Linnaeus' classification system preceded Darwin's theory by 100 years, so it lacked the insights regarding genetic and evolutionary relationships that we have now. It is, nevertheless, a useful system for the simple reason that organisms that are related to each other usually bear obvious similarities in appearance. Other important features that suggest genetic relationships between organisms that are important factors in their classification are their **life histories** (including the presence or absence of distinct **larval stages**), their patterns of **embryonic development** (growth patterns following fertilization) and, for animals, the types of **body cavities** they have. Many of these factors are taken into account to assign an organism to one or another taxonomic group. Modern biochemical science has added **molecular genetic** analyses as a tool to determine evolutionary relationships. Such analyses provide specific details of the DNA and proteins within an organism. Statisticians have also worked on systematics through comparative mathematical analyses of numerous body characteristics (**numerical taxonomy**).

The framework developed by Linnaeus and refined since then for the classification of organisms is shown below. It is a hierarchical system consisting of groups within groups and specifies a taxonomic breakdown of the largest taxon (the kingdom) into smaller and more specific ones, with species being the smallest taxon. It also represents a dynamic system in which, even though the categories usually stay the same, the animals and plants assigned to them might shift from one group to another as we gather more biological information.

Kingdom
    Phylum (or Division, for plants)
        Class
           Order
               Family
                  Genus
                     Species

The last two categories in this hierarchy, **genus** and **species**, are expressed together to constitute the **scientific name** of a species. Scientific names are generally derived from Latin or Greek roots and therefore are always either underlined or italicized when written. Also, for a given scientific name, like ours—*Homo sapiens*—the genus always comes first and is capitalized, while the species comes second and is not capitalized. If more than one species within a genus is being discussed, it is common to spell out the genus the first time it is given but then give only the first initial of the genus when it is mentioned shortly afterward. For example, all modern people belong to the species *Homo sapiens,* but fossil remains of extinct hominids, like *H. erectus* and *H. habilis,* have been found in Africa and, according to some anthropologists, lived as long as 3.5 million years ago. When more than one species within a genus is referred to, but not specified individually, you will see the name of the genus followed by **spp.** (e.g., by *Homo* spp.). When a single, nonspecified species is implicated it is written as *Homo* **sp**.

Now we will use our taxonomic framework as an example of how a great white shark is classified:

Kingdom   Animalia
   Phylum   Chordata
      Class   Chondrichthyes
         Order   Squaliformes
            Family   Lamnidae
               Genus   *Carcharodon*
                  Species   *carcharias*
                     Scientific name—*Carcharodon carcharias*

The white shark was originally described by Linnaeus. Sometimes you will see the scientific name of a species followed by the last name of the person who described it, either with or without parentheses around the person's name. If the species was originally described under a different genus than is currently in use, as is the case with the white shark, then the authority's name is in parentheses, e.g., *Carcharodon carcharias* (Linnaeus). More descriptions of common species are attributed to Linnaeus than to any other individual. Linnaeus believed his taxonomic system reflected the pattern of creation of living things.

Most phyla, particularly the most diverse phylum, Arthropoda, include so many different kinds of animals that the categories given above are not enough to organize them adequately. That kind of situation is addressed by establishing sub- and super-categories such as **subphylum**, superorder, subclass and the like. The phylum Chordata also uses some of these additional categories. For example, the white shark mentioned above is grouped under the chordate subphylum Vertebrata.

Phyla (the plural form of phylum) are the major categories that are used to group different kinds of animals together. Plants are commonly grouped into **divisions**, instead of phyla, but the two categories are essentially equivalent.

Within each taxonomic category, there may be many members or only a few. Many living species have yet to be described and classified, particularly small ones like **nematodes** (roundworms), which have few distinguishing characteristics. The phylum Arthropoda includes over 800,000 contemporary species, most of which are insects. Many additional extinct species are known from the fossil record. The invertebrate phylum Priapulida has only two **genera** (plural form of genus) and eight species. Thus, some groups are very diverse and others are not, which leads us to our next topic.

Species Diversity

Species diversity simply refers to the number of different species in a given area, or within a taxon. The species diversity of the world, in terms of those living species that are known and classified, is given in Figure 4.1. It is clear from the figure that invertebrates have by far the greatest number of species, followed by land plants. Only about 3% of the known species are vertebrates, despite their obvious ecological successes. About half of the vertebrate species are fishes. Since many living species remain undescribed, estimates of the actual number of contemporary species place the total well above the 1,390,992 given in the figure. Recent estimates place the actual number of species on earth anywhere from 5 to 30 million.

We are now ready to take a look at the individual taxonomic groups themselves, particularly as they relate to fisheries. They are summarized in Table 4.1. We will begin with the kingdoms.

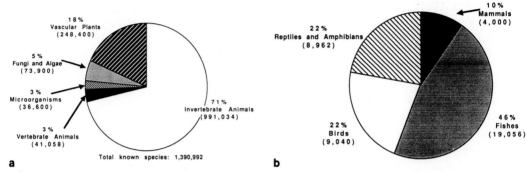

Figure 4.1. (a) Diversity of contemporary known species showing their distribution among major taxonomic groups. (b) Diversity of contemporary vertebrate species.

Table 4.1. Taxonomic groups of importance to fisheries

KINGDOM MONERA:	These are prokaryotes (before the nucleus); includes bacteria and blue-green algae.
KINGDOM FUNGI:	Decomposing heterotrophs. This and the following kingdoms are eukaryotes (have cells with a true nucleus).
KINGDOM PROTISTA:	Most are single-celled eukaryotes, but some are multicellular. May have characteristics of either plants or animals.
DIVISION CHLOROPHYTA:	green algae
DIVISION CHRYSOPHYTA:	includes diatoms
DIVISION PYRROPHYTA:	includes dinoflagellates
DIVISION PHAEOPHYTA:	brown algae and kelps
DIVISION RHODOPHYTA:	red algae
KINGDOM PLANTAE:	Multicellular photosynthetic plants, most live on land.
DIVISION TRACHEOPHYTA:	vascular plants
KINGDOM ANIMALIA:	Multicellular animals.
PHYLUM PORIFERA:	sponges
PHYLUM COELENTERATA:	jellyfish, corals
PHYLUM MOLLUSCA:	molluscs
CLASS GASTROPODA:	snails, abalones
CLASS BIVALVIA:	also called Pelecypoda, includes clams, mussels oysters, scallops
CLASS CEPHALOPODA:	octopi and squid
PHYLUM ARTHROPODA: arthropods	
CLASS CRUSTACEA:	Includes crabs, shrimps, lobsters, crayfish, krill and copepods
PHYLUM ECHINODERMATA:	includes sea urchins and sea cucumbers
PHYLUM CHORDATA:	chordates
SUBPHYLUM UROCHORDATA:	tunicates

Table 4.1.—*Continued*

SUBPHYLUM VERTEBRATA:	vertebrates
CLASS AGNATHA:	jawless fishes; lampreys and hagfishes
CLASS CHONDRICHTHYES:	cartilagenous fishes; sharks, rays, ratfishes
CLASS OSTEICHTHYES:	bony fishes
FAMILY ACIPENSERIDAE:	sturgeons
FAMILY CLUPEIDAE:	herrings, sardines and shad
FAMILY ENGRAULIDAE:	anchovies
FAMILY SALMONIDAE:	salmons and trouts
FAMILY CYPRINIDAE:	carps, goldfishes and minnows
FAMILY ICTALURIDAE and	
FAMILY SILURIDAE:	catfishes
FAMILY CLARIIDAE:	walking catfishes
FAMILY GADIDAE:	cods, hakes and pollocks
FAMILY SCIENIAE	drums, croakers and seatrouts
FAMILY SCORPAENIDAE:	rockfishes
FAMILY CENTRARCHIDAE:	sunfishes and freshwater basses
FAMILY SCOMBRIDAE:	tunas and mackerels
FAMILY PLEURONECTIDAE and	
FAMILY BOTHIDAE:	flatfishes
CLASS AMPHIBIA:	frogs
CLASS REPTILIA:	turtles and alligators
CLASS MAMMALIA:	mammals
ORDER CETACEA:	whales, dolphins and porpoises
ORDER CARNIVORA:	otters, seals and sea lions

The Kingdoms of Life

Kingdoms are such large and broad taxa that it is difficult to assure common ancestry among their members, so more than any other taxonomic category, kingdoms tend to be categories of functional convenience. Various biologists have advocated lumping species into as few kingdoms as two, or splitting them into as many as five. A two-kingdom system has one for animals and the other for plants. A three-kingdom system has one for animals, one for plants and one for prokaryotes. We will use the five-kingdom system because it is the most descriptive option. The five kingdoms in this system are **Monera**, **Fungi**, **Protista**, **Plantae** and **Animalia**. We will consider them in that order.

Monera

The kingdom Monera includes the prokaryotes—bacteria and blue-green algae. This is the most ancient kingdom. The oldest series of fossilized life forms discovered so far dates back to about 3.8 billion years ago. Another, the Onverwacht series from Swaziland, Southeast Africa, suggests that prokaryotes were alive between 3.2 and 3.5 billion years ago. About 2,700 contemporary species of organisms in the kingdom Monera have been described, although many of them are difficult to distinguish. They are also the world's smallest organisms.

Bacteria can be very important to fisheries for several reasons. Most bacteria are **decomposers**, obtaining their nutrition from the remains of dead animals and plants, and for that reason they are important for the recycling of biological materials. Species of both bacteria and blue-green

Bacteria can be very important to fisheries for several reasons. Most bacteria are **decomposers**, obtaining their nutrition from the remains of dead animals and plants, and for that reason they are important for the recycling of biological materials. Species of both bacteria and blue-green algae play another critical ecological role through their place in the nitrogen cycle. Prokaryotes are the only organisms that can take inorganic nitrogen gas (N_2) and change it into compounds like ammonia (NH_3) that make the nitrogen available for synthesis of larger organic molecules. This process is called **nitrogen fixation** and is extremely important for the growth and productivity of plants.

Bacteria are also important because some species cause diseases, attacking fishes and other living things rather than decomposing dead ones. Many of these diseases are **infectious** and can spread rapidly through a population, causing significant losses. In some cases, bacterial diseases can be treated with **antibiotics**. Bacteria are successful in part because they are adaptable to a wide range of living conditions, including the insides of other organisms and places with extreme temperatures, and also because they reproduce rapidly. Bacteria reproduce by simple cell division, or **binary fission,** and can double their population size as quickly as every 7 minutes. Bacteria living within the bodies of other organisms may also be beneficial, including those that normally inhabit the digestive tracts of animals. What bacteria lack in morphological complexity they more than make up for by their adaptability to living in extreme and changing environments.

Most bacteria are heterotrophs, but some are autotrophic. A few are able to photosynthesize, but a more interesting group of autotrophic bacteria obtain their energy from inorganic chemical compounds rather than sunlight; these bacteria are called **chemoautotrophs**. Recently, some scientists have speculated that the first organisms were chemoautotrophic, perhaps living near sites of volcanic activity like the **hydrothermal vents** at the bottom of the ocean.

Blue-green algae are thought to have promoted the change in the early atmosphere by releasing oxygen (O_2), a by-product of their photosynthesis. This transition of the atmosphere was well underway 2.5 billion years ago and presumably made conditions more suitable for later life forms. Today's blue-green algae are common nuisances in potable water supplies by giving water a bad taste. Their photosynthetic pigments, which allow the plants to capture energy from sunlight, include **chlorophyll a**. Unlike the photosynthetic eukaryotes, however, blue-green algae have their chlorophyll scattered along membranes within the prokaryotic cell rather than encapsulated within organelles called **chloroplasts**. In fact, blue-green algae are thought to have been the ancestral progenitors of chloroplasts, and bacteria were perhaps the progenitors of **mitochondria.**

Prokaryotes, as a group, are morphologically simple but biochemically diverse. Their biochemical abilities allow them to exist where eukaryotic organisms cannot. For some, their rapid rates of **mutation** allow them to adapt quickly to changing circumstances, including our efforts to combat them with antibiotics (mutations and other features of cells are discussed in Chapter 6). Their structure includes a cell membrane that lacks cholesterol and is usually surrounded by a cell wall. The DNA of prokaryotes is not contained within a cell nucleus and it does not form chromosomes, but instead occurs as a single, circular molecule. Some prokaryotes are able to move by gliding or "swimming" with **flagella**, the slender hairlike extensions of the cells (flagella is the plural form of **flagellum**). The bacterial flagella are not constructed like the flagella of eukaryotes, but they do allow motile bacteria to move either toward or away from chemical stimuli. This behavior is known as **chemotaxis.**

Fungi

The kingdom Fungi and all of the remaining kingdoms consist of organisms with eukaryotic cells. Like the bacteria, fungi can be important to fisheries since many fungal species cause diseases of fish and shellfish. Some cause particular problems by attacking egg masses. All fungi are heterotrophic and most species share the ecological role of decomposers with bacteria. There are about 80,000 species of fungi.

Most fungi grow as masses of thin filaments, but a few, including the yeasts, are unicellular. Each filament is called a **hypha** and the **hyphae** grow into a mass called the **mycelium**. (When you pick a mushroom, most of the mycelium is left in the ground.) The hyphae are surrounded by a cell wall that is made of **chitin**, a substance that also forms the shells of crustaceans. To obtain nutrition, the hyphae usually secrete digestive enzymes and then absorb the partially digested organic materials.

People are probably most familiar with this group due to mushrooms, which are the reproductive structures of terrestrial fungi. Yeasts are important to the fermenting industries, and some antibiotics, including penicillin, are produced by fungi.

Protista

The kingdom Protista is a large and diverse one. While this group includes primarily unicellular organisms, it also includes the **kelps**, which are multicellular and reach large sizes. The protists include **algae**, which are photosynthetic autotrophs, and the **protozoa** ("first animals"), which are unicellular heterotrophs. While the protozoans are sometimes important as pathogens, we are going to pay most of our attention to algal protists. The five important algal divisions are **Chlorophyta, Chrysophyta, Pyrrophyta, Phaeophyta** and **Rhodophyta.** These divisions are somewhat easier to remember when you realize that their root meanings refer to different colors.

Chlorophyta means "green plant," and members of this division are commonly called **green algae.** The division includes both unicellular and multicellular plants. The unicellular ones are typically flagellated phytoplankton. Green algae are most common in freshwater, and some live in terrestrial habitats. Among the marine species, some are gathered for food. There are approximately 7,000 species of chlorophytes.

Chlorophytes have chlorophyll a and chlorophyll b within their chloroplasts, and they are thought to have been the progenitors of land plants. (The different forms of chlorophyll allow botanists to understand the affinities between different groups of plants.) Also like the land plants, chlorophytes have cell walls made with **cellulose**, and they store their food reserves as starch. The reproductive cycles of green algae can be complicated and may involve both **asexual** (without sexual recombination) and **sexual** modes. Sexual reproduction usually involves **fertilization** by the union of **haploid gametes** (eggs and sperm). Haploid means only one set of chromosomes are carried, so when fertilization occurs, the two sets of chromosomes combine to make a **diploid zygote.** Sexual reproduction of vertebrate animals also involves union of haploid gametes. However, the variability in reproductive patterns is far greater among green algae than vertebrates.

Chlorophyta is one of the three algal divisions that include **seaweeds.** Seaweeds are simply marine algae that are multicellular and **macroscopic,** or big enough to see easily with the unaided eye. Macroscopic plants are also sometimes called **macrophytes.** The chlorophytic seaweed that you are probably most familiar with is sea lettuce (genus *Ulva*). It is a bright green plant with a single, very thin blade perhaps as long and wide as a lettuce leaf. Algae don't have leaves like land plants; their parts are called **thalli.** *Ulva* is only two cell diameters thick and is sometimes used as a condiment in salads and soups. Another green alga, *Enteromorpha*, is shaped like fine tubes with walls only one cell thick (*Enteromorpha* means "gut-shaped"). Both *Enteromorpha* and *Ulva* are found in the **intertidal** zone, or that area of the beach that is alternately submerged and exposed by the tides (Figure 4.2).

Green algae are often found living in association with other organisms. Perhaps the best examples of this arrangement are the lichens. Lichens are actually composite organisms with one part being a green alga and the other a fungus. Such a biological relationship, where two species live in close association, is called **symbiosis.** There are three patterns of symbiosis—**mutualism, commensalism** and **parasitism.** The lichen example represents mutualism, where both species benefit. Green algae are also mutualistic symbionts within sea anemones, corals, giant clams and other animals, where the algae provide some of the products of photosynthesis to the animals and the

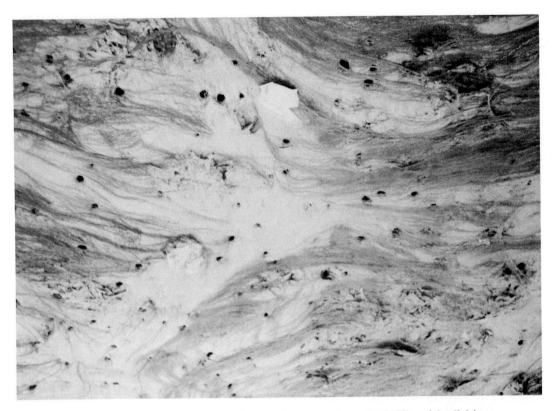

Figure 4.2. A sandy intertidal area showing filamentous *Enteromorpha* and leafy *Ulva* of the division Chlorophyta. The holes in the substrate are the burrows of ghost shrimp. (Courtesy of Ken Chew.)

animals provide refuge and nutrients to the algae. Parasitic symbiosis occurs among the **pathogenic** bacteria and fungi that attack living organisms and cause disease. In parasitism, the parasitic species benefits and the host species is harmed. The situation where one species benefits and the other is neither helped nor harmed by the association is called commensalism.

The division Chrysophyta means ''golden plants.'' These are unicellular members of the phytoplankton and form the base of the world's largest food chains. This division includes the **diatoms**. Since most of the earth is covered with water, the diatoms and other phytoplankton play the greatest role in converting some of the energy in sunlight into organic compounds. Diatoms store their food reserves as oil droplets. Sometimes, when zooplankton are feeding heavily on dense concentrations, or blooms of diatoms, enough of these oil droplets are released to form a natural oil ''slick'' at the surface.

Chrysophytes owe their coloration to a combination of pigments, including chlorophyll *a* and fucoxanthin. There are about 12,000 species of chrysophytes, and nearly 10,000 of them are diatoms. Most are microscopic, but what they lack in size they more than make up for in ecological importance. Some diatoms are shown in Figure 4.3. An interesting feature of diatoms is that they are encased within a close-fitting pair of **valves** impregnated with **silica** (SiO_2). Glass is made of the same substance, and the valves of many diatoms fit together much like the halves of a Petri dish or a pillbox. Diatomaceous earth, a sedimentary deposit used for industrial filtration and polishing compounds, is made of the remains of diatoms that piled up on the bottoms of lakes or seas. The presence of the rigid valves makes the process of asexual reproduction by cell division a bit more

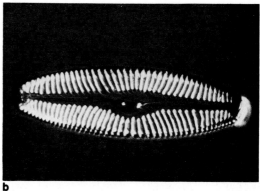

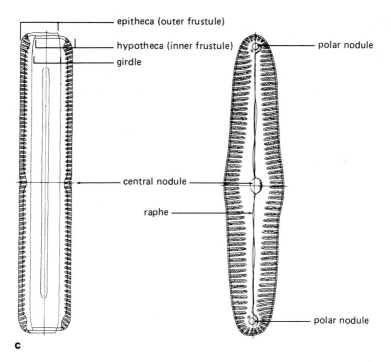

Figure 4.3. (a) A centric datom, division Chrysophyta. (b, c) Another diatom, showing the valve structure. (Courtesy of Ken Adkins and Kendall/Hunt.)

complicated for the diatoms than for other cells. Each new daughter cell takes one valve with it and then forms a new valve to fit the old one.

The next division, Pyrrophyta, translates as "fire plants," which refers to their orange-red color. Members of this division are also important constituents of phytoplankton communities, and like the diatoms, pyrrophytes are unicellular and small. The common name for this group is the **dinoflagellates.** The name of the division probably relates to the "red tides" that sometimes result from massive dinoflagellate blooms. There are about 1,100 species of dinoflagellates and most are marine. Each dinoflagellate has a transverse groove (Figure 4.4) with two flagella that allow the

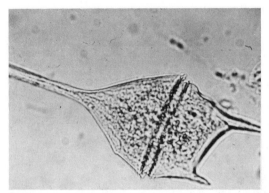

a

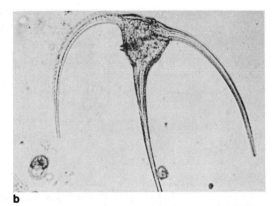

b

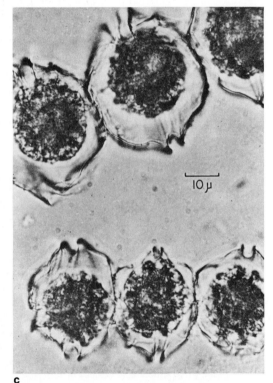

c

cells to rotate or move weakly in the water. Dinoflagellates often have bizarre shapes with heavy cellulose-containing cell walls and long spikes or other "armor." Presumably, the armoring of dinoflagellates and the siliceous valves of diatoms are adaptations that discourage zooplanktonic herbivores from feeding on them.

Dinoflagellates get their reddish color from a mixture of pigments known as xanthophylls, plus beta-carotene and chlorophylls *a* and *c*. Red tides have given dinoflagellates a notorious reputation in many coastal areas, and for good reason. Fish kills and even human mortalities have been caused by dinoflagellate blooms. A number of species produce **toxins,** or poisonous substances, that can harm other animals. In many cases these poisons are **neurotoxins,** which act directly on the nervous system. The toxins are synthesized by dinoflagellates and are then either stored within cells as an **endotoxin,** or released into the water as an **exotoxin.** The release of exotoxins is what causes widespread fish kills during blooms of certain dinoflagellates, such as *Gymnodinium breve,* which occurs on the Atlantic Coast and Gulf of Mexico.

Even massive blooms of nontoxic dinoflagellates can cause fish kills, particularly during the summer, if the bloom results in a loss of oxygen in the water. Along the Pacific Coast, from California to Alaska, another dinoflagellate, *Protogonyaulax catenella,* produces an endotoxin that affects vertebrate nervous systems and builds up in clams, mussels and other filter feeding shellfish that ingest them. When people eat

Figure 4.4. (a) A dinoflagellate, showing the transverse groove. (b) Another dinoflagellate armored with long spines. (c) *Protogonyaulax catenella,* a dinoflagellate that causes toxic red tides in the Pacific Ocean. (Courtesy of Ken Adkins and Ken Chew.)

the contaminated shellfish they may become ill or die. Local public health departments monitor shellfish beds and close off the harvests during toxic episodes. The association of toxic conditions with the visual appearance of a red tide is misleading. Some species of pyrrophytes cause red tides that are nontoxic, while other species produce toxic conditions without coloring the water. If you are unsure about the edibility of shellfish in your area, it is best not to take chances. Contact your public health department and ask if there is a **paralytic shellfish poisoning,** or **PSP** alert in the area. Persons who eat contaminated shellfish may notice a tingling of the lips and tongue, followed by progressive numbness in the extremities. In extreme cases, death results from cessation of breathing. There is no antidote for the toxin, known as **saxitoxin,** but those that survive the poisoning will recover fully. The toxin does not seem to harm shellfish.

Another interesting feature of dinoflagellates concerns the ability of some species, like *Noctiluca scintillans,* to produce their own light. This is known as **bioluminescence,** and many different types of organisms, including the familiar fireflies, have this ability. At night during the warmer months of the year, it is common to see bluish or greenish sparkles of light in the waves as they strike a beach, or near your hand as you draw it through the water. The light is produced by chemical reactions within the dinoflagellate cells as they are agitated. Scientists are unsure of the benefits of bioluminescence to planktonic organisms even though many kinds of phytoplankton and zooplankton display the response.

The division Phaeophyta includes kelps, the largest of the seaweeds. Phaeophyta means "dusky plant" and the common name for this group is **brown algae.** There are about 1,500 species of brown algae, and most are found in temperate, cold water marine habitats. Like the chrysophytes, brown algae contain the pigments chlorophyll *a* and *c* and fucoxanthin, but unlike the chrysophytes, phaeophytes are multicellular macrophytes. Phaeophytes store their food reserves chiefly as laminarin, a carbohydrate, and have cell walls that contain cellulose.

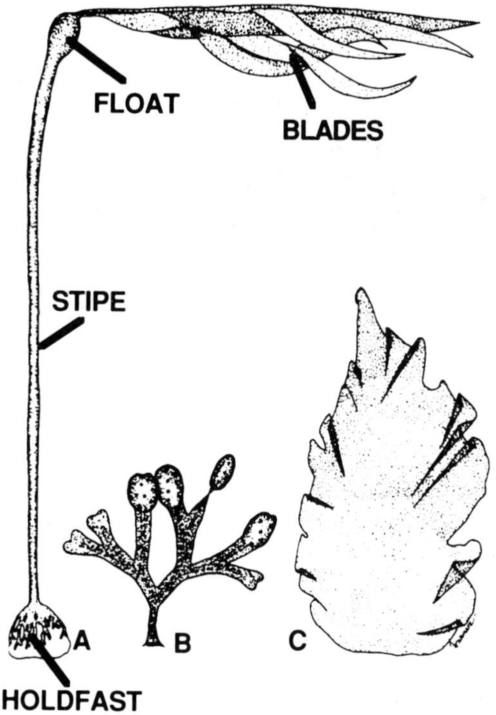

Brown algae that form **kelp beds** are the fastest-growing plants on earth. Some species can grow up to 2 feet in height per day and reach lengths well over 100 feet. These plants have an enormous ecological importance in that diverse communities of fish and shellfish are associated with the kelp beds in much the same sense as forest ecosystems and wildlife are associated on land.

The body parts, or thalli of the brown algae are often specialized into a **holdfast,** which provides anchorage to a rock or other hard substrate, a **stipe,** which extends like a stalk upward toward the surface where there is more light available, and **blades,** which are the primary sites of photosynthesis. In addition, there may be one or more **gas bladders** that act as floats to keep the blades in a good position to receive sunlight. Examples of "giant" kelps are *Macrocystis* and *Nereocystis* (Figure 4.5). Both of these genera are harvested. Kelps are usually harvested from the beds for the purpose of extracting thickening agents from them (e.g., algin). Some people eat *Nereocystis,* and pieces of the stipe can be pickled or candied. Kelp is also harvested in the North Pacific after herring have spawned on it and is

FLOAT

BLADES

STIPE

A   **B**   **C**

HOLDFAST

Figure 4.5. (a) *Nereocystis,* a giant kelp, division Phaeophyta. (b) *Fucus,* a rockweed, division Phaeophyta. (c) *Porphyra,* a red alga, division Rhodophyta. (Courtesy of Eric Warner.)

shipped eggs and all to Japan, where herring roe is a delicacy.

Another species, *Laminaria saccharina,* provides a seaweed product called **kombu.** Kombu is used in a wide variety of oriental soups, stews and prepared food products. Species of the genus *Laminaria* are smaller than the giant kelps. Brown algae of many kinds have also been used as feed supplements for livestock and as a fertilizing mulch for agriculture.

Most people who live near the seashore in temperate climates are familiar with **rockweed,** a common brown alga that grows attached to rocks in the upper intertidal zone. Rockweed is a smaller plant than the kelps and belongs to the genus *Fucus.* When the plant matures, the tips of each thallus become swollen and bumpy. The small bumps contain the gametes, either eggs or sperm. When the tide goes out and the gametes are ready for release, the plant dries out somewhat and, as the tide comes back in, the gametes are ejected into the water. Then a chemical substance released from the eggs begins to attract the sperm by **chemotaxis** (chemical attraction), and fertilization results.

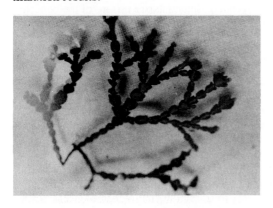

Figure 4.6. A filamentous coralline alga, division Rhodophyta. (Courtesy of Lynwood Smith.)

The last division, **Rhodophyta,** means "red plants," and this division includes the seaweeds commonly known as red algae. Rhodophyta is the most diverse group of seaweeds and includes about 4,000 species. Most red algae are marine, and they tend to grow in warmer waters than brown algae. The pigments that give them their distinctive colors are chlorophyll *a* and *d,* carotenoids and phycobilins, including the red pigment phycoerythrin. Red algae tend to be smaller macrophytes than the browns, and they also store their food products as carbohydrates (starch). Their cell walls have cellulose and are sometimes also impregnated with calcium carbonate ($CaCO_3$). Calcified red algae are called coralline algae (Figure 4.6). The growth patterns of red algae tend to fall into one of three categories: **filamentous** (branching), **membranous** (flat, with blades) or **encrusting** (flat against a hard substrate).

The uses of red algae rival the uses of kelp in importance and character. Thickening agents, including agar and carrageenin, are extracted from red algae. Agar is extracted from various genera including *Gracilaria* and *Gelidium.* Carrageenin is extracted from Irish moss, *Chondrus crispus,* which grows on both sides of the North Atlantic. Red algae of many species are also used directly as food. The most popular genus is *Porphyra,* which is used to make **nori.** Nori is the most widely used of the edible seaweeds, and it is often used in soups, crackers and sushi. Nori can also be dried, seasoned and eaten alone. Demand for nori has made *Porphyra* a valuable resource, and aquaculture of *Porphyra* is common, particularly in Japan, to meet that demand.

The various divisions of algae are sometimes grouped within the plant kingdom rather than with the protists. Under the classification system used here, the plant kingdom is reserved for the land plants.

Plantae

The land plants and rooted aquatic plants are sometimes considered under a single division, **Tracheophyta,** which means "windpipe plants," or **vascular** plants. Otherwise, the kingdom is grouped into ten separate divisions. We will keep it simple and consider them together. There are 248,400 species of tracheophytes.

Land plants evolved and radiated over roughly the same geological time period as fishes, beginning about 500 million years ago. Since that time, a few have taken up an aquatic existence while retaining most of the major features of the land plants. The larger algae, the seaweeds, make good use of the water that surrounds them for structural support, supply of nutrients and dispersal of reproductive stages. Land plants had to solve these problems differently, and they developed woody components and roots for structural support, a vascular system of internal tubes for transport of water and nutrients, and flowers and seeds for reproductive dispersal. The tracheophytes that have returned to the water have vascularization (**xylem** and **phloem**), roots, flowers and seeds, and these plants serve important ecological roles in nearshore marine and freshwater areas. Aquatic tracheophytes are much more common in freshwater environments than they are in marine environments. They frequently occur in shallow areas of lakes, marshes and wetlands, and along riverbanks. Many fisheries depend upon the productivity and presence of these plants. A number of fishery stocks, including salmon and shrimp, also depend upon vascular plant communities because the juvenile stages spend critical periods foraging and finding refuge among the plants.

An example of a marine tracheophyte is eelgrass, *Zostera marina* (Figure 4.7). Eelgrass and other marine grasses provide very important shallow water refuges for fish and shellfish, particularly those that spend some of their time as juveniles within **estuaries**. Estuaries are nearshore areas where freshwater from rivers meets the ocean (considered in Chapter 10). Besides the advantages that accrue from their physical presence in creating habitats and refuges, and stabilizing the substrate with their roots, these plants also introduce a great deal of food into the ecosystem. They typically grow up from buried roots during the warmer months and then die back or get swept away during winter storms. The remains of these plants then break down into tiny pieces and begin to

Figure 4.7. Eelgrass, *Zostera marina,* division Tracheophyta, shown in the upper left. (Courtesy of Ken Chew.)

decompose, to become an important food source known as **detritus**. Tracheophytes are important sources of detritus in rivers, lakes, estuaries and other nearshore waters having sand or mud substrates. Phaeophytes and other algae provide a similar function in exposed coastal areas that usually have rocky or hard substrates. Both sources of detritus are critical elements of food chains that support important fisheries.

The last kingdom we will consider, the animals, is covered in the following chapter. A summary of some of the major features of organisms in the other four kingdoms is presented in Table 4.2.

Table 4.2. Summary of general features of the major taxonomic groups.

Taxonomic group	Common names	Examples	Cellular features	Macroscopic features	Mode of Nutrition	Importance
Kingdom Monera	Bacteria	*Escherichia coli*, an intestinal symbiont in humans	Prokaryotes—no organelles, cell walls with non-cellulose polysaccharide some with flagella	(Unicellular)	Heterotrophic (some chemosynthetic or photosynthetic autotrophs)	Decomposers, pathogens, digestive symbionts
	Blue-green algae	*Oscillatoria*, an occasional nuisance in freshwater	Prokaryotes, *chlorophyll a* (not bound in chloroplasts), cell walls	Unicellular and filament-forming	Photosynthetic	Nitrogen fixation, fouling of water supplies and noxious blooms
Kingdom Fungi	Fungi, mushrooms, yeast	*Saprolegnia*, a fish pathogen	Eukaryotes, filamentous hyphae, cell walls with chitin	Multicellular, hyphae form a mycelium	Heterotrophic, absorb dissolved substances, often associated with plant	Decomposers, pathogens, fermentation
Kingdom Protista						
Division Chlorophyta	Green algae	*Ulva*, sea lettuce; *Enteromorpha*	Eukaryotes, *chlorophyll a* and *b* within chloroplasts, cell walls with cellulose, many with flagella	Unicellular and multicellular (seaweeds)	Photosynthetic	Some are harvested for food, many are symbionts with other organisms
Division Chrysophyta	Diatoms, golden-brown algae	*Coscinodiscus*, a centric diatom	Eukaryotes, siliceous capsules, *chlorophyll a* and *c*, fucoxanthin	(Unicellular phytoplankton)	Photosynthetic	Base of aquatic food chains

Table 4.2.—*Continued*

Taxonomic group	Common names	Examples	Cellular features	Macroscopic features	Mode of Nutrition	Importance
Division Pyrrophyta	Dinoflagellates	*Protogonyaulax catenella*, causes toxic "red tides"	Eukaryotes, two flagella in transverse groove, armored, *chlorophyll a* and *c*, carotenoids	(Unicellular phytoplankton)	Photosynthetic	Aquatic food chains, noxious blooms, toxic red tides
Division Phaeophyta	Brown algae, kelps	*Laminaria*, source of edible kombu; *Nereocystis*, bull kelp	Eukaryotes, cell walls with cellulose and algin, *chlorophyll a, and c,* fucoxanthin	Multicellular, kelps have holdfast, stipe and blades; some have gas bladder floats	Photosynthetic	Kelp bed ecosystems, source of industrial chemicals, harvested as food
Division Rhodophyta	Red algae	*Porphyra*, source of nori; *Chondrus crispus*, source of carrageenin	Eukaryotes, cell walls with cellulose, some with calcium carbonate, *chlorophyll a* and *d,* phycoerythrin	Multicellular seaweeds	Photosynthetic	Harvested as food, source of industrial chemicals
Kingdom Plantae Division Tracheophyta	Vascular plants, land plants	*Zostera marina* eelgrass	Eukaryotes, *chlorophyll a* and *b,*	Multicellular, roots, stems, leaves, flowers seeds, vascular system	Photosynthetic	Sea grass ecosystems

5
Forms of Aquatic Life II

Frederick G. Johnson

Introduction

The kingdom Animalia includes over 1 million described species divided among 30 phyla. We will only concern ourselves with six phyla, as indicated in Table 4.1. Three phyla support the majority of the world's fisheries. The first two, **Mollusca** and **Arthropoda**, include the shellfishes. The last phylum, **Chordata**, includes the fishes and is the only phylum that includes vertebrate animals; the other phyla include only invertebrates.

Members of the kingdom Animalia are multicellular heterotrophs. That definition alone separates them from all of the other kingdoms except Fungi, but animals and fungi differ in their means of obtaining nutrition. We learned earlier that fungi take up dissolved organic substances from their environment, often after secreting digestive enzymes to break these substances down. Animals, for the most part, tend to ingest their food in whole pieces, store it in a body cavity and digest it later. In other words animals "eat" their food while fungi "soak up" theirs. Also, since the food does not often go to the animals, the animals must go to the food. Consequently, most animals have muscle tissues and body structures that give them mobility and the ability to capture or gather food. Animals have cells of the eukaryotic type, but their cells lack cell walls and chloroplasts. Animal cells contain mitochondria and cell nuclei with chromosomes, as well as other organelles. The animals are thought to have evolved from protozoan members of the kingdom Protista. In fact, the structure of sponges is very similar to a simple aggregation of protozoan cells.

Figure 5.1. A marine sponge, phylum Porifera. The larger openings are for water outflow, while the tiny openings on the sides are incurrent openings. (Courtesy of Ken Adkins.)

Porifera

The phylum Porifera includes the sponges, of which there are about 5,000 species. All but 150 of the species are marine; the rest are found in freshwater. The phylum is divided into four classes, and one of those, the Demospongiae, includes sponges that are harvested for human use. The skeletons of the sponges are used after the living tissues have been dried and cleaned away. Fibers made of spongin, a proteinaceous material, form the flexible portion of the skeleton. Another part of the skeleton consists of needle-like spicules of silica, and these must be removed before the 'natural' sponges are used.

Sponges are very simple in structure (Figure 5.1). Their bodies consist of a skeleton and only a few different kinds of cells. There are no tissues, organs, or organ systems. Sponges are **sessile**, meaning they live attached to a substrate and don't move, apart from the dispersal of their planktonic gametes. The cells act together in a coordinated fashion without centralized control by a nervous system. If you separate the cells of a sponge by passing it through a fine-mesh sieve, the cells will reassort themselves on their own and form tiny new sponges.

Sponges feed on tiny particles of food in the water and, therefore, are **filter feeders**. Flagellated cells within the sponge create water currents that bring water and food in through small openings and then pass the water out through larger openings after the food has been removed. Some sponges are conspicuous in coral reefs and form associations with symbiotic algae that live inside them. Sponges are not eaten by people.

Coelenterata

The next phylum, **Coelenterata** (the C in Coelenterata is pronounced as an s, and the o is silent), includes jellyfishes, corals and sea anemones. A few of the jellyfishes are eaten in a prepared state as food, primarily in the Orient. There are about 9,000 species of coelenterates, and most are marine. **Hydra** is a familiar, though inedible, freshwater example.

Coelenterates have a body plan known as **radial symmetry**. That means you can take a jellyfish, for example, lay it down flat and slice it through the middle at any angle. No matter which way you slice it, both halves will look the same. If you try that with a human you will get into a lot of trouble and besides that, only one slice will yield mirror-image halves. Humans, fishes and most shellfishes have **bilateral symmetry**, but the bodies of coelenterates radiate outward from a central point like the spokes of a wheel. Individual coelenterates are either sessile, as **polyps**, or free-living, as **medusae** (jellyfishes, Figure 5.2). In either case their bodies are structured in much the same way.

In some textbooks, the phylum coelenterata is called **Cnidaria** (the C is silent in Cnidaria). An interesting feature of this group is that nearly all are predators, but none have any teeth or other movable hard parts, and those species that can move at all do so only very feebly. Jellyfishes are planktonic, while sea anemones and corals are sessile. The distinguishing feature of the coelenterates that allows them to be carnivores is that they have specially modified organs called **nematocysts**. Nematocysts are tiny "stinging capsules" that are formed by specialized cells known as **cnidocytes**. Nematocysts appear on the tentacles of coelenterates and on other structures associated with feeding or defense. There is a microscopic trigger on each nematocyst that causes the discharge of a harpoon-like structure and venom when a tentacle comes into contact with prey (Figure 5.3). One nematocyst is produced by each cnidocyte. Each nematocyst is used only once, and then more are produced by new cnidocytes. Many nematocysts will be discharged during a typical feeding encounter. Some coelenterates like fire coral, the Portuguese Man-of-War, sea nettles and others have nematocysts that are potent enough to be dangerous to swimmers. Other species can be handled by people without harm because the skin of our hands is too thick for the nematocysts to reach our sensitive nerve endings. Never try to eat a coelenterate raw, however, because the skin of your tongue and lips is thin enough to be penetrated.

The importance of coelenterates as human food is minor at best. Jellyfishes are probably eaten more for their unusual texture and appearance than for any other reason. Coelenterates have considerable ecological importance in tropical marine habitats, however, because of their occurrence in coral reefs. Coral reefs are the most diverse marine communities, despite a general lack of dissolved nutrients necessary for the growth and productivity of phytoplankton and seaweeds (Figure 5.4). The lack of phytoplankton makes the water clear and leads to a scarcity of zooplankton, the normal prey of the corals. The corals in tropical reefs owe their considerable productivity to the presence of symbiotic green algae within their tissues, which provide the corals with nutrition. Coral reefs, islands and atolls are formed from the accretion of coral skeletons made of calcium carbonate. These

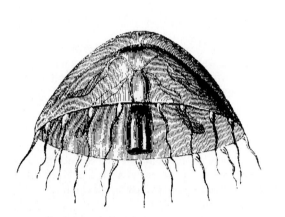

Figure 5.2. A medusa, phylum Coelenterata. Medusae are free-floating members of the planktonic community.

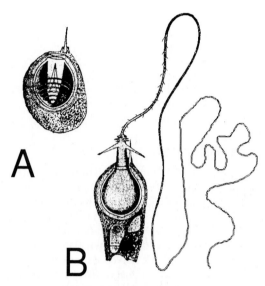

Figure 5.3. (a) An undischarged nematocyst showing a trigger at the top and the coiled thread inside. (b) A nematocyst after it has discharged.

communities would not be present without the mutualistic association between the corals and the algae.

Mollusca

The phylum Mollusca includes about 100,000 species of soft-bodied invertebrates that are usually protected by one or more hard shells. The root word *mollus* means soft in Latin. These animals, commonly called molluscs (mollusks), are widely distributed in freshwater, marine and terrestrial habitats. The shells are composed of calcium carbonate and a proteinaceous component called the **periostracum**, both of which are secreted by a flap of tissue called the **mantle**. Molluscs have three general body regions—the **head**, with sensory and feeding organs, the **foot**, a muscular organ for attachment or movement and a **visceral** hump that contains organs of the digestive and reproductive systems. Most molluscs have gills called **ctenidia** (the c is silent) that lie within a cavity between the body and the mantle. This cavity is called the **mantle cavity. Cilia**, the microscopic structures on the ctenidia that are similar to flagella, but are smaller and more numerous, beat in unison to create currents that draw water through the mantle cavity. Three classes of molluscs currently support fisheries.

Gastropoda

The first class, **Gastropoda**, means ''stomach foot.'' The term refers to the tendency of these animals to lie ''belly-side'' down on the substrate, using the muscular foot to adhere and creep slowly along. The gastropods include snails, slugs, abalones, limpets and **nudibranchs** (sea slugs).

Slugs and nudibranchs have no shells, but the rest of the gastropods have a single shell. Gastropods have some interesting variations in body symmetry. Molluscs in general are bilaterally symmetrical, and so are gastropods when they begin life as larvae. Later, during one of the larval stages, gastropods undergo a transformation called **torsion**, where some parts of the body twist

Figure 5.4. A coral reef community in the Caribbean Sea showing many corals but no seaweeds. The branching structures are coelenterates called gorgonians. Polyps of the gorgonians on the right side of the photo are visible. The fish in the center is a squirrel fish.

around into new positions, 180° from where they had been, and the bilateral symmetry is lost. A further variation in body form is produced by the coiled growth patterns of some of the gastropods. The larval stages include the **trochophore** and **veliger** stages (Figure 5.6). The larvae ordinarily lead a planktonic existence, while most of the adults are benthic.

The part of a gastropod that is eaten as food by humans is the foot, which can be very valuable. Species of abalone (*Haliotis* spp.), for example, have a large muscular foot that is highly prized as food (Figure 5.5). Gastropods tend to grow rather slowly, and they are easily overfished. These factors, together with their delicious flavor, combine to keep market prices high. Sometimes the flesh of gastropods is tough, so it needs to be pounded to break up the muscle fibers during preparation. (Terrestrial gastropods like *Helix* are also eaten as escargot, but they are not products of a fishery.) Most of the gastropods eat plants and are, therefore, **herbivores**.

Gastropods feed with a unique rasping organ called a **radula**. The radula consists of a fairly long belt-like structure across which rows of hooked teeth are arranged. The teeth are very hard and are sometimes impregnated with iron compounds to make them even harder. You can pick up the radula of some species with a magnet, or if the animal is intact, it will often turn toward a magnet that is placed near it. The radula is designed to scrape food from surfaces and pass the food on to the digestive tract. As the radula becomes worn away at the working end near the mouth, it is continuously replaced by growth from within. Limpets use the radula to scrape algal films from rocks,

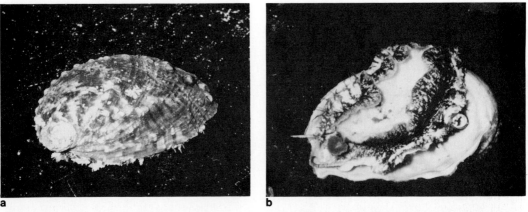

a   b

Figure 5.5. (a) Top view of an abalone, *Haliotis kamtschatkana,* showing the single shell. (b) View of the underside, showing the muscular foot, and at the left, the mouth and the head. (Courtesy of Lynwood Smith.)

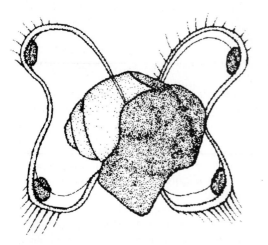

Figure 5.6. A planktonic veliger larva of a gastropod.

while abalone rasp away at kelp fronds and other seaweeds. Some gastropods are meat eaters (carnivores) and drill their way through the shells of other molluscs to eat them. Others, like tropical snails of the genus *Conus,* are able to capture fishes and can even kill humans with an eversible feeding structure that shoots out and injects venom. Some gastropods are filter feeders. There is a wide diversity of forms, habits and habitats within this group.

An ecological generalization that is usually safe to make is this—if an organism goes to the trouble of producing and protecting itself with a hard shell, it probably tastes good. That is certainly true of the gastropods. The nudibranchs, which live in the presence of predators but lack shells, either have bad-tasting flesh or protect themselves with undigested nematocysts obtained from their coelenterate prey (Figure 5.7). Abalones, due to their high value, are now produced by aquaculture as a supplement to wild harvests. Other gastropod snails of various kinds are gathered from the wild for food, or by shell collectors.

Bivalvia

The next class, **Bivalvia**, represents a group of organisms that has long been grown through aquacultural practices throughout the world. The term bivalve refers to the two shells these animals have. The shells generally cover the whole animal and they are joined by a flexible **hinge** that is made of the same substance that constitutes the outer shell layer, the periostracum. Many bivalves have **siphons** that can extend beyond the opening or gape of the shells (Figure 5.8). The shells or valves can be closed by the contraction of the adductor muscles. If you have eaten scallops you are familiar with adductor muscles, as they are usually the only part of a scallop that is eaten. Scallops have a single adductor, but many bivalves have two (Figure 5.9). Bivalves also include clams, mussels, oysters and cockles. Members of this class occur in both marine and freshwater habitats. Some species, particularly the giant clam, *Tridacna,* reach large sizes and attain weights in the hundreds of pounds.

Bivalves are filter feeders and draw water into their mantle cavity by the ciliary action of the ctenidia. The ctenidia also help remove food, mostly phytoplankton, from the water. The cilia sort the food and pass it toward the mouth. It is ironic that the largest bivalves, the giant clams, live in habitats having the lowest abundance of phytoplankton. Giant clams occur in tropical reefs and their growth is largely due to symbiotic algae within their mantle. The giant clams expose the mantle tissue to sunlight and receive food products from the photosynthesis that results.

While some bore into rock or wood, most clams live buried within soft substrates like sand or mud and use their foot to dig. They extend their siphons to the substrate surface to draw in water for feeding and respiration. Another name for the class Bivalvia is **Pelecypoda**, which means "hatchet-foot," and refers to the unique shape of the muscular foot that is adapted for burrowing. Most bivalves are filter feeders, and that is why this group is a major source of paralytic shellfish poisoning derived from toxic dinoflagellates. One exception to the filter feeding mode is an obscure group of predatory clams called septibranchs, not the giant clams. Rumors that giant clams clamp down on divers or swimmers that step into them and eat their legs are untrue. The septibranchs use their siphons like a "slurp gun" to ingest unsuspecting crustaceans that get too close by creating sudden inward jets of water. Other bivalves use their siphon like a suction hose to bring in materials that have already settled on the bottom, mostly organic detritus. *Macoma* is an example of a clam that is a deposit feeder rather than a filter feeder.

Figure 5.7. A nudibranch, a gastropod which has no shell. The feathery structures are tipped with nematocysts obtained from coelenterate prey. (Courtesy of Ken Adkins.)

Figure 5.8. Examples of bivalves with long siphons. The two on the left are geoducks, *Panope generosa,* and on the right are horse clams, *Tresus* sp. (Courtesy of Ken Chew.)

Figure 5.9. A rock scallop, *Hinnites,* with one valve removed, showing the white adductor muscle in the center. (Courtesy of Ken Chew.)

a   b

Figure 5.10. (a) Mussels of the genus *Mytilus,* an important edible bivalve. (b) A mussel with one valve removed showing the byssus in the lower center. (Courtesy of Ken Chew.)

A number of bivalves live attached to substrates instead of buried within them. These include mussels, which form **byssal threads** for attachment to rocks, pilings, culture ropes, or to each other. Byssal threads are made by the extrusion of a liquid substance through a groove in the foot. The thread soon hardens into a tough but flexible fiber as the foot is removed from the spot where it placed the thread. The byssal threads form the **byssus,** the so-called "beard" of the mussels, which is removed before the mussels are eaten (Figure 5.10). Oysters and some scallops cement one of their shells to a rock or other hard surface, leaving them sessile as adults. They disperse to new areas through their larvae, which change from veligers to **spat** before selecting a place to settle (Figure 5.11). Other species of scallops are free-living and can "swim" away from predators by clapping their shells together like castanets.

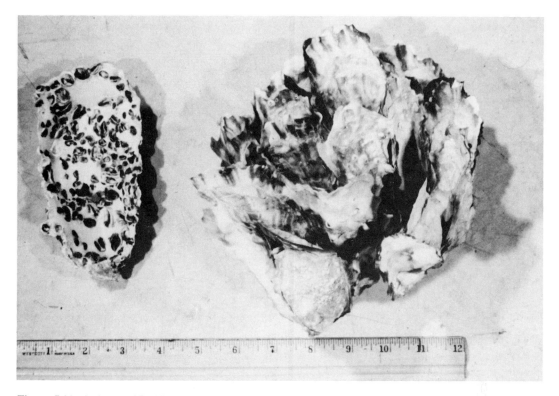

Figure 5.11. A cluster of Pacific oysters, *Crassostrea gigas,* resulting from the settlement of spat on a single mother shell. A piece of mother shell with over 100 spat is shown on the left. (Courtesy of Ken Chew.)

Some bivalves cause significant economic damage. Foremost among them are the **ship-worms,** which are actually clams that bore into wooden boat hulls, pilings and bulkheads (Figure 5.12). *Teredo* is an example of a shipworm genus. People have been trying to find ways to poison or discourage the settlement of these animals for years, but without much success. The many existing **antifouling** measures have yet to include an effective means to kill only the **fouling organisms** and not other aquatic species. This is a particularly serious problem with marine aquaculture systems. Aquaculture structures, cages, ropes and nets must be kept free of the many kinds of fouling organisms in order to protect the well-being of harvestable ones. Even if an antifouling poison is found that is selective enough in its toxic action to leave the target species unharmed, the possibility still remains that the substance will be present in the seafood products consumed by humans.

The following is a list of some of the most important bivalve shellfishes:

- *Crassostrea gigas,* the Pacific oyster (Figure 5.13). It is produced by various aquacultural practices. It originated in Asia and is now grown in the United States, Canada and Europe. Pacific oyster larvae for stocking the oyster beds are now produced by hatcheries. Where introduced, Pacific oysters have displaced many native oyster species in commercial importance.
- *Ostrea edulis,* the European oyster. This oyster has now been displaced by *C. gigas.*
- *Crassostrea virginica,* the American oyster. This oyster is widely harvested along the Atlantic and Gulf Coasts of the United States.

Figure 5.12. Pieces of wood showing damage caused by shipworms. (Courtesy of Ken Chew.)

- *Mytilus edulis,* the bay mussel. This mussel is commonly grown by hanging culture techniques in countries bordering the North Atlantic and North Pacific oceans.
- *Mya arenaria,* the Eastern softshell clam. This species has been introduced on the Pacific coast and is harvested in the wild by hydraulic escalator dredges.
- *Mercenaria mercenaria,* the northern quahog. This is a hard-shell clam of the North Atlantic, and it is now produced with the aid of hatcheries.
- *Tapes philippinarum (Venerupis japonica)* the Manila clam (Figure 5.14). Also a hard-shell clam, it is originally from Asia but was introduced on the west coast of North America. This clam is now grown through aquaculture.
- *Saxidomus giganteus,* the butter clam. This is a native west coast species that is harvested commercially in the wild.
- *Panope generosa,* the geoduck. This large clam is harvested by hand by divers assisted with water jets and is found along the west coast of North America. It burrows to a depth of 3 feet.
- *Siliqua patula,* the razor clam (Figure 5.15). This clam is a very rapid burrower and supports a popular recreational fishery off the west coast of North America.
- *Spisula,* a genus of surf clams. These are harvested in commercial quantities off the Atlantic coast of North America.
- *Pecten* spp., along with other genera (Figure 5.16). These are scallops, most of which are harvested in the wild by boats dragging gear over the bottom. However, some are grown by hanging line and cage culture.
- *Pinctada* spp., pearl oysters. These are also grown by suspended culture in Japan, but for jewelry rather than food. Freshwater bivalves are also used for pearl culture.

Figure 5.13. Internal view of a Pacific oyster. The flaps of tissue in the lower center are the gills, or ctenidia. The darker flap of tissue around the margin is the mantle. The large light colored mass at the upper right is gonadal tissue. (Courtesy of Ken Chew.)

Figure 5.14. Hard-shell clams of the Pacific. All but the clam at the lower left, with foot and siphon exposed, are Manila clams (*Tapes philippinarum*). The clam that is preparing to bury itself with its extended foot is *Protothaca staminea*. Both clams are commonly called littleneck or steamer clams. (Courtesy of Ken Adkins.)

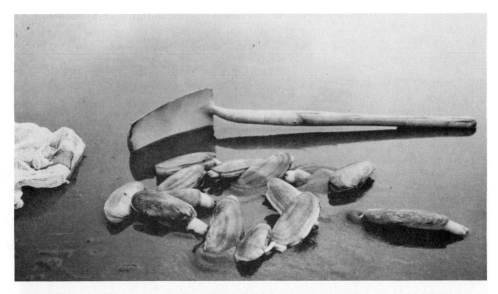

Figure 5.15. Razor clams, *Siliqua patula*, provide important recreational opportunities along the U.S. Pacific Coast. Razor clams are rapid burrowers and provide exciting sport. (Courtesy of Ken Chew.)

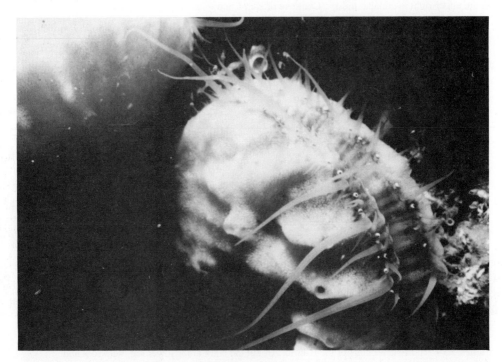

Figure 5.16. This scallop is encrusted with a sponge and shows the sensory tentacles and eyes at its mantle margin. (Courtesy of Ken Adkins.)

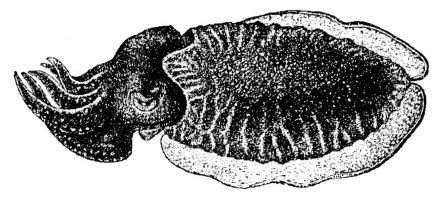

Figure 5.17. The cuttlefish, *Sepia,* belongs to the class Cephalopoda.

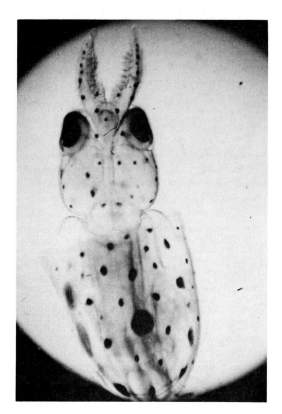

Cephalopoda

The last class of commercially important molluscs is **Cephalopoda**, which translates as "head-foot." Cephalopods are highly modified for active predation and have excellent eyes, tentacles with sucking discs and a modified radula that serves as a "beak" of sorts, all located around the head area. Cephalopods are carnivores and are exclusively marine. The cephalopod body plan is in contrast to that of the filter feeding bivalves, which have less centralization of function

Figure 5.18. A common market squid, Loligo, shown just after hatching from the egg case. The dark spots are chromatophores that allow the animal to change color. (Courtesy of Lynwood Smith.)

Figure 5.19. A SCUBA diver behind a large octopus, class Cephalopoda. (Courtesy of William High.)

around the head and are therefore considered to be **decephalized.**

The shell may be reduced or absent among the cephalopods. *Nautilus* has a full, coiled external shell. Cuttlefishes have an internal shell. *Sepia* is an example of a cuttlefish genus (Figure 5.17). The common market squid, *Loligo* (Figure 5.18), has a reduced internal shell that looks like a clear, flexible piece of plastic, and this structure is called a **pen.** The pen is made of the same substance that forms the periostracum. *Octopus* has no shell (Figure 5.19). It is easy to remember that squid have pens if you also remember that squid, along with other cephalopods, have **ink sacs.** The ink is discharged to confuse predators, such as eels, by producing a dark cloud as a visual screen and interfering with the predator's sense of smell as well.

Cephalopods are able to move rapidly in pursuit of their prey or away from predators. The cephalopod foot is highly modified as a muscular funnel that directs water jets for swimming or sudden darting movements. Water that enters the funnel is ejected from the mantle cavity.

The tentacles are muscular and have excellent **tactile** abilities (the ability to discriminate by the sense of touch). Octopoids have eight tentacles, squids have ten and *Nautilus* has over 90. The tentacles are arranged in a ring around the head. The nervous system and sensory organs of cephalopods are the most elaborate of all the invertebrates, and the abilities they confer to this group rival those of vertebrates. Cephalopods have an unusual ability to perceive and react to their perceptions.

Cephalopods also have a remarkable ability to change the color of their bodies. They can use this ability for camouflage or to signal each other. The color changes are produced by very rapid expansions or contractions of pigment-filled skin cells called **chromatophores.** It is believed that a cuttlefish, for example, is able to send visual messages to more than one other cuttlefish at a time by flashing chromatophore patterns across different parts of its body. Octopi not only change the colors of their skin, but can change their textures as well to blend even more "artfully" into their backgrounds.

All of the cephalopod genera mentioned so far are good to eat. Substantial fisheries exist for squids, octopi and cuttlefishes in various parts of the world. Some of these animals get very large and are, perhaps, the largest of invertebrate animals. Giant squid of the genus *Architeuthis* have been found with a 12-foot body and 60-foot tentacles, weighing about two tons. Giant squid live deep in the ocean and are preyed upon by sperm whales. Biologists speculate that giant squid reach lengths well in excess of 70 feet, judging from the size of scars left on whales by squid tentacles.

Arthropoda

The phylum **Arthropoda** is the most diverse one, encompassing over 800,000 living species, including insects, crustaceans, spiders and mites. The phylum is subdivided into countless subgroups of bilaterally symmetric organisms having jointed appendages, segmented bodies and **exoskeletons.** An exoskeleton is a skeleton that covers the outside of the body. Humans and fish have **endoskeletons,** which are located internally. (An octopus has no skeleton at all and can modify its body form by the action of its muscles and by moving its body fluids from one region to another. That is why a large octopus can slip through a very small opening.) Arthropod exoskeletons cover the entire body surface and a portion of the lining of the digestive tract. There is a complete lack of cilia among the arthropods, which is unusual for a major animal phylum. The extensive exoskeleton may be the reason for the lack of ciliated surfaces.

The exoskeleton offers both advantages and disadvantages to the arthropods. Obviously, the advantages are most significant, since the arthropods are clearly the most ecologically successful group of animals. Arthropods occur in nearly every habitat where animal life is possible. One of the advantages of the exoskeleton is that it provides an effective barrier between the body and the environment, thus it can "seal" the body off from its environment to a certain extent. This can be par-

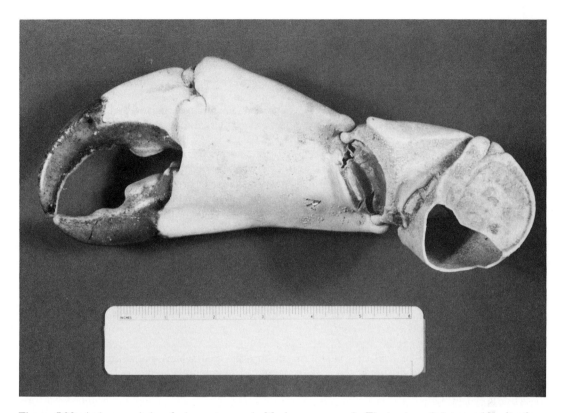

Figure 5.20. A claw, or chela, of a large stone crab, *Menippe mercenaria.* The hard exoskeleton and its sites for muscle attachment give the claw sufficient power to crush prey. In the stone crab fishery, the animals are released alive after the claws are removed. The claw shown weighed over one pound. (Courtesy of Ken Adkins.)

ticularly valuable for resisting water loss, or **desiccation**, and partially explains why the insects have been so successful at colonizing terrestrial habitats.

Arthropod exoskeletons are composed principally of **chitin** and proteins, and many are hardened with calcium. If the exoskeleton is hardened, it offers greater protection from predators. Mobility is provided by jointed appendages and flexibility is provided by joints in the exoskeleton between the body segments. The range and power of movement that the exoskeleton makes possible are also distinct advantages. The inner surface of an exoskeleton provides sites for muscle attachment that allow for a wide range of movements, including those that involve a considerable amount of power, speed, or leverage. The appendages have special internal muscle attachment sites called **apodemes** that further enhance the range of muscular movements. These are the white flaps of material you find when you crack a crab leg or claw and pull the meat away. Crabs with larger claws (Figure 5.20) develop enough crushing power to crack the shells of their bivalve prey (or damage the fingers of unwary humans).

A major disadvantage of the arthropod exoskeleton is that it restricts growth. In order for an arthropod to grow it must shed its old exoskeleton and form a new, larger one. This process is called **molting**, or **ecdysis**. Sometimes large numbers of molted exoskeletons (**exuvia**) wash up on the beach, giving the erroneous impression that a shellfish kill has occurred. Molting is a complicated process that involves replacement of all the body coverings—the gill linings, the surfaces of the

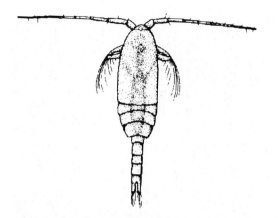

Figure 5.21. A copepod of the genus *Calanus*. These crustaceans are the earth's most numerous animals and are the major herbivores in planktonic food chains.

eyes, antennae and other sensory structures, and portions of the gut lining. Besides the delicate nature of the process itself, it also leaves the animals very vulnerable to attack. It takes some time for the new exoskeleton to harden and for mobility to be regained. For these reasons, molting generally takes place in a protected area. A hard exoskeleton also gets in the way of sexual intercourse. Female crabs, for example, mate just after they molt, and they are accompanied by a hard-shelled male. When a female gets ready to molt she releases a chemical substance in her urine that attracts the male, who remains and protects her while she is vulnerable.

Crustacea

The class **Crustacea** includes aquatic arthropods that feed with appendages called **mandibles**. This group includes crabs, lobsters, shrimps, crayfishes, barnacles, krill and copepods. There are about 25,000 species of crustaceans, and many support important fisheries. Others, like krill and copepods, occupy very important positions in aquatic food chains, but are not common in the human diet. **Krill** are planktonic shrimp-like crustaceans that can be very abundant in colder marine waters where whales go to feed upon them. **Copepods**, extremely abundant in most aquatic habitats, are planktonic herbivores and feed on diatoms and other phytoplankton. Much of the world's animal biomass consists of a single marine copepod genus—*Calanus* (Figure 5.21). An individual *Calanus* is rather small, about the size of a rice grain. It is said that if you took all the *Calanus* in the world and put them in a big pile and then took all the other animals on earth and made another pile, the *Calanus* pile would be the larger of the two. Planktonic crustaceans provide the major links in the very productive aquatic food chains between phytoplankton and fishes, and these arthropods are the most numerous of all animals.

Some of the planktonic crustaceans are larval stages of animals that will be benthic as adults. Crustaceans have diverse larval life histories, and most begin with the **nauplius** stage. Each successive molt is often accompanied by a **metamorphosis**, and each larval stage in the series may bear little resemblance to the adult (Figure 5.22). Often the adults have a number of body segments fused together under a rigid **carapace** that covers the head and thoracic regions. Among crabs, the abdomen is folded beneath the carapace, but the segments of the abdomen are not fused together. The abdomens of shrimps, crayfishes and lobsters extend behind the carapace. There will typically be one pair of appendages per body segment, whether the segments are fused or not (Figure 5.23). The appendages may be specialized as antennae, mouthparts, walking or swimming legs and claws. Large claws are called **chelae**. Other appendages are used as "gill bailers" to ventilate the gill chambers (located just underneath and to either side of the carapace). It is the beating of the gill bailers that makes a crab sound like it is "ticking" when you take it out of the water. The gill bailers are needed to ventilate the gills because the gills are not ciliated like the ctenidia of molluscs.

Most benthic crustaceans are omnivores and feed on both animals and plants. They use chelae to grasp and crush, while the mandibles and other mouthparts are used to grind up the food and pass it to the mouth. Some benthic crustaceans are filter feeders, most notably the barnacles. Barnacles are sessile as adults and feed with feathery appendages that strain food from the water. Like most

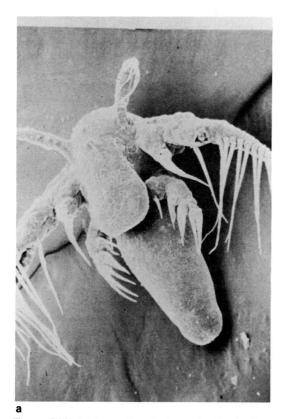
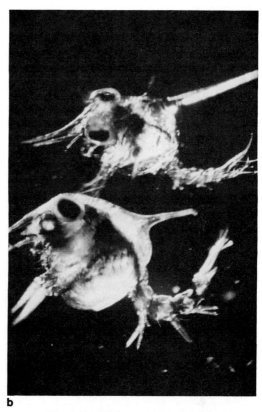

a **b**

Figure 5.22. (a) A scanning electron micrograph of a crustacean nauplius larva. (b) Later stages of crab larvae, called zoea. (Courtesy of Wally Clark and Dave Armstrong.)

Figure 5.23. A pandalid shrimp, *Pandalus danae,* showing paired appendages near the head area, compound eyes and the spiny rostrum between the eyes. (Courtesy of Lynwood Smith.)

Figure 5.24. Jan Armstrong, a fishery scientist, holding a red king crab (*Paralithodes camtschatica*) from the North Pacific. (Courtesy of Dave Armstrong.)

Figure 5.25. Dungeness crab, *Cancer magister,* shown within a crab pot. (Courtesy of Ken Chew.)

crustaceans, barnacles are edible. As you might expect from such a diverse group, just about every conceivable feeding type is found among the crustaceans, including parasitic copepods that can be found clinging to the skin of fishes.

The following are examples of crustaceans that support major commercial fisheries:

- *Paralithodes* and *Lithodes,* the king crabs. King crabs, harvested in the North Pacific and Bering Sea, include three species, commonly known as red, brown and blue king crab (Figure 5.24). This valuable fishery began after World War II and crashed in the early 1980s.
- *Chionoecetes,* the tanner or snow crab. These are smaller than the kings and are also fished in the North Pacific and Bering Sea. The tanner crab fishery became more important as the king crab fishery went into decline.
- *Cancer magister,* the Dungeness crab (Figure 5.25). This medium-sized crab is fished in the North Pacific from California to the Aleutian Islands.
- *Callinectes sapidus,* the blue crab (Figure 5.26). This crab is fished on the east coast, particularly in Chesapeake Bay and in the Gulf of Mexico. Both the hard-shelled and soft-shelled phases are marketed.
- *Menippe mercenaria,* the stone crab. This is also an east coast and Gulf crab and is known for its massive claws. Only the claws are harvested, and the live crab are tossed back into the water to regenerate new claws.

a   b

Figure 5.26. (a) Top and (b) bottom of a blue crab. The last pair of walking legs is adapted for swimming. The specimen shown is a male, as indicated by the narrow abdomen on the underside. (Courtesy of Ken Chew.)

- *Homarus americanus,* the Maine lobster (Figure 5.27a). This lobster has been fished in the North Atlantic for many years and supports a very valuable fishery.
- *Panulirus* spp., the spiny lobster (Figure 5.27b). There are a number of species of spiny lobsters that support fisheries in various parts of the world, including the Central Pacific and Australia. They differ from Maine lobsters in that they lack the large claws, and they get their common name from the large, spiny antennae.
- *Procambarus* is but one of the many crayfish genera. Freshwater crayfishes support fisheries in North America, Europe and other regions as well. They bear a morphological resemblance to the Maine lobster but are smaller in size (Figure 5.27a). Crayfishes are now commonly produced by aquaculture in the southern United States, often in combination with agricultural production.
- *Macrobrachium,* the giant river shrimp, or freshwater prawn (Figure 5.28). This shrimp and the crayfish are the only freshwater crustaceans listed here. Like the crayfish, it is also produced by aquaculture in various parts of the world.
- *Penaeus monodon,* the tiger shrimp. This large shrimp, and others like it, have become some of the world's greatest aquacultural cash crops (Figure 5.29). These shrimps are being produced through saltwater pond culture in many countries. Two other penaeids, the white shrimp, *P. setiferus* and the brown shrimp, *P. aztecus,* provide the second most valuable U.S. fishery. These two species are harvested wild in the Gulf of Mexico. Large shrimps are also called prawns.
- *Pandalus* spp., the Pacific shrimps (Figure 5.30). There are several species of pandalids that support capture fisheries in the North Pacific. They differ from penaeids in having a visible hump near the middle of their abdomen, whereas shrimp of the genus *Penaeus* have abdominal segments that form a gradual curve to the tail.

Insecta

Another class of arthropods that feeds with mandibles deserves mention, even though it is primarily a terrestrial group. It is the class **Insecta**, the most diverse class of organisms on earth. A single order of insects, Coleoptera (beetles), makes up one fourth of all known animal species. Insects are very important to inland fisheries, where many freshwater fishes feed on adult insects and insect larvae. In lakes, rivers, marshes and estuaries, insects are important constituents of food chains.

a   **b**

Figure 5.27 (a) An American lobster, *Homarus americanus,* on the right, next to a crayfish, *Pacifastacus.* Both the claws and tails are eaten. *H. americanus,* also called Maine lobster, is a marine species. Crayfishes live in freshwater. (b) A spiny lobster, *Panulirus,* from Hawaii. This lobster has no claws and only the tail is eaten. (Courtesy of Ken Adkins.)

Echinodermata

Echinodermata means "spiny skin." This phylum is exclusively marine and includes starfishes, sand dollars, sea urchins and sea cucumbers. Only sea urchins and sea cucumbers are harvested for food. Other echinoderms are taken for ornamental purposes. There are about 6,000 living species of echinoderms.

These animals have an interesting pattern of body symmetry. Their primary symmetry, or that pattern present early in life, is bilateral. One of the typical echinoderm larval stages is called the **pluteus,** and the pluteus larvae are bilaterally symmetric. As with most invertebrate phyla, the larval stages of echinoderms tend to be planktonic, and the larvae metamorphose and later settle to the bottom where the adults lead a benthic existence. Metamorphosis toward the adult stages, however, is accompanied by a change in symmetry to a secondary radial pattern. Moreover, the adult symmetry is organized into five parts and is therefore called **pentamerous radial symmetry.** The five-part symmetry is obvious among the starfishes because most starfishes have five arms or rays.

Another unique feature of this phylum is the presence of a **water vascular** system and **tube feet.** Echinoderms pump seawater throughout their bodies in order to move their body parts and activate the tube feet, which are used for attachment and locomotion. This intimate contact with seawater presumably has made adaptation to freshwater or other non-marine habitats impossible for the echinoderms. The water vascular system provides one source of internal structure, a hydraulic

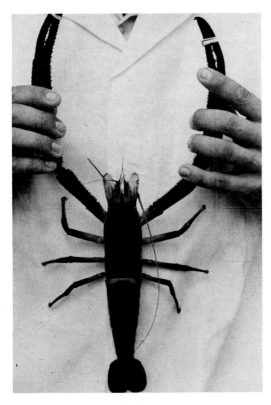

Figure 5.29. Penaeid prawns grown through aquaculture in Japan. These are locally known as karuma ebi (*Penaeus japonicus*). Note the smooth outline of the top of the abdomen. (Courtesy of Conrad Mahnken.)

Figure 5.28. The freshwater prawn, *Macrobrachium*. (Courtesy of Ken Chew.)

one, while a calcareous endoskeleton provides another. The endoskeleton is most obvious among the sea urchins and sand dollars because it is usually left intact after they die and commonly washes up on the beach. The endoskeleton, also called a **test**, shows patterns of pores arranged in five rows where the tube feet and other parts of the water vascular system penetrate. The spines of echinoderms are also parts of the endoskeleton, since each spine is covered by a very thin layer of living tissue. Getting pierced by sea urchin spines can be a painful and troublesome experience for humans because the spines have a tendency to break off and cause infections. The tropical genus *Diadema* has long, needle-sharp spines and is particularly troublesome to swimmers who blunder into them.

Sea urchins are harvested for their roe (uni), which is removed from within the test and eaten raw with sushi rice. The roe is soft, has fine texture and is orange or cream-colored. Examples of species that are important in this regard are the North Pacific red urchin, *Strongylocentrotus franciscanus* (Figure 5.31) and the cosmopolitan green urchin *S. droebachiensis*. Sea urchins, which belong to the class **Echinoidea**, are often harvested by divers. Sea urchins are herbivores and are typically found in great abundance around kelp beds where they exert a major influence on the kelp bed community. Kelp is a preferred item of the sea urchin diet, and sea urchins use a five-toothed structure called **Aristotle's lantern** to break up and ingest the kelp fronds. Kelp bed ecosystems are dramatically affected by abalones and sea urchins, which eat the kelp, and starfishes and sea otters, which eat the sea urchins and abalones. Removal of a major predator, like the sea otter (which was almost harvested to extinction for its hide) can cause entire communities to change in character. Top predators that maintain a delicate ecological balance in these communities are known as **keystone predators**, and fishery practices that affect keystone species need to be carefully studied and conscientiously managed.

Figure 5.30. Pacific prawns of the genus *Pandalus.* Note the sharp curve or hump on the abdomen. (Courtesy of Conrad Mahnken.)

Sea cucumbers don't look like other echinoderms at all, nor are they used in salads. They belong to the echinoderm class **Holothuroidea**. They look somewhat like large slugs with tube feet that lie about in crevices or creep slowly along the sea bottom. They lack the fused endoskeleton and hard spines typical of sea urchins. Their endoskeleton is reduced to tiny calcareous **ossicles** that are scattered loosely throughout the soft body wall. If you handle a live sea cucumber, it will often contract its body wall musculature and assume the turgid shape of a football. If you keep handling one, it is likely to spew its entire complement of internal organs forth, forming a very obnoxious sticky mass that will keep you occupied for some time trying to remove it from your hands. This is a protective response among sea cucumbers that is called **evisceration**. A sea cucumber that eviscerates on a predator will creep off and later regenerate its internal organs. Either the band of muscles that line the body wall, or the entire body wall is eaten as food. The muscles by themselves can be fried, while the body walls are often dried and later cut into pieces for soups, stews or pickling. Both sea cucumbers and sea urchins are popular to oriental tastes. An example of an edible North Pacific sea cucumber is *Parastichopus* (Figure 5.32).

Sea cucumbers eat detritus. They have a whorl of sticky tentacles around the mouth that they either extend up toward the water surface or lay down on the substrate. Those that hold their tentacles up like the limbs of a tree are **suspension** feeders; they trap detritus that is settling down from the water above (Figure 5.33). Those that lay their tentacles down to grope along the

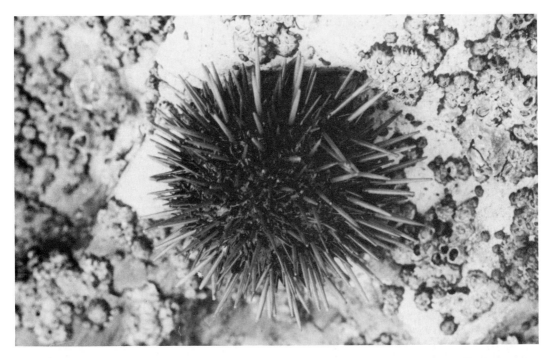

Figure 5.31. The red sea urchin, *Strongylocentrotus franciscanus,* which belongs to the phylum Echinodermata. The top side with its movable spines is shown. (Courtesy of Dave Armstrong.)

Figure 5.32. End view of a sea cucumber (*Parastichopus californicus*) showing tube feet at lower left and sensory papillae on top. (Courtesy of Ken Adkins.)

Figure 5.33. The feeding tentacles of a suspension feeding sea cucumber. Note that there are ten tentacles and that one is inserted into the mouth to remove food from its sticky surface. (Courtesy of Ken Adkins.)

Figure 5.34. An edible tunicate, *Halocynthia raretzi*, grown by hanging line culture in Japan. It is eaten raw and has a spicy flavor. (Courtesy of Ken Chew.)

substrate are deposit feeders. Detritus that sticks to the tentacles is removed when the food-laden tentacle is passed into the mouth and then pulled away clean.

Chordata

Our last phylum, and the one to which we belong, is **Chordata**. The distinguishing feature of chordates is that they have a structure called a **notochord** at least at some point in their life history. A notochord is a flexible rod that runs along the body's main axis. Chordates are bilaterally symmetric, so if you slice a chordate from top to bottom, passing through the midline of the notochord, you will be left with mirror image halves. There are about 45,000 living species of chordates. All vertebrates are chordates, but not all chordates are vertebrates. The phylum is divided into three subphyla—**Urochordata**, **Cephalochordata** and **Vertebrata**. The first two subphyla consist of invertebrates.

Urochordata includes the tunicates, or sea squirts. These animals look like sponges when they are adults. They begin life as planktonic larvae that look much like tadpoles, and then they undergo a dramatic and very rapid metamorphosis to become sessile, filter-feeding blobs on rocks or other substrates. They feed by drawing water through siphons by the action of a ciliated **pharynx with gill slits**, or pharyngeal basket, and strain food particles out of the water that passes through the basket. The notochord is present only in the larval phase of the life history. Tunicates are grown in the Orient by suspended line aquaculture and are eaten raw (Figure 5.34). There are about 1,300 species of Urochordates, and all are marine. The tunicates derive their common name from the cellulose-containing tunic that covers the body of the adults. Some of the urochordates remain planktonic for their entire lives—these include the salps and the larvaceans.

The subphylum Cephalochordata includes only about 20 species of small, eel-shaped chordates known as **lancelets**. Lancelets are free-living filter feeders that are anatomically similar to tunicate larvae. They have a notochord and a nerve cord that runs along the top side of the notochord, a pharynx with ciliated gill slits and a tail. The notochord forms a flexible axis that allows lancelets to swim with undulating movements or to burrow quickly through sand. A small fishery for lancelets off the coast of China has existed for many years. Lancelets are not very important to fisheries, but they are thought to be similar to the ancestral chordates that led to the very successful adaptive radiation of the vertebrates.

Some of the major features of the invertebrates are summarized in Table 5.1, while harvest methods and locations of major commercial shellfishes are presented in Table 5.2. Before we discuss the various classes and families of vertebrates, most of which are fishes, we should first consider some general features of fish anatomy and some anatomical terms. We will begin with the external anatomy of fish, as shown in Figure 5.35, starting with the head and working our way toward the tail.

Fish Anatomy

Visible on the head is the mouth, surrounded by **jaws**, the **eyes**, the **nares** (nostrils) and the gill covers, or **opercles**. In locational terms, the head is **anterior** on the body, and the tail is **posterior**. The top side of the body is **dorsal**, while the bottom or belly side is **ventral**. Structures that lie along the body's midline, along the plane that divides the body into mirror-image halves, are **medial. Lateral** is the term that contrasts with medial and refers to structures that are located away from the midline, or more to the right and left sides of the body. Structures that are laterally located, like your arms and legs, often occur in pairs, while those that are medially located, like your nose, usually do not occur in pairs. The mouth is medial, the eyes are lateral, and the opercles are posterior to the mouth and are lateral.

There are several kinds of **fins**. Most of the fins are supported by **fin rays**, which may either be **spiny** or **soft**. Fishes that have fins with only soft rays are sometimes called **soft-rayed fishes** while those that have at least some spiny supporting rays are often called **spiny-rayed fishes**. Some fins lack supporting fin rays and these are called **adipose fins**. Adipose fins are dorsally located, and an adipose fin is shown in Figure 5.36. Different kinds of fishes can generally be grouped into fish families on the basis of the presence and placement of the fins. The fins and other anatomical features are used to identify unknown specimens with the aid of **dichotomous keys**. These keys appear in guide books and distinguish between features in a step-by-step fashion to arrive at the proper taxonomic group of a given specimen.

There are usually two pairs of lateral fins, the **pectorals** and the **pelvics**. Pectoral fins are homologous with our arms, while pelvic fins are homologous with our legs. (The term **homologous** means the structures are similar in origin and construction, and homology implies that there is ancestral affinity. Homologous structures may be quite different in appearance and function, however, as are the wings of a bat, the flippers of a porpoise and the arms of a human. A contrasting term, **analogous**, refers to structures that are similar in form or function but bear no common origin or ancestral affinity. The wings of a bat and the wings of a butterfly are examples of analogous structures.) Pectoral fins are located just posterior to the opercles, whereas pelvic fins are ventral and may either be anteriorly located, quite near the pectorals as shown in Figure 5.35, or they may be posteriorly located as shown in Figure 5.36. In general, pelvic fins are posteriorly located among sharks and soft-rayed fishes and are anteriorly located among spiny-rayed fishes.

Table 5.1. Major Features of Invertebrates.

Taxonomic Group	Common Names	Example	Body Symmetry	Habitats
Phylum Porifera	sponges	*Spongia, Hippospongia*—commercial sponges	most are asymmetrical	all aquatic, most marine
Phylum Coelenterata	jellyfishes, corals, sea anemones, hydroids	*Physalia*—the Portuguese Man-of-War	radial	all aquatic, most marine
Phylum Mollusca Class Gastropoda	snails, slugs, abalones, limpets, nudibranchs	*Haliotis*—abalone, *Strombus,* conch	primary bilateral may be modified later by torsion or coiling	all habitats
Class Bivalvia	clams, oysters, scallops, mussels	*Crassostrea gigas*—the Pacific oyster	bilateral	all aquatic, most marine
Class Cephalopoda	octopus, squid	*Loligo opalescens*—the common market squid	bilateral	all marine
Phylum Arthropoda Class Crustacea	shrimps, crabs, lobsters, crayfish, barnacles, krill, copepods	*Paralithodes camtschatica*—red king crab	bilateral	all habitats
Phylum Echinodermata	sea urchins, starfish, sea cucumbers	*Strongylocentrotus franciscanus*—the red sea urchin	primary bilateral, secondary pentamerous radial	marine only
Phylum Chordata Subphylum Urochordata	tunicates, sea squirts	*Ascidia*—common sea squirt	bilateral	marine only
Subphylum Cephalochordata	lancelets	*Branchiostoma*—the lancelet	bilateral	marine only

Table 5.1.—*Continued*

Body Features	Larval Stages	Principal Feeding Modes	Importance
cellular level of organization, create water currents with flagellated cells, sessile	present	filter feeders	source of natural bath sponges—skeletons made of spongin fibers
have tentacles, stinging cells with nematocysts, corals are sessile and secrete calcareous endoskeleton	present	carnivores, many have mutualist symbioses with green algae	some jellyfish taken as food, corals very significant ecologically, some species hazardous to swimmers
most protected by single shell, attachment and locomotion with muscular foot	veliger, others	feed with radula, most are herbivores but all modes represented	many commercial species, most are marine, some terrestrial
have 2 shells joined by hinge, foot used for digging or attachment, decephalized	veliger, others	filter feeders, create water currents with ciliated ctenidia	many commercial species, most are marine, important in aquaculture, some used for jewelry
have tentacles with sucking discs, shell may be absent, excellent sensory and swimming capabilities, chromatophores	most development takes place within the eggs	carnivores	many commercial species, large fisheries for squid
body covered by exoskeleton, segmental bodies, jointed appendages	nauplius, zoea, others	feed with mandibles, all feeding types represented, many are omnivores	support some of the world's most valuable fisheries, also becoming important in aquaculture, planktonic forms extemely important in food chains
calcareous spines and endoskeleton, water vascular system, tube feet	pluteus, others	sea urchins are herbivores and feed with Aristotle's lantern, sea cucumbers are detritivores	sea cucumbers are harvested for food, sea urchins for their roe
adults, sessile covered with tunic, larvae have a notochord	present "tadpole" larvae	filter feeders with ciliated pharynx	taken as food and grown by aquaculture in the Orient
free-living, has notochord, nerve cord, tail, eel-like appearance	development is direct	filter feeders with ciliated pharynx	little commercial value, but taken as food in China

Table 5.2. Commercially Important Invertebrates of North America.

Common name	Region	Capture gear
1. Abalone	Pacific Coast	diver
2. Clam		
Hard shell	Atlantic, Pacific	escalator dredge, hydraulic dredge, diver, kicking, by hand
Geoduck	North Pacific	diver
Soft shell	mid-Atlantic	by hand, escalator dredge
3. Crab		
Blue	mid-Atlantic, Forida, Gulf of Mexico	mechanical dredge, pot, trawl, trotline
Dungeness	Pacific	pot
King	North Pacific, Bering Sea	pot
Snow (tanner)	North Pacific, Bering Sea	pot
Stone	Florida	pot
4. Crawfish	most states	trap
5. Lobster		
American	North Atlantic	trap, trawl
Spiny	Florida, California	trap
6. Mussel	North Atlantic, Pacific	diver, aquaculture, rake
7. Octopus	North Pacific	pot, trawl
8. Oyster	Atlantic, Pacific	tongs, dredge, diver, aquaculture
9. Scallop	Atlantic, North Pacific	trawl, dredge (drag) diver
10. Sea cucumber	North Pacific	diver, trawl
11. Sea urchin	Pacific	diver
12. Sea weed	Pacific, Atlantic	by hand
13. Shrimp	Atlantic, Gulf Mexico, Pacific	aquaculture, trawl pot
14. Squid	Atlantic, Pacific	trawl, purse seine, pound net, jig
15. Whelk	North Atlantic	pot

The medial fins include the **dorsal fins**, of which there are typically one, two or three, and one or two **anal fins**. The anal fins are ventrally located and occur just posterior to the anus. The **caudal fin** (tail fin) and adipose fin are also medial fins. The narrow region of the body just anterior to the caudal fin is called the **caudal peduncle**. Some very rapid-swimming fishes like tunas have accessory fins in the caudal peduncle area called **finlets**.

Along the sides of the body are the **lateral line** and the **scales**. The lateral line is a sensory organ that detects vibrations. We will learn more about the lateral line in Chapter 7. The scales form a protective covering for the body. There are four kinds of scales—**placoid, ganoid, cycloid** and **ctenoid** (the c is silent in ctenoid). Placoid scales are found on sharks and rays, and are homologous with vertebrate teeth. They are composed of an outer layer of hard enamel and an inner matrix of dentine (Figure 5.37). Shark teeth are actually enlarged placoid scales that grow continuously around the jaws. Ganoid scales are large plate-like scales and are found on sturgeons. Cycloid scales are typical of soft-rayed bony fishes, while ctenoid scales are commonly found on spiny-rayed bony fishes, although there are exceptions to this. Ctenoid scales differ from cycloid scales in that their posterior, exposed margins are toothed or comblike rather than smooth. The anterior portions of these scales are embedded within the skin. As a bony fish continues to grow past the juvenile stage, it enlarges the margins of the scales it already has rather than creating new scales. Both cycloid and ctenoid scales have incremental

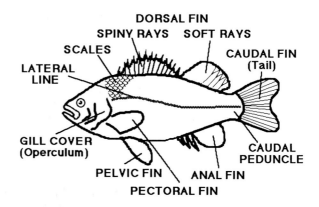

Figure 5.35. External anatomy of a spiny-rayed fish.

growth rings called **circuli**, which can be studied much like the growth rings of a tree and are valuable to fishery scientists for establishing age and growth rates.

The internal anatomy of a soft-rayed fish, the rainbow trout, is shown in Figure 5.36. Apart from the gills, the form of the kidney, the pyloric caeca and the swimbladder, the internal organs of this fish are quite similar to our own in both form and function. The **gills** which lie beneath the opercles serve the same function as our lungs. The digestive pathway is called the **alimentary** canal and includes the **mouth, esophagus, stomach, pyloric caeca, intestine** and **anus**. The pyloric caeca are digestive diverticula that look like spaghetti. A long **kidney** lies just ventral to the backbone and is dark reddish-brown in color. A swimbladder filled with air is present in most fishes and lies ventral to the kidney; its main function is to maintain buoyancy.

Vertebrates

It is convenient to group vertebrates into seven classes: **Agnatha, Chondrichthyes, Osteichthyes, Amphibia, Reptilia, Aves** and **Mammalia**. All but Aves (birds) are harvested through fisheries and will be considered in this section. The first three classes are fishes. There is often confusion about the correct plural term of the word "fish." **Fish** refers to one or more individuals of a given kind of fish, while **fishes** refers to two or more different kinds, or species, of fish.

Agnatha

The class **Agnatha** includes a few dozen species that reflect features of the earth's most primitive vertebrate. The term agnatha means "without jaws," and these eel-shaped fishes lack not only jaws but pectoral fins, pelvic fins and scales as well. Their gill chambers are vented by circular pores rather than opercles. Their bodies are elongate and cylindrical and are supported by a cartilaginous endoskeleton rather than a bony one. Their mouths are circular and modified for sucking. This class includes the **lampreys** (Figure 5.38) and **hagfishes** (Figure 5.39). Their impact on fisheries tends to be negative.

Hagfishes are marine fishes with habits that most of us would consider disgusting. For starters, hagfishes produce unusual amounts of **mucus**, and that is why they are sometimes called "slime eels." It only takes about four hagfish (a foot or so long) to turn the water in a five-gallon bucket into a solid gell. Hagfish feeding habits range from scavenging for food along muddy sea bottoms to parasitizing larger fishes. They enter the mouth or anus of another fish to eat it from the inside. After this they slither, well fed, out of a nearly empty bag of skin and scattered bones. Hag-

Rainbow trout *Oncorhynchus mykiss*

(formerly *Salmo gairdneri*)

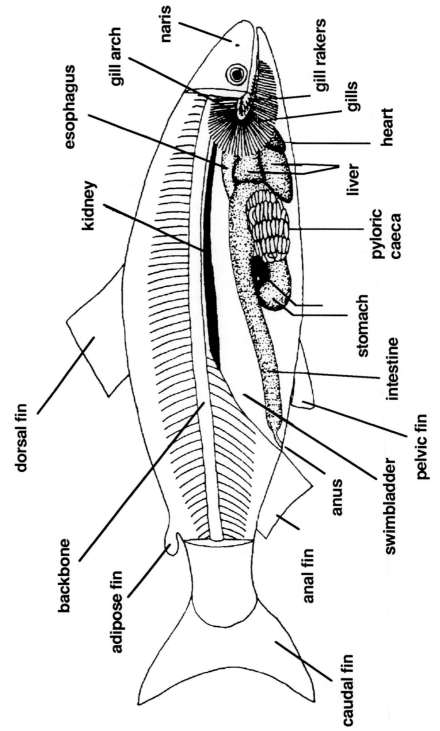

Figure 5.36. Internal anatomy of a soft-rayed fish, the rainbow trout. (Courtesy of Eric Warner.)

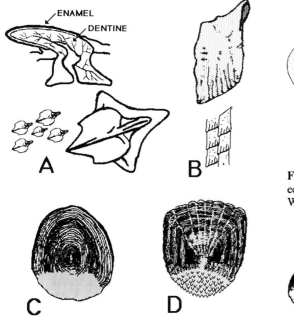

ENAMEL
DENTINE

A

B

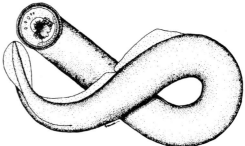

Figure 5.38. A lamprey showing the circular mouth equipped with hook-like teeth. (Courtesy of Eric Warner.)

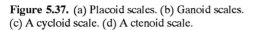

C

D

Figure 5.37. (a) Placoid scales. (b) Ganoid scales. (c) A cycloid scale. (d) A ctenoid scale.

Figure 5.39. A hagfish, class Agnatha.

fishes often attack other fishes that have been caught by longlines, traps, or nets. The sight of hagfish slithering out of fishes that were harvested for food is not welcome to most commercial fishermen. Other commercial fishermen are now harvesting hagfishes for their skins, needed for manufacturing "eel-skin" leather goods.

Lampreys have filter feeding larvae, but some of the adults are parasitic on larger fishes. Their mouths are equipped with hook-like teeth that are used to attach to the outside of a fish. The lamprey can then rasp away at the flesh of its host or suck away its body fluids. These fishes can be serious nuisances to salmonid and other important fisheries. Many lampreys spend time in both freshwater and salt water during their lifetimes. They hatch in freshwater and spend their early lives as wormlike larvae in river bottoms. The parasitic species then migrate to the ocean and return to freshwater later to spawn. Other species remain in freshwater throughout their lives. Those that return from the sea to spawn are called **anadromous**.

Chondrichthyes

The class Chondrichthyes is a relatively primitive group of cartilaginous fishes that have retained many of their general features for millions of years. Chondrichthyes means "cartilage fishes." This class consists of primarily marine fishes with jaws, placoid scales, and paired pelvic and pectoral fins that lack fin rays. The pelvic fins are positioned posteriorly on the body. Males have copulatory organs called **claspers** at the pelvic fins. Many of these fishes produce large eggs called "mermaid's purses," while others bear live young. This group includes fishes commonly known as **sharks**, **rays** and **ratfishes**.

Everyone is familiar with sharks. There are about 800 species of sharks; most have very sharp teeth, and all are predators in one way or another, with many being scavengers. While their

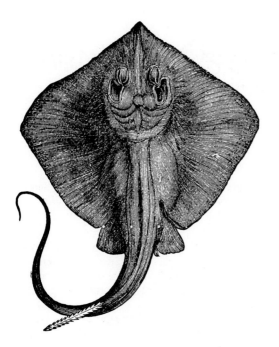

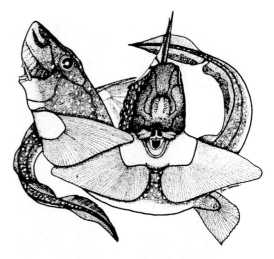

Figure 5.41. Two ratfish (*Hydrolagus collei*), one shown head on and the other from the side. Ratfishes belong to the class Chondrichthyes. (Courtesy of Eric Warner.)

Figure 5.40. A stingray, class Chondrichthyes. Note the barbed stinger on the tail.

notorious reputations are overblown, they do tend to be good at biting things. They can also be very large, reaching lengths of 60 feet, and that makes them the largest extant fishes. The largest fish, the **whale shark** (*Rhincodon typus*), is a filter feeder that feeds on zooplankton. The basking shark (*Cetorhinus maximus*) is another large one (45 feet long), and it also is a filter feeder. The white shark (*Carcharodon carcharias*) is an active predator and eats just about anything it wants to. It is the most dangerous shark to swimmers and reaches lengths of 20 feet or more. Even though the teeth and jaws of humans are smaller than those of the great white shark, humans have eaten more white sharks than white sharks have eaten humans. White sharks attack whales, seals, sea lions, other sharks, other fishes, birds, crabs, sea turtles and an occasional surf board. Fossilized teeth suggest members of the white shark group once reached lengths of about 50 feet. Most of the other species of sharks are smaller and less threatening.

Sharks (and other members of this class) lack swimbladders. Pelagic sharks need to swim continuously to keep from sinking and to keep water flowing past their gills. They have gill slits rather than opercles. Many sharks are taken by humans for food. Dogfish sharks, *Squalus acanthias,* have been used for "fish and chips" and also as a source of vitamin A (from the livers). Sharks are also targeted by fisheries for their fins, which are used in soups.

Rays are modified for a more benthic existence with dorso-ventrally flattened bodies and greatly enlarged pectoral fins (Figure 5.40). The skates are in this group, as are stingrays (which have a poisonous barb on the tail) and the electric rays (which can generate bursts of electric current for feeding and protection). The largest ray, the manta, feeds on pelagic zooplankton and reaches a width of 22 feet from wing tip to wing tip. The pectoral fins, or "wings" of skates are often consumed by humans and have a flavor similar to that of scallops. This group tends to forage along soft substrates for invertebrates.

Ratfishes are also known as **chimaeras**. The ratfish (*Hydrolagus*) gets its common name from its teeth, which look like the incisors of rodents (Figure 5.41). These are very strange looking fishes

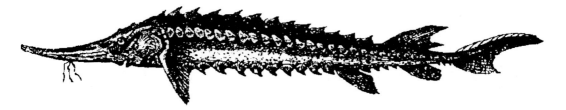

Figure 5.42. A sturgeon, family Acipenseridae, showing large ganoid scales.

that feed mostly on invertebrates. They have little importance to fisheries apart from some industrial uses (machine oil and fertilizer).

Osteichthyes

Osteichthyes means "bony fishes." This class includes the great majority of freshwater and marine animals that support fisheries. The general features of this group include a bony endoskeleton, a backbone divided into segments called **vertebrae**, jaws, fins with fin rays, scales, opercles and swimbladders in most members. Cycloid and ctenoid scales of bony fishes usually overlap each other like shingles. The opercles cover a single pair of openings to the gill chamber. Water enters the mouth, flows over the gills and then exits posterior to the gills between the opercle and each side of the body. A bony fish may either swim with its mouth open to ventilate the gill chamber, or it can fan the opercles while the body is stationary to achieve ventilation.

About 40% of the 20,000 bony fish species are freshwater fishes; the rest are marine. There are 38 orders and 425 families in this class, but we will consider only about a dozen of the more important families here. Bony fishes, the largest subgroup of which are called **teleosts**, are known in almost all of the world's aquatic habitats. They can be found in freshwater environments as much as 3 miles above sea level and in marine environments as much as 7 miles below sea level, at the bottom of the deepest ocean trenches.

Acipenseridae

It is convenient to discuss different groups of fishes at the family level. Taxonomic family names are easy to distinguish because they end in the suffix -**idae.** The first family we will cover, **Acipenseridae**, includes sturgeons (Figure 5.42). This is a primitive group with features that have not changed appreciably from those of its distant ancestors. Sturgeons have mostly cartilaginous skeletons but the heads are bony. They have five rows of large ganoid scales called **scutes** that run longitudinally along the body. These are large fishes, and adults range in size from 3 to 20 feet long and weigh up to 3,200 lbs. Sturgeons are important food fish throughout their range in North America and Eurasia. They are also valued for their eggs, which enter the market as very expensive caviar. Sturgeons feed on fishes and invertebrates near the bottom, where they forage over soft substrates with a long snout and sensory **barbels** located near the mouth. Sturgeons are often taken in large rivers by recreational anglers using smelt, herring, or pieces of lamprey as bait. Commercial fisheries for sturgeons are often threatened by the high value of the meat and roe, and also because the fish do not become sexually mature until they are about 10 years old. Sturgeons are long-lived, and the older females produce millions of eggs if given the chance. All sturgeons spawn in freshwater, but some are anadromous and feed in bays and estuaries. An example of this family is *Acipenser transmontanus,* the white sturgeon.

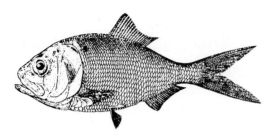

Figure 5.43. A menhaden, family Clupeidae.

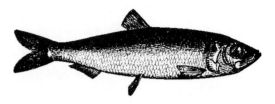

Figure 5.44. A herring, genus *Clupea.*

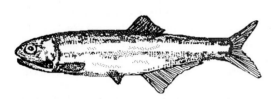

Figure 5.45. An anchovy, family Engraulidae. Note the large jaw.

Clupeidae

The next family, **Clupeidae**, is very important commercially, and it is also important because these fishes provide forage for other valuable piscivorous fishes like salmon and tuna. This family includes **herring, sardines, shad, alewives** and **menhaden** (Figure 5.43). These are small, schooling fishes that can be identified by the single dorsal fin, cycloid scales, soft fin rays, deeply forked caudal fin and the posteriorly positioned pelvic fins. They look somewhat like anchovies but have smaller mouths. Most species are marine, but some, including shad, are anadromous, and a few are found exclusively in freshwater. These fishes are visual predators upon zooplankton. Members of this group are sold in canned, dried, pickled and fresh forms throughout the world. Many clupeid species are also used for roe, bait, fish meal, fertilizer and other industrial purposes. Specific examples of this family include *Sardinops sagax,* the Pacific sardine, *Clupea harengus,* the Atlantic and Pacific herring (Figure 5.44), *Alosa pseudoharengus,* the alewife, *Brevoortia tyrannus,* the Atlantic menhaden and *B. patronus,* the Gulf menhaden.

Engraulidae

Anchovies belong to the family **Engraulidae**. These fishes are filter feeders on zooplankton; they feed while swimming forward with open mouths and flared opercles, straining zooplankton out of the water with the aid of long, comb-like gill rakers that line the gill arches. They have mouths and jaws that extend well past the eye (Figure 5.45). Otherwise, their appearance and habits are similar to those of the clupeids. Anchovies are commercially harvested primarily for reduction to fish meal and fish oil. The world's most important engraulid is *Cetengraulis mysticetus,* the Peruvian anchoveta. The northern anchovy, *Engraulis mordax,* is also an important commercial species. The small fishes such as those included in the last two families are increasingly used in aquacultural ventures. In Japan, for example, miscellaneous catches of small wild fishes are eagerly bought up by net pen aquaculturists for feeding whole to valuable piscivorous species like **yellowtail** (jacks—family Carangidae). Aquacultural production of fishes is placing ever greater demands on both fresh and processed forms of lower-value fishes.

Salmonidae

The family **Salmonidae** includes the familiar **salmons** and **trouts,** and also **whitefishes** and **chars.** These are soft-rayed fishes with cycloid scales and posteriorly positioned pelvic fins (Figure

5.46). There is a single dorsal fin and a rayless adipose fin posterior to the dorsal fin. The adipose fin is often clipped (removed) in hatchery-produced fish to indicate their origin or to indicate the presence of an unseen tag within the snout. Most salmons and some trouts and chars are anadromous. Members of this family spawn in freshwater, usually in streams, and have large eggs that are fertilized externally. Eggs are usually deposited in shal-

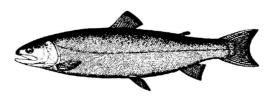

Figure 5.46. A rainbow trout, family Salmonidae.

low depressions called **redds**. The redds are excavated by the fish in the gravel of streambeds. Anadromous species include the five species of **Pacific salmon**: *Oncorhynchus tshawytscha*, the **king** or **chinook** salmon, *O. kisutch*, the **silver** or **coho** salmon, *O. nerka*, the **sockeye** or **red** salmon *O. keta*, the **chum** or **dog** salmon and *O. gorbuscha*, the **pink** or **humpback** salmon. The **rainbow trout**, *O. mykiss,* used to have the scientific name *Salmo gairdneri*. Anadromous individuals of this species are called **steelhead**. Another anadromous species, *Salmo salar,* is found throughout the North Atlantic and goes by the somewhat misleading (because it is a trout) common name **Atlantic salmon**. Other important salmonids include the cutthroat trout, *O. clarki*, the brown trout, *S. trutta,* the brook trout, *Salvelinus fontinalis* and the lake trout, *Salvelinus namaycush*. Brook trout and lake trout are actually chars.

Collectively, the salmonids provide some of the world's most important recreational and commercial fisheries. Most of the commercial fisheries are either marine or **terminal fisheries** (those which take place at the mouth of the river system where the fishes spawn). Most of the recreational fisheries are inland fisheries and take place in lakes, ponds, rivers and nearshore marine areas. Landlocked forms of some anadromous species also provide recreational fisheries in lakes. Production of these valuable fishes has been enhanced by hatcheries for nearly a century. In "put-and-take" fisheries, the fish are caught shortly after their release from a hatchery. Some of the recreational fisheries for salmonids are lucrative operations in countries bordering the North Atlantic. Fees of $7,000 are sometimes paid for the chance to catch an Atlantic salmon in a private waterway. Allocation of salmonid fishes among different user groups has been more contentious than for any other family of fishes.

Cyprinidae

The next family, **Cyprinidae**, is the most important in terms of global aquaculture production and also is the largest fish family, having over 1,600 species. This family includes **carps**, which are produced for food in pond culture. It also includes ornamental **goldfishes**, and **minnows**, which are

often used as bait, along with **shiners, chubs, squawfish** and **dace**. The scientific name of the common carp is *Cyprinus carpio*. The cyprinids are freshwater fishes, but some enter brackish water. There is no adipose fin and they lack teeth on the jaws, but do have teeth in the throat. The scales are usually large and cycloid (Figure 5.47). Some cyprinids have small barbels, often at the sides of the mouth. These fishes have small mouths and typically feed upon a variety of plants, plankton, benthic invertebrates or sometimes small fishes.

Figure 5.47. A carp, family Cyprinidae, showing the large scales and barbels around the mouth.

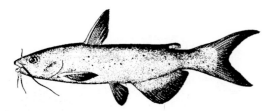

Figure 5.48. The channel catfish, showing barbels and an adipose fin.

Figure 5.49. A cod, family Gadidae. The three dorsal fins and two anal fins are distinguishing features of this family.

Ictaluridae, Siluridae and Clariidae

We have included three families of catfishes, **Ictaluridae (North American** or **bullhead catfishes)**, **Siluridae (European catfishes)** and **Clariidae (walking catfishes)**. These fishes can be distinguished by their sensory barbels around the mouth and the fleshy adipose fin. Some have no scales, and most live in freshwater ponds, lakes, or rivers. An ictalurid, the **channel catfish** *Ictalurus punctatus* (Figure 5.48), has become the most important aquaculture species in the United States. (crayfishes are second in importance, followed by rainbow trout). Silurid and clariid catfishes are also grown in aquaculture in many parts of the world. Most of these fishes are omnivores and forage along the bottom for food. Clariid catfishes are difficult to keep within their enclosures and are able to ''walk'' over land for considerable distances. Some catfishes, particularly those in major rivers, attain large sizes.

Gadidae

The family **Gadidae** includes **cods, hakes, haddock** and **pollocks**. These are easy to recognize by the large number of medial rayed fins. They usually have three dorsal fins and two anal fins. The scales are small and cycloid, and all of the fin rays are soft. The pelvic fins are anteriorly positioned and some species have medial chin barbels (Figure 5.49). These **demersal** (living near the bottom) fishes are found primarily in cold marine waters and support important commercial fisheries on both coasts. They are typically captured in large amounts by trawls. They have large mouths and are active predators, feeding on small fishes and invertebrates. Important commercial species in the Pacific include the **Pacific** or **gray cod** (*Gadus macrocephalus*), the **Pacific hake** (*Merluccius productus*) and the **Alaska** or **walleye pollock** (*Theragra chalcogramma*). Important species in the Atlantic include **Atlantic cod** (*Gadus morhua*), **pollock** (*Pollachius virens*) and **haddock** (*Melanogrammus aeglefinus*).

Sciaenidae

The family **Sciaenidae** supports important commercial and sport fisheries in the eastern United States and Gulf of Mexico. Some species are caught by recreational anglers in California. This family includes **drums, croakers** and **seatrouts**; these primarily nearshore marine fishes are found in warm waters. Pelvic fins are anterior in these fishes, and scales may be cycloid or ctenoid, but most are ctenoid. The fin rays are usually spinous anteriorly and soft posteriorly. There is no adipose fin. Croakers and drums have modified and muscular swimbladders that can produce sounds. An example of this group is the redfish, or red drum, *Sciaenops ocellatus* (Figure 5.50). This is the fish that set off all the smoke alarms after a chef from New Orleans developed the recipe for blackened redfish. The popularity of the recipe added so much demand for the fish that the commercial red drum fishery in Louisiana had to be temporarily shut down in 1988.

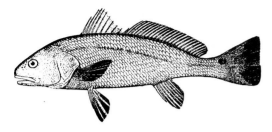

Figure 5.50. The red drum, family Sciaenidae.

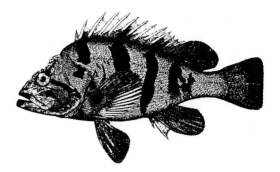

Figure 5.51. *Sebastes,* a genus of the family Scorpaenidae.

Scorpaenidae

The next family, **Scorpaenidae**, includes the **rockfishes** and **scorpionfishes**. These are marine fishes with spines on the head and fins. The spines on the dorsal and anal fins are venomous, so great care should be taken when handling them. Rockfishes are important sport and commercial species along the Pacific coast, where there are 65 species of a single rockfish genus, *Sebastes* (Figure 5.51). These fishes are often found near rocky outcrops, reefs and drop-offs. The most important commercial species is *S. alutus,* the Pacific ocean perch. Rockfish fillets are often misleadingly

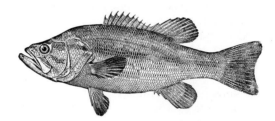

Figure 5.52. *Micropterus,* a freshwater bass of the family Centrarchidae.

marketed as **red snapper**. The Pacific ocean perch is red in color but is not a snapper. Snappers belong to the family Lutjanidae and are common along the Atlantic and Gulf Coasts, but not the Pacific Coast. The scientific name of the "real" red snapper is *Lutjanus campechanus*. Rockfishes are somewhat similar to snappers in appearance. Both have large mouths, large eyes, are active carnivores and have anteriorly placed pelvic fins. Snappers lack venomous spines. Both families are important commercially and recreationally. Rockfishes bear live young and some species live over 100 years.

Centrarchidae

Centrarchidae is the family of freshwater fishes known as the **sunfishes**. Members of this family include **smallmouth bass** (*Micropterus dolomieui*), **largemouth bass** (*M. salmoides*), **bluegill** (*Lepomis macrochirus*), **pumpkinseed** (*L. gibbosus*) and various other species of sunfishes and **crappie**. These are the most important gamefishes in the United States (Figure 5.52) These fishes are indigenous (native) to the warmer freshwater habitats of North America east of the Rocky Mountains, but they have been introduced widely outside their native range. They are active carnivores that prey upon a variety of insects, aquatic invertebrates and small fishes. The anterior portions of the dorsal and anal fins are spiny, the pelvic fins are anterior, and the single dorsal fin may have a notch between the spiny and soft-rayed portions.

Scombridae

Scombridae includes **tunas, mackerels, bonitos** and **albacore**, all of which are fast-swimming oceanic fishes. This family is built for speed. Scombrids are torpedo-shaped and have a very

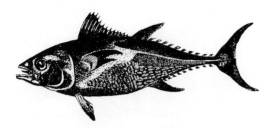

Figure 5.53. A tuna, family Scombridae.

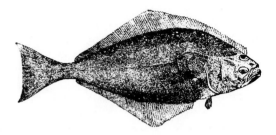

Figure 5.54. A halibut, family Pleuronectidae, showing both eyes on the right side of the body.

Figure 5.55. A flounder, family Bothidae, showing both eyes on the left side of the body.

narrow caudal peduncle, often with accessory finlets and lateral **keels,** and two dorsal fins that slip into grooves in the body when depressed (Figure 5.53). These are pelagic schooling fishes, and many migrate long distances. Some species have scales on only a portion of the head and lateral line. All are active carnivores, and the large tunas are piscivores. The bluefin tuna, *Thunnus thynnus,* reaches a length of 10 feet and weight of 1,500 lbs. This family is very important to commercial fisheries in warmer marine waters throughout the world, often far offshore. The United States consumes nearly half of the world's tuna harvest, most of it in canned form. Raw tuna brings high prices in Japan for sashimi.

Pleuronectidae and Bothidae

The last two families of fishes that we will cover are **flatfishes.** The family **Pleuronectidae** includes the so-called **righteye flounders** (Figure 5.54), while the family **Bothidae** includes **lefteye flounders** (Figure 5.55). These families include demersal fishes that are variously called **flounder, sole, halibut, dab** or **turbot.** They all undergo a dramatic metamorphosis as juveniles. They begin life as planktonic larvae, similar in appearance to larvae of other fishes. Soon one of the eyes begins to migrate to the other side of the head until both eyes are on either the right or left side of the body. Pleuronectidae includes flatfishes with both eyes on the right side of the body while Bothidae includes those with their eyes on the left side. When the eyes have reached their final position, the fish assume a benthic existence, eyed-side up. They become laterally compressed, light in color on the underside and dark in color on the eyed side. These fishes are predators of invertebrates and small fishes. They can mimic the color of the bottom or cover themselves with loose sediment to further conceal themselves. These fishes support important commercial fisheries in temperate marine waters throughout the world. The largest of the flatfishes are the halibuts. Pleuronectid halibuts include **Pacific halibut,** *Hippoglossus stenolepis,* **Atlantic halibut,** *H. hippoglossus* and **Greenland halibut,** *Reinhardtius hippoglossoides.* The **California halibut,** *Paralichthys californicus,* is a bothid. Halibuts are usually fished commercially by longlining. Sportfisheries for these fishes are also popular. Just about all of the flatfish species are good to eat. Smaller flatfishes include the **English sole,** *Parophrys vetulus,* which is another pleuronectid. Flatfishes generally live on soft seabottoms.

Table 5.3 lists many of the important commercial species of fish by common name, together with the primary places they are fished and the primary methods by which they are caught. The relative importance of commercially harvested fishes and shellfishes (in terms of weight of landings) is given in Figure 5.56. Descriptions of the commercial fishing methods themselves are given in Chapter 15, and descriptions of sportfishing methods are covered in the last four chapters.

Table 5.3. Commercially Important Fish Species of North America.

Name	Region	Capture gear
1. Albacore	Pacific	troll
2. Anchovy	California	purse seine
3. Blackcod	Pacific, Bering Sea	longline, traps, trawl
4. Bluefish	mid-Atlantic	encircling gill net
5. Bonito	California	purse seine
6. Butterfish	Gulf of Mexico	trawl
7. Carp	most states	each seine
8. Catfish	southern states	pond culture
9. Cod	North Pacific, Bering Sea, North Atlantic	trawl, gill net, longline
10. Dogfish	North Atlantic, North Pacific	gill net, longline, trawl
11. Eel	mid-Atlantic	electrofishing, pot
12. Flounder	Atlantic, Pacific, Bering Sea	trawl
13. Grouper	Florida, Gulf of Mexico	longline, trap
14. Halibut	Pacific, Bering Sea	longline
15. Herring (sardine)	North Atlantic, Pacific, Bering Sea	purse seine, gill net
Herring roe	North Pacific	diver
16. Yellow perch	Great Lakes	gill net
17. King mackerel	Florida	troll, gill net
18. Spanish mackerel	mid-Atlantic	pound net, trap net
19. Menhaden	Atlantic, Gulf of Mexico	purse seine, pound net
20. Mullet	Florida	gill net
21. Pacific sardine	California	purse seine
22. Pacific whiting (hake)	Pacific	trawl
23. Pollock	North Pacific	trawl
24. Red drum (redfish)	Gulf of Mexico	hook and line
25. Rockfish	Pacific	trawl, longline, auto jig
26. Salmon		
Chum	Pacific	purse seine, gill net
Coho	Pacific	troll, purse seine, gill net
King	Pacific, Bering Sea	troll, purse seine, gill net, fishwheel
Pink	North Pacific	purse seine, troll, gill net
Sockeye	North Pacific	gill net, purse seine

Table 5.3.—*Continued.*

Name	Region	Capture gear
27. Sardine (juvenile herring)	Maine	stop seine, purse seine
28. Shad	mid-Atlantic, Pacific	gill net
29. Shark	Atlantic, Pacific, Gulf of Mexico	longline, drift, gill net
30. Snapper (red)	Gulf of Mexico	hook and line, traps
31. Striped bass	Atlantic, Pacific	beach seine, gill net, longline
32. Swordfish	Atlantic, Pacific	harpoon, longline
33. Thread herring	Florida, Gulf of Mexico	purse seine
34. Tilefish	Atlantic	longline
35. Tuna		
Albacore	West Coast	troll
Bluefin	Atlantic	harpoon, longline
Bonito	California	purse seine
Yellowfin	Eastern Pacific, Atlantic	purse seine, longline
36. Weakfish	Atlantic	pound net
37. White sturgeon	Columbia River	longline
38. Yellow perch	inland	gill net

Amphibia

Apart from the **mammals**, which are covered in Chapter 17, we will not cover the remaining classes of vertebrates in much detail. The class **Amphibia** includes the frogs. *Rana* is a common frog genus. Frogs have aquatic larvae called tadpoles. The jumping legs of the adults are prized as food, particularly in Europe, and wild frogs in some places have disappeared as a result. Unfortunately, frogs die after the legs are removed, unlike crabs which can regenerate new legs. Aquaculture of frogs has begun to supply the demand for frog legs and also provides frogs as biological specimens for education and research.

Reptilia

The class **Reptilia** includes the turtles and alligators. These animals have been harvested heavily at times for food and for hides and shells. The green sea turtle, *Chelonia mydas,* is a preferred food item, and populations of this turtle have been seriously depleted. Both the adult turtles and the eggs, which are deposited in beach sand, are eaten. Some freshwater turtles, including the snapping turtle, are also harvested for food. Tortoise shell combs and other accessories come from the hawksbill turtle, a marine species. Alligators were once endangered because of their valuable hides, but are now recovering under a protected status, and regulated hunts are taking place in some states. Aquaculture industries in the southern United States now produce alligators for both meat and hides.

Conclusion

Amphibians are thought to have descended from air-breathing bony fishes. Some modern fishes have lungs, and the swimbladder may have arisen as an early adaptation for life in freshwater environments that were low in oxygen content rather than as an organ for controlling bouyancy. An-

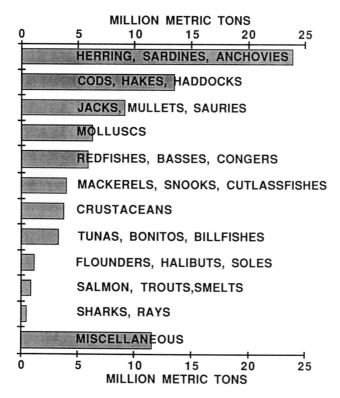

Figure 5.56. Relative contribution of major groups to the world catch by weight of landings (1986).

cient lobe-fin fishes called **coelacanths** are known from the fossil record to have stalked or lobed lateral fins, suggesting an affinity with the limbs of the early amphibians (lungfishes also have lobed fins and are thought to be the direct progenitors of amphibians). Coelacanths were common in the Devonian period, 400 million years ago. In a remarkable scientific discovery, a living coelacanth (*Latimeria chalumnae*) was discovered in deep waters off the coast of South Africa in 1938. Scientists are still studying this so-called "living fossil" to learn more about the origin of terrestrial vertebrates. Amphibians remain dependent upon aquatic environments during their larval stage, but reptiles were the first vertebrates to become truly terrestrial by encapsulating the early aquatic existence within their eggs. Reptiles are thought to have evolved from amphibians and to have given rise to the mammals. Recent findings suggest that birds are the descendents of dinosaurs, which were warm-blooded, rather than the cold-blooded reptiles. That makes birds the closest living relatives of the once-dominant dinosaurs.

We began and ended these last two chapters on the forms of aquatic life with a discussion of the ancestral affinities that bind their classification into a meaningful order. In the following chapters, we will look inside these organisms to learn more about how they lead their lives.

6
Life Functions

Frederick G. Johnson

Introduction

Life functions refer to the processes that keep living things alive and able to reproduce their kind. These processes are grouped within the scientific discipline of **physiology**. In a very real sense, the primary reason that fisheries exist is to provide the biochemical energy and raw materials (like amino acids) that allow for growth and maintenance of human life functions. To understand physiological processes fully, scientists often need to deal with events that take place at the molecular level. In other words, to understand physiology, we need to break life functions down to their smallest parts—the biochemical reactions that sustain living activities—and find out how the parts work. This is often an intimidating endeavor for students of biology, but it need not be. Living processes are fundamentally simple and elegant; otherwise, they would not have worked so well for so long.

We will examine a number of physiological processes at the molecular level, including how energy is processed, how information is stored and processed, how organisms move, how they respire, how their cells and organs communicate and how aquatic organisms maintain their salt and water balance. We will answer some very fundamental questions in this chapter, including why animals eat and why they breathe.

Genes and Proteins

The key to understanding biological processes at the molecular level lies in understanding the actions of proteins. We were introduced to their importance to a limited extent in Chapter 3. In this section, we will look more closely at how proteins are formed and discuss the diversity of their functions.

Proteins are polymers of amino acids. A **polymer** is a large molecule formed by the bonding together of smaller subunit molecules called **monomers**. For proteins, the monomer subunits are **amino acids**. There are only about 20 different organic amino acids, and the specific order in which they are linked together by covalent chemical bonds determines the unique features of each resulting protein. **Covalent bonds** are those in which electrons are shared by two or more atoms within the bonded molecule. Energy is stored in chemical bonds, and bonds of the covalent type store a relatively large amount of energy. Forming chemical bonds (referred to as **synthesis**, or **anabolism**) requires input of energy, while breaking chemical bonds (**catabolism**) releases energy.

Most biologically functional proteins consist of dozens or hundreds of amino acids. While energy is stored in the chemical bonds of proteins, the utility of proteins to living systems far exceeds the energy storage role. The functions of proteins stem largely from the molecular diversity that results from the bonding of amino acids in unique sequences. When a protein is formed, the peptide bonds first connect the amino acids together in a linear sequence. This is known as the **primary structure** of the protein (Figure 6.1). The amino acid chain may then fold or twist into spiral or pleated shapes along some or all of its length, and that process confers the **secondary structure** of the protein. **Tertiary structure** tends to be the most important element of protein function and results from the folding of the primary and secondary structures into a complicated, unique, and usually globular molecule that has specific biochemical properties. The functions of proteins in living systems include tissue **structure**, **regulation** of biochemical reactions, **movement** by muscle contraction, **transport** of respiratory gases and other substances, **immunity** to diseases and **recognition** and control of events at cell membranes. Before we discuss these various functions (which will occupy much of this chapter), we will briefly consider the information-storage system that produces the proteins in living organisms—the nucleic acids.

Nucleic acids (DNA, deoxyribonucleic acid, and RNA, ribonucleic acid) are also polymers of smaller monomer subunits, the **nucleotides**. There are only four different nucleotides, but the DNA molecules that make up chromosomes are so large that thousands of nucleotides are included in each chromosome. Each nucleotide is composed of a sugar component (ribose), a phosphate component and one of the four **nitrogenous bases**. The nitrogenous bases pair together in a specific fashion opposite each other in the DNA (Figure 6.2).

The four nucleotides are specified by the nitrogenous bases **adenine** (A), **guanine** (G), **thymine** (T) and **cytosine** (C). The bases form relatively weak bonds with each other, called **hydrogen bonds**, when paired together across the middle of the two strands of the DNA. Adenine can only pair with thymine, and cytosine can only pair with guanine to form the "steps" of the DNA "ladder." In humans there are 46 chromosomes in each cell, 23 from the mother and 23 from the father. Each chromosome has roughly 20,000 genes. Each gene consists of about 1,000 nucleotides. The specific order of nucleotides in the gene con-

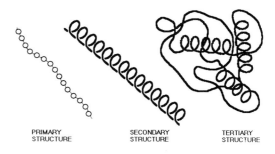

PRIMARY STRUCTURE   SECONDARY STRUCTURE   TERTIARY STRUCTURE

Figure 6.1. A diagram showing how the primary structure of proteins, which results from the sequential linkage of amino acids, becomes incorporated into secondary and tertiary structural patterns.

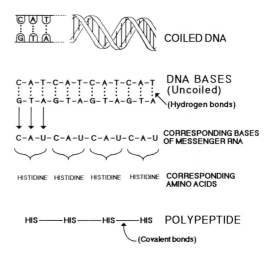

Figure 6.2. A sequence of steps showing how genes direct the synthesis of proteins. A polypeptide consists of a small protein segment.

stitutes what is known as the **genetic code**. There are four "letters" in the code (the four nitrogenous bases), and each "word" is three letters long, giving 64 possible words. These words are called coding triplets, or codons. Sixty-one of those words specify amino acids. Since there are only 20 amino acids for the 61 triplets to code for, many of the amino acids are specified by more than one nitrogenous base triplet. The remaining three triplets are signals that act to stop the protein synthesis process at the appropriate point.

Considering what it accomplishes, the biological information storage process is actually quite simple. It is analogous to dealing innumerable hands from the same deck of cards. The role of genes can be reduced essentially to that of "blueprints" for the order of amino acids that form specific proteins. Some genes may act directly in regulatory capacities, and other portions of DNA may have no function at all (the so-called "junk" DNA). For the most part, however, an individual gene specifies the synthesis of an individual protein. It is the specificity of the proteins that directs the elaboration of life.

In mechanistic terms, eukaryotic cells and organisms reproduce themselves by first making a new copy of their blueprints. The nitrogenous base pairs separate when the DNA strands unravel, and new nucleotides are appropriately matched on both sides of the exposed single strands to make a new copy of the double-stranded chromosomes. This process is called **replication**. In order for the structure and activities of the cells to be established, specific genes that are required at given times unravel from the double-stranded state, and the exposed genes then direct the synthesis of the other type of nucleic acid, RNA. The RNA forms with the correct sequence of nitrogenous bases to produce the protein for which the gene serves as a blueprint. This process is called **transcription**. Finally, the RNA segment travels to another part of the cell where it directs the synthesis of the protein that was encoded by the gene, and this final step is called **translation**. When chromosomes replicate in the process of cell division, it is known as **mitosis**. Replication and segregation of chromosomes that leads to the formation of sex cells is called **meiosis**. A **mutation** results in a change in the order of the nucleotides that make up the genetic code. These mistakes will be carried forward during mitosis and meiosis and usually result in proteins having an improper sequence of amino acids after the mutated gene is activated. A mutation may or may not improve the effectiveness of a protein produced from expression of the mutated gene, but only rarely is it improved. Mutations in important genes are usually harmful and are often fatal. The sum of the rare exceptions to that trend, however, accounts for much of what we call evolution.

One of the functions of genes is to code for the production of structural proteins. Structural proteins are made of either fibrous or globular subunits. Fibrous proteins consist of repeated sequences of amino acids in which secondary structure often dominates. Examples of fibrous proteins include **collagen**, which forms skin, connective tissue and tendons, and **keratin**, which makes up hair, wool, scales and nails. Globular proteins make up the filaments in the internal structure of eukaryotic cilia and flagella. These filaments, called **microtubules,** are actually spiral assemblages of globular proteins.

Another gene function is the production of regulatory proteins. This is the function of most genes. Most regulatory proteins regulate the progress of biochemical reactions in cells. Such reactions constitute **cellular metabolism**. The proteins that regulate metabolism are called **enzymes**, and they greatly increase the rates of chemical reactions (the enzymes act as **catalysts**). Often the enzymes work together with nonprotein **enzyme cofactors** to control cellular metabolism. These cofactors may be required in the diet, where they are known as **vitamins**. Enzymes owe their activity to the tertiary structure of the amino acid chains, which creates pockets where the reacting molecules fit together perfectly. Such pockets are called the **active sites** of the enzymes. After the reactant molecules are seated within the active sites, the tertiary structure of the enzyme flexes slightly, in what is known as a **conformational change**, and this causes the chemical reaction to be completed. Enzymes are not the only regulatory proteins in living systems. Other regulatory proteins act as **hormones** by activating or deactivating components of the **endocrine system**.

Transport proteins include **hemoglobin** and **hemocyanin**, which serve to transport respiratory gases, oxygen and carbon dioxide through the circulatory system. Hemoglobin is a reddish protein found in the blood of fishes and other vertebrates, and it contains iron. Hemocyanin is a bluish-colored respiratory pigment that contains copper and is found in the blood of molluscs and crustaceans. Many other substances, including various nutrients and some toxicants, are transported in living systems by binding onto transport proteins.

Proteins that confer immunity against foreign molecules include the **antibodies**. Antibodies are produced by white blood cells to recognize foreign cells, attach to their surfaces and render them inactive (recently discovered viruses that attack antibody-producing cells in humans cause the disease known as AIDS). Other proteins that serve a recognition function are located in cell membranes, where they act as receptors and transporters of substances that reach the membrane. For example, these membrane-bound proteins may act to move selected substances into or out of a cell, or to initiate a sequence of cellular events once a chemical message from outside the cell has been received by a receptor.

Another major function of proteins is to produce movement. The globular structural proteins that make up the microtubules of cilia and flagella serve this purpose. This is also the function of the contractile proteins, **actin** and **myosin**, which make up muscle tissue.

Movement

Swimming, walking, grasping, chewing, crawling and other displays of overt movement all involve the contractility of protein units. These units are visibly obvious when one looks at **striated muscle** tissue through a microscope. The striations are actually bundles of the proteins actin and myosin, which are aligned parallel to each other in contractile sets of filaments (Figure 6.3). The thick filaments are made of myosin and the thin filaments are made of actin. When a muscle is relaxed the thick and thin filaments do not overlap very much, but when the muscle contracts the thick and thin filaments move toward each other to produce greater overlap. The trigger that initiates the contraction sequence is generally an electrical impulse from the nervous system.

The actual mechanism that causes sliding of the thick and thin filaments against each other is not fully understood. We do know that the process requires an input of chemical energy. This energy is provided by a compound called **ATP (adenosine triphosphate)**, about which we will learn more in the following section. Some biochemists envision the contraction process as a ''ratchet-type'' mechanism, where ATP is attached to the filaments; as

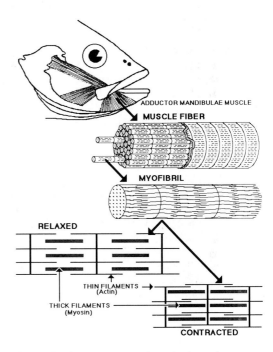

Figure 6.3. The structure and mechanism of muscular contraction.

some of the energy in ATP is released by the breaking of one of its chemical bonds, that energy is translated into another notch of mechanical motion inward. When the contraction is finished, the filaments disengage from each other and return to the relaxed state. This relaxation can be assisted by the contraction of an **antagonistic** muscle that opposes the action of the first one. Meanwhile, the chemical bond in ATP that was broken to yield the energy used for contraction is reformed and the muscle is made ready to contract again.

Assemblages of contractile proteins are present in just about all cells, whether those cells are components of muscle tissue or not. Some cells move through their medium, even without cilia or flagella, and most cells undergo internal movements. All cells that divide need to pinch themselves off at the middle before producing the two ''daughter'' cells. Segregation of chromosomes to opposite ends of cells during mitosis and meiosis also involves intracellular movements along protein microtubules. While the exact mechanisms of these contractile protein movements and the movements of cilia and flagella are not fully known, we do know that they are very similar in widely dissimilar organisms in terms of the apparent mechanisms, the proteins involved and the source of energy that drives the systems.

Energy

While matter tends to move through ecosystems in cyclic patterns (e.g., the **carbon cycle**, the **nitrogen cycle**, the **phosphorous cycle**, the **water cycle**), energy flows through ecosystems in a **one-way** fashion. In this section, we will follow the pathway that energy takes from its source, the sun, until it is used in biological systems. This energy occurs in various forms along the way, but these forms tend to be much the same among different organisms.

We already learned about one category of energy—**kinetic energy**—the energy that is reflected in heat or the movement of molecules. When you put a pan of water over a flame on the stove, the water molecules move more and more rapidly as the temperature rises. That is because the kinetic energy of the water molecules increases as the metal atoms of the pan move faster and collide with them, transferring the energy of their movement to the water molecules. Eventually, the kinetic energy will become great enough for the water molecules to overcome the mutual attraction that holds them together as a liquid, and they will enter the gaseous state when the water boils. The other category of energy is **potential energy**. Potential energy is **stored energy**. It can be stored in many forms, such as water held at elevation behind an hydroelectric dam or in the covalent chemical bonds of organic molecules. The first law of thermodynamics tells us that energy can change from one form to another, but it cannot be created or destroyed.

The energy that drives biological systems comes from sunlight. It strikes the earth in the form of **radiant energy**, which can be measured in **wavelengths** to reflect the quality of radiant energy, or in **photons** to reflect the quantity of energy. Radiant energy is produced by nuclear fusion reactions in the core of the sun, where hydrogen nuclei strike each other with such force that they fuse into helium nuclei (that is also how a hydrogen bomb works). In the process, a great deal of energy is released, and this energy comes from the conversion of a small amount of matter (according to the classic formula, $E = mc^2$, for which Albert Einstein is famous). Radiant energy that leaves the sun travels outward in all directions at the speed of light, and some of it strikes the earth.

The solar energy that reaches the earth amounts to about 10^{24} calories per year. A calorie is the amount of energy required to increase the temperature of one milliliter of water by one degree centigrade. About 30% of that energy is reflected back out into space by the earth's atmosphere. That is why the earth appears luminous when viewed from space. Another 20% is absorbed by the atmosphere itself, thereby increasing the kinetic energy of the atmospheric molecules. Only a small portion (1% to 3%) is absorbed by oxygen (O_2) and ozone (O_3) in the upper layers of the atmosphere, but that portion includes the potentially harmful ultraviolet wavelengths.

About half of the incoming solar radiation reaches the surface of the earth. Most of it is absorbed by water or land, thereby increasing water and land temperatures. This heating causes some of the water to evaporate and drives the earth's weather and water cycle. Some of the energy that heats the earth's surface radiates back up toward the atmosphere in longer (infrared) wavelengths than the visible light wavelengths that came from the sun. These longer wavelengths cannot pass through the atmosphere as easily as the shorter incoming wavelengths did, so they become trapped. Atmospheric carbon dioxide helps to trap the infrared heat, and this phenomenon is known as the **greenhouse effect**. Humans have increased the amount of carbon dioxide in the atmosphere by burning forests and fossil fuels, and this issue has raised the concern that these activities are contributing to a relatively sudden **global warming**.

Less than 1% of the incoming solar radiation is captured by organisms. The overall average is probably closer to 0.1%, but of the sunlight that actually strikes a plant, a maximum of 3% may be transformed into biological production. In terms of the total production of new biomass, however, this translates into about 120 billion metric tons of organic production on earth each year. This production of organic matter by plants is called **primary production.**

Most of this primary production takes place in water and enables production at other levels in food chains. The annual fishery harvest stands at about 80 million metric tons, or somewhat less than one thousandth of the total world amount of primary production. As we will see, this does not mean that there is an overabundance of untapped primary production available. For one thing, we eat relatively few aquatic plants directly. Moreover, those animals that do eat aquatic plants retain relatively little of the energy produced by the plants within their own tissues. People utilize a much more significant proportion of land plant production than aquatic plant production. Humans now use 20% to 40% of the energy captured by terrestrial plants. How much of the energy we use from aquatic production hinges in part upon the length of the food chains that supply the items of seafood in our diets.

Energy in new biological material is stored in the chemical bonds of organic molecules. These molecules include sugars, larger carbohydrates, lipids, proteins and nucleic acids. The energy in these molecules represents a pool of potential energy that may be put to biological use by the plants that originally synthesized the molecules, or by the heterotrophs that consume the plants. Animals eat to acquire the energy in these molecules. Each time energy is used, or converted to another form as it "works" its way through the ecosystem, much of it ends up in forms that cannot be used by organisms. In other words, each time the pool of potential energy moves up a link in a food chain or food web, the pool shrinks considerably.

The chemical equation for photosynthesis proceeds like this:

$$6H_2O + 6CO_2 + \text{light energy} \rightarrow C_6H_{12}O_6 + 6O_2$$

This means that plants require water, carbon dioxide and sunlight to produce **glucose** (a simple sugar) and oxygen. Glucose molecules can then be linked together into polymers (polysaccharides or starches), modified and changed into other types of organic molecules, or used directly as a source of potential energy. We will keep our biochemistry simple and consider glucose to be our prototype energy-source molecule, which it tends to be in living systems. Glucose is the form in which potential energy is generally transported within or-

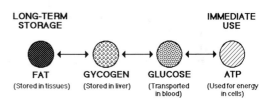

Figure 6.4. The forms and locations of chemical energy sources within animals.

ganisms. Potential energy that is stored in organisms tends to be incorporated into larger molecules (sugar polymers and fats), while potential energy for immediate use is in the form of ATP. Figure 6.4 shows a typical spectrum of the forms of potential energy in a fish or a mammal.

Anabolic reactions mediated by certain enzymes allow glucose to be synthesized into glycogen, other starches and fats. The enzymes are produced by the activation of appropriate genes and can be turned on and off at appropriate times. We also encountered an example of how ATP is used, where it provided the biochemical means for movement. In the next section, we find out how ATP is regenerated from glucose.

Respiration

Respiration can be defined in two ways. The first definition deals with the flow of energy in biochemical reactions that yields ATP, and the second involves the flux of respiratory gases.

Definition #1: Respiration is the oxidation of organic molecules, using oxygen as an elec-

tron acceptor.

Definition #2: Respiration is the process of taking oxygen from the environment and returning carbon dioxide to the environment.

Let us first consider respiration in the context of definition #1. Remember that a substance is oxidized when it loses electrons, and it is reduced when it gains electrons. The organic compound we are going to oxidize is glucose. Oxygen (O_2) will be reduced in the process and, with the addition of hydrogen, will be converted to water (H_2O). The energy that will be captured by ATP comes from electrons, which have high energy levels in glucose, and move to lower energy levels when they reach oxygen (as we will see, electrons, e^-, are followed by protons, H^+, and that is why O_2 becomes H_2O when it receives two electrons). Here is the chemical equation for oxidation of glucose:

$$C_6H_{12}O_6 + 6O_2 \rightarrow 6CO_2 + 6H_2O + energy$$

First of all, note that this equation is the reverse of the one for photosynthesis. Photosynthesis and respiration are complementary and mutually perpetuating processes. Secondly, we find that the energy at the right of the equation is in two forms. About 60% of the energy (which comes from the electrons that form the covalent bonds in glucose) is given up as heat, or kinetic energy. The remaining 40% is stored as potential energy in a covalent bond in ATP. The 40% efficiency level of this reaction is equal to or better than the efficiency of the power plants that generate our electricity. The biological oxidation of glucose can regenerate 38 ATP molecules per glucose molecule.

Energy is released from ATP during cellular activity when a covalent bond is broken between the last two phosphate groups in the ATP molecule. The remaining molecules are **ADP** (adenosine diphosphate) and phosphate. The ATP is regenerated by rebonding a phosphate to ADP. (The P here stands for a phosphate group, or chemically, PH_2O_3, and adenosine is the combination of the sugar monomer ribose with the nitrogenous base adenine). There is a **coupled reaction**, or a linkage between the chemical reaction given above and the following reaction:

$$38ADP + 38P + energy \rightarrow 38\ ATP$$

For each molecule of glucose oxidized, 38 ATP molecules is the maximum number that can be recharged by these coupled reactions. The covalent bonds that link the terminal phosphates to the ATP molecule are relatively easily broken and re-formed and, therefore, are well-suited for their role as providers of the ''currency'' for energetic transactions within cells. The energy-yielding reaction that breaks ATP down into ADP plus phosphate requires water and the assistance of an enzyme.

Before we discuss the individual steps involved in the oxidation of glucose, we should get a better understanding of how energy can be released when electrons move from positions of high energy to positions having lower energy. We can use the analogy of a fluorescent light fixture in this regard. Fluorescent lamp bulbs contain inert gas atoms sealed within a tube. When you plug a fluorescent fixture into an electrical power source, the electrical energy excites the atoms in the tube, causing the electrons of the gas atoms to absorb some of the energy provided by the electricity. In doing so, the electrons assume higher-energy orbitals around the nuclei of the excited atoms. The electrons then spontaneously revert back to lower-energy orbitals and, in doing so, they give up the difference in energy between the higher and lower energy states. The energy given up is in the form of light, rather than the electrical energy that was provided originally. After the electrons revert to their lower-energy state, they can become re-excited, allowing the light fixtures to emit a constant level of radiant energy as this occurs again and again (as long as they receive electrical energy).

The cellular metabolic pathway for the complete oxidation of glucose also operates in a manner that takes advantage of electrons moving from higher energy levels to lower energy levels. The pathway makes use of enzymes and coupled reactions, as already noted, where reactions that yield energy are coupled with those that require energy. This pathway also makes use of **electron carriers** (NADH and FADH$_2$) that assist in the movement of the electrons. The metabolic pathway has three stages—**glycolysis**, the **Krebs cycle** and **electron transport**.

Glycolysis liberates a relatively small amount of energy for use by the cell, and it can take place throughout the cell's **cytoplasm** (the semi-fluid substance that fills the cell). This stage requires no oxygen and is therefore **anaerobic**. The last two stages (which sometimes are considered together as respiration in the sense of definition #1) do involve the respiratory gases and are therefore **aerobic**, and only occur in the mitochondria among eukaryotic cells.

In the glycolysis stage, the 6-carbon glucose molecule is broken down into two 3-carbon molecules of **pyruvic acid** (Figure 6.5). In this process, two ATP molecules are recharged and two electrons are carried off by NADH. The Krebs cycle produces carbon dioxide and involves the further breakdown and cyclic processing of carbon-containing units, with the primary result being the liberation of electrons to electron carriers. During the Krebs cycle, two pyruvic acid molecules yield six electrons to 6NADH and four to 2FADH$_2$, in addition to recharging 2ATP. In the last stage, electron transport, the electron carriers provide electrons to a cascading series of enzymes called the **electron transport chain**, that couple the energy given up by the electrons to the synthesis of ATP from ADP. In this stage, 34 ATP molecules are recharged for each glucose molecule that originally entered the metabolic pathway. This is also the stage where oxygen is used as the ultimate electron acceptor and is reduced to water. Figure 6.6 summarizes the major features of the oxidation of glucose.

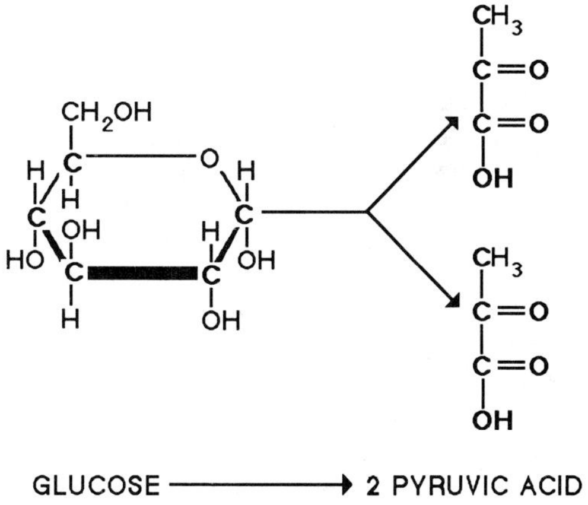

Figure 6.5. A chemical representation of glycolysis.

Even though these events take place continuously within the cells of fishes, invertebrates and even ourselves, they may be hard to relate to since they are so far removed from our scales of

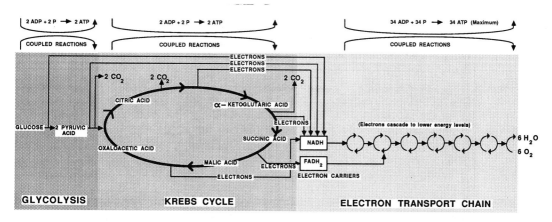

Figure 6.6. The three stages involved in the oxidation of glucose.

awareness. We can feel these processes, however, when it comes to strenuous muscular activity. During a very demanding series of muscular contractions, for example, it is common for much of the ATP that is immediately available to the muscles to be used up. This is true of a human running a race, a fish swimming to escape a predator, or a fish fighting a fisherman trying to reel it in. What happens in these situations is that an immediate demand for recharged ATP is created, and that demand is satisfied by glycolysis. Glycolysis is a relatively rapid process and it does not require oxygen, but it is also relatively inefficient in that only about 2% of the energy available in glucose is provided by the ATP molecules recharged by glycolysis. Glycolysis also creates an accumulation of pyruvic acid, which then converts to **lactic acid**. These acids, particularly lactic acid, can be toxic to the tissues at high concentrations, and their buildup during strenuous exercise makes our muscles ache. The level of these acids is reduced by processing them through the Krebs cycle and electron transport, but the latter stages often are not rapid enough to keep up with their production. In this situation, an animal generates an **oxygen debt** by proceeding with energy-yielding glycolysis at a faster rate than the subsequent steps can proceed. After the initial burst of activity is finished and, if the animal survives, the lactic acid that was produced is converted to pyruvic acid, which then enters the Krebs cycle until its concentration diminishes. Since the rest of the process is aerobic and requires oxygen, the animal ventilates or breathes heavily until the oxygen debt is repaid. Fish can die from generating an oxygen debt too great to repay (probably from acidosis), but humans rarely suffer a penalty that severe for strenuous exertion.

The topic of panting brings us neatly to the second definition of respiration—the exchange of respiratory gases. We have seen how carbon dioxide is liberated and oxygen is required during respiration, according to our first definition. But we still need to consider how to get those gases into and out of the cells, as our second definition of respiration suggests. There are four basic steps that serve to move these gases into and out of an animal's body.

We will consider the movement of respiratory gases using oxygen as our example. Movement of carbon dioxide utilizes the same steps that we will describe for oxygen, but in the reverse order. For aquatic organisms such as fishes, crustaceans and molluscs, the oxygen is present in a dissolved state in the water. The first step in the process involves the **bulk flow** of water past the animal's gills. Bulk flow means that the water is moving past the gills in volume, bringing the dissolved oxygen into contact with the cell membranes of the gills. The second step involves the diffusion of oxygen molecules across the gill membranes and into the blood. Diffusion proceeds without the input of energy from ATP, and it is therefore a **passive** process. In addition, **diffusion** occurs only from areas having a relatively high concentration of the diffusing substance to areas having a lower concentration of that substance, or **down a concentration gradient**. Oxygen diffuses into the body of an aquatic animal because the animal has used up its oxygen, creating a lower concentration inside than outside (movement of a substance **up a concentration gradient** is an **active** process that requires energy from ATP). The third step involves the bulk flow of oxygen-containing blood from the gills to the target tissues that need the oxygen.

While in the blood, the oxygen may either be dissolved in solution or bound to carrier molecules such as hemoglobin or hemocyanin. Also, the carrier molecules may either be free in the blood or within blood cells. Among fishes and mammals, the carrier molecule, hemoglobin, occurs within the red blood cells, while hemocyanin typically occurs free in the blood plasma. Movement of blood through the **circulatory system** is promoted by a pumping action of a blood vessel or a heart, which is a muscular contraction process that requires energy from ATP. Fishes and mammals have a **closed circulatory system**, where the blood is always contained within circulatory vessels. Crustaceans and most molluscs have an **open circulatory system**, where the blood may enter **open sinuses** within the tissues.

The fourth step of respiratory gaseous exchange takes place when the oxygen diffuses from the blood to target cells where cellular metabolism has created a low concentration of available oxygen. The same cells liberate carbon dioxide, creating a higher concentration of that gas, which diffuses out of the cells into the blood and is carried away to the gills. That finishes our look at energy and respiration. We hope that if you are asked why you eat and why you breathe, you will now have a lengthy answer.

Digestion and Excretion

Now we know something about how the intake of organic molecules in food is useful to animals. These compounds provide energy for body maintenance, growth, reproduction and movement. They also provide the raw materials (like the amino acids that make up proteins) for the synthesis of molecules needed for physiology and elaboration of new tissue. When animals feed, the organic molecules in their food are usually too large to be **assimilated,** or taken into the blood after crossing cell membranes of the digestive tract. The breakdown of organic molecules from food into smaller subunits that can be assimilated and used is accomplished by **digestion.**

Digestion tends to proceed in three stages—mechanical breakdown, chemical breakdown and assimilation. Mechanical breakdown serves to break the food up physically into smaller parts that are more easily processed by the digestive tract. The digestive tract is also called the **alimentary canal** (shown in Figure 5.36), and in a fish or a mammal it includes the following parts (given in the same order that food takes in passing through): mouth—esophagus—stomach—intestine—anus. In humans the mechanical breakdown of food (**trituration**) is accomplished by chewing. Some animals have accessory organs (such as crops) anterior to the stomach that accomplish mechanical breakdown. Mechanical breakdown exposes more surface area to the action of chemical breakdown. Most fish swallow their food whole and utilize little or no mechanical breakdown before the commencement of chemical breakdown.

Chemical breakdown results in the solubilization of food into smaller organic compounds. In humans, chemical breakdown actually begins in the mouth through the release of salivary amylase, an enzyme that breaks down starches, but most chemical breakdown among animals takes place in the stomach. This process is mediated by **digestive enzymes** and is accentuated by acidic conditions within the stomach. The chemical reactions themselves are hydrolysis reactions (hydrolysis means "to break with water") because water is consumed in the process. A series of enzymes are produced by glands and each enzyme breaks down (hydrolyzes) a particular class of food compounds. Proteins are hydrolyzed into peptides and amino acids, starches into monosaccharides, and so forth. Portions of the digestive tract are lined with specialized mucus so the tissues lining the digestive tract itself are not digested by the enzymes. The mucus also provides lubrication, while muscular action called **peristalsis** assists in the movement of material along the alimentary canal. Chemical breakdown yields nutritive compounds that are small enough to be assimilated. Portions of the food that are not assimilated are passed on through the alimentary canal and are eliminated as **feces.**

Assimilation takes place in the intestine. In mammals, the **small intestine** is for assimilation and it is a very long organ, while the **large intestine** is shorter and it is involved with the formation of the feces. Key features of the assimilation process include a great deal of available surface area and close proximity to the circulatory system (after a meal the supply of blood to your digestive tract is increased, drawing blood and oxygen away from the brain; that is why you feel sleepy after a meal). Assimilated substances cross cell membranes of the intestine either passively, by diffusion, or by active transport with the aid of membrane proteins and energy from ATP. The nutritive substances then enter the blood and circulate to the tissues.

Solid wastes are eliminated through the anus as feces, but for many animals indigestible solids may also be regurgitated. Symbiotic bacteria or other microorganisms often reside within the digestive tracts of animals, where they assist with various digestive processes, including the formation of feces. Bivalves have two kinds of feces—real ones and pseudofeces. Pseudofeces are pellets of material taken in by filter feeding, which are separated and ejected before entering the digestive tract. In aquatic ecosystems, feces are important food sources for many organisms.

Liquid wastes are eliminated by the kidneys, or the **nephridia**, depending upon the organism in question. Liquid wastes tend to include water, salts and nitrogenous wastes. Nitrogenous wastes are, for the most part, products of protein catabolism, and the excreted forms of these wastes among aquatic organisms are **ammonia** or **urea**. As we will learn in the next section, some of these processes and substances are very important to the salt and water balance of aquatic organisms.

Salt and Water Balance

This section addresses **osmosis** and **osmoregulation**. Osmoregulation concerns the maintenance of fluid balance within organisms—or the level of dissolved substances (**salts** and other **solutes**) within body fluids. Requirements for solute concentrations within body fluids tend to be rather specific, while the levels of solutes in aquatic media may vary widely. Osmoregulation, then, is a process that helps most aquatic organisms maintain an internal environment that is quite different from the chemical properties of their external environment.

Some of the physical processes that govern the relative distribution of water and solutes between the inside and the outside of a cell or an organism are similar to processes we encountered for gaseous exchange. For example, if you put some salt or sugar in a container of pure water and leave it alone, it will eventually dissolve and diffuse throughout the volume of water in the container until the concentration of solute is even throughout the water. This will happen even if the water is not stirred, because the solute will keep diffusing from areas of high solute concentration to areas of lower solute concentration (down a concentration gradient) until the concentration is even throughout (and there is no longer a gradient). Heat, a reflection of the kinetic energy or degree of motion of the water molecules, enhances the rate of diffusion. With osmosis, however, we introduce a new variable—the semipermeable membrane. Cell membranes are **semipermeable**, meaning they are permeable to some substances, like water, which can move freely across the membranes, but they are impermeable to others, such as salts and other solutes. That is why cells can maintain differential concentrations of solutes on their insides versus their outsides. Consequently, if there is more salt on one side of a cell membrane than the other, the salt cannot diffuse across the membrane to even out the concentration difference, *but the water can.* That is what osmosis is all about.

Osmosis can be defined as follows: the movement of water from an area of low solute concentration (and therefore high water concentration) to an area of high solute concentration across a semipermeable membrane that restricts the passage of solutes. That definition is quite a mouthful, but it is perhaps easier to remember when put this way—**water moves towards salt across a cell membrane**. In osmosis, water will keep diluting the salt until one of the following happens (for the sake of simplicity, we are going to refer to the sum of the various solutes as salt):

a)   the relative concentrations are evened out,
b)   the cell bursts, or
c)   the **osmotic pressure** causing water to move across the membrane is equalled by the physical (hydrostatic) pressure on the other side of the cell membrane.

The last case is common for plant cells since they have cell walls, which provide a rigid boundary to the cell membrane. When your house plant loses water, for example, the salt concentration of the cells increases and the plant wilts. When you rewater the plant, water enters the cells by osmosis until the cell membrane is forced against the cell wall and the hydrostatic pressure inside the cell increases (until it opposes the osmotic pressure of the water trying to get in). This physical pressure is called **turgor pressure**, and it is the reason why plants regain their stature after they are watered.

Now we will consider an animal cell and place it into three different osmotic situations. We will use red blood cells as an example, placed within beakers of water having various salt con-

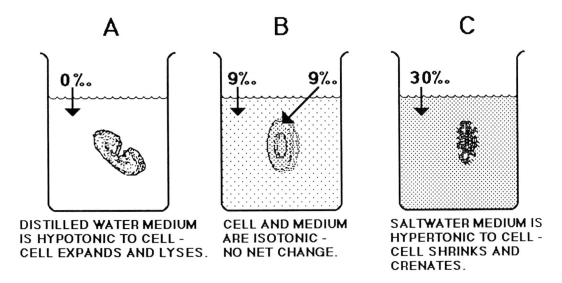

A   **B**   **C**

DISTILLED WATER MEDIUM IS HYPOTONIC TO CELL - CELL EXPANDS AND LYSES.

CELL AND MEDIUM ARE ISOTONIC - NO NET CHANGE.

SALTWATER MEDIUM IS HYPERTONIC TO CELL - CELL SHRINKS AND CRENATES.

Figure 6.7. Three possible osmotic scenarios that result when a cell (human red blood cell) is placed into media having different salinities.

centrations (Figure 6.7). Our experimental cells have an initial salt concentration of 0.9%, or $9^o/_{oo}$ (nine **parts per thousand**); they lack cell walls and have a semipermeable membrane. First, we put one cell into a beaker of distilled water. Since there is no salt in the water but there is some salt in the cell, water moves into the cell until **lysis** occurs (it bursts). The appropriate terms for the relative solute concentrations in this case are these—the cell is **hypertonic** (oversupplied with salt) to the medium, while the medium is **hypotonic** (undersupplied with salt) to the cell. In the second case, we place a cell into $9^o/_{oo}$ brackish water. Water moves across the cell membrane (because the membrane is permeable to water), but the amount that enters equals the amount that leaves; the cell does not change its shape because the cell is **isotonic** (same) with its medium. Finally, we place a cell in seawater, which has a **salinity** of $30^o/_{oo}$. Now the medium is hypertonic to the cell, and the cell is hypotonic to the medium. In this case, the cell **crenates** (shrinks like a raisin), since water will "follow the salt" and move to the outside of the cell.

    Unless an aquatic organism actively intervenes when the salinity of its medium changes, it will face changes similar to those undergone by our experimental blood cells. The extent to which organisms intervene is a measure of their osmoregulation. Aquatic organisms can be categorized as either osmoconformers or osmoregulators. **Osmoconformers** do not resist changes in the salinity of their environment and allow their internal condition to vary as their external conditions vary. **Osmoregulators** resist changes in the salinity of their environment and keep their internal conditions relatively stable. Most fish are osmoregulators, while invertebrates may either be osmoregulators or osmoconformers. Don't confuse these two terms with euryhaline and stenohaline. The latter terms refer to tolerance (survival) levels, rather than physiological response strategies. **Euryhaline** organisms can tolerate widely variable salinities while **stenohaline** organisms can exist only within a narrow range of environmental salinities. Estuarine and intertidal organisms are euryhaline, while lacustrine (lake-dwelling) and deep-sea organisms are stenohaline.

    Marine invertebrates tend to be isotonic with their environments. Some intertidal invertebrates, notably the bivalves, can simply shut themselves off from the environment or behaviorally reduce their exposure to unsuitable extremes in salinity for brief periods. Most cartilaginous fishes live in saline environments, and they are also nearly isotonic. These fishes retain **urea** (a nitrogenous waste product) in their blood to achieve an internal solute concentration near that of seawater. Bony fishes face a more complicated situation, since they are either hypertonic when in freshwater or hypotonic in seawater.

Bony fishes in freshwater are hypertonic to their medium. Water will move into their cells continuously, particularly across the thin membranes of the gills. To eliminate the excess water, these fishes urinate profusely. In addition, they retain most of their salts through the action of the kidney, they get some salts from their diet, and they have specialized cells in the gills that "pump" salts into the body by active transport, a process that requires ATP.

Bony fishes in seawater are hypotonic to their medium. Consequently, they constantly lose water through the gills and skin. They make up for the loss of water by swallowing seawater. They then eliminate the salts taken in with the seawater by active transport of the salts up a concentration gradient, using the chloride cells of the gills. Again, this process requires energy from ATP. Salts are also lost with the feces and urine, but the urine volume is minimal.

Anadromous and catadromous fishes travel between fresh- and saltwater environments, and undergo elaborate physiological changes that prepare them for the different salinities they face. For young salmonids preparing to enter salt water as they migrate downstream, this process is known as **smoltification**.

Temperature

Organisms live within a fairly narrow range of temperatures, and this is dictated in large part by the thermal properties of water. Apart from some unusual prokaryotes found within hot springs and hydrothermal vents, the upper temperature limit for living organisms is about 50°C. Above that, proteins **denature**, or lose their specific tertiary structures (that is what happens when a cooked egg begins to harden and turn white). The lower temperature limit for aquatic organisms is slightly below the freezing point of water, at –1°C or –2°C. The **tolerance limits** for individual species fall within this range.

Aquatic organisms that can survive a relatively wide range of **ambient temperatures** (the temperatures of their environment) are considered **eurythermal**. Intertidal marine organisms, and organisms that live in shallow lakes, ponds and streams can be expected to be eurythermal. **Stenothermal** organisms can withstand only a narrow range of ambient temperatures. Animals that live only in deeper ocean waters or near the bottoms of larger, deep lakes can be expected to be stenothermal. The terms described above only relate to tolerance limits and should not be confused with poikilothermic and homeothermic, which refer to body temperatures.

Poikilothermic animals have internal body temperatures that vary as ambient temperatures vary. Poikilothermic animals are sometimes erroneously called "cold-blooded," although if they are in a warm environment, they will usually have a warm body temperature. **Homeothermic** animals have nearly constant internal body temperatures, regardless of variation in ambient temperature, and homeothermic animals are sometimes referred to as "warm-blooded." To keep their body temperature nearly constant, homeothermic animals produce heat through their metabolic chemical reactions. They often have specialized circulatory systems that either centralize or dissipate heat, insulative systems for retaining heat and evaporative systems for eliminating heat. Fishes and invertebrates are poikilotherms, while mammals are homeotherms.

Since water has such a high heat capacity, it is an excellent buffer for sudden changes in atmospheric temperature. Large bodies of water have relatively constant temperatures. Heat transfer between an aquatic animal and the water it lives in is very rapid though, and that is why most aquatic animals are poikilotherms. That is also why water only a few degrees cooler than the air temperature feels so cold when you plunge into it. Only highly metabolic and highly insulated animals can be homeothermic in water. It also helps if they are large, so they have a relatively small ratio of surface area to body volume. Heat production of a homeotherm is proportional to its body volume, while heat loss to the water is proportional to its surface area. Some of the largest and more active fishes are actually semi-homeothermic, because they produce more heat by their metabolic activity than they lose to the water. These fishes include the great white shark, some of the larger tunas and billfishes. All of these are active predators, and they take advantage of the warming of their blood and shunt it into the head area to keep their senses and responsiveness as acute as possible. The advantage of homeothermy lies in the ability to react quickly regardless of ambient temperature and to maintain internal chemical reactions at relatively constant rates.

Besides its direct impact on survival (in terms of tolerance), temperature also bears important metabolic consequences. Relatively small changes in temperature produce relatively large changes in the rates of chemical reactions. In general, an increase of only 10°C is enough to double the rate of a chemical reaction. The temperature dependence of a process is typically denoted by its Q_{10}, which stands for the rate of a process at one temperature divided by its rate at a temperature ten degrees lower (for example, the rate at 20° divided by the rate at 10°). The Q_{10} of most biological processes falls in the range of 2.0 to 2.5. If the ambient temperature is too low for normal biochemical reactions, many poikilotherms are unable to move, or may move only feebly. Experienced recreational fishermen tend to appreciate the significance of water temperatures in determining the likelihood that fish will be "on the bite."

A number of other strategies allow aquatic animals to cope with the temperatures encountered in their environment. Some of these strategies are behaviorally mediated and include habitat selection, migration, torpor or hibernation, adjustment of muscular activity, timing of reproduction and life cycles, and vertical migration for the purpose of selecting optimum water temperatures. Temperature and oxygen content are two of the major physical factors that determine the distribution of aquatic animals, while the presence or absence of food and predators are the major biological factors.

Internal Communication

There are two major systems for internal communication, control of body functions and integration of internal information. They are the **nervous system** and the **endocrine system**. The nervous system is a rapid-response system, while the endocrine system is slower to respond but more prolonged in its actions. Another fundamental difference between the two systems concerns the mode of information transfer. Endocrine signals (**hormones**) are carried in the blood, and **nerve impulses** are conducted along cells of the nervous system called **neurons**. We will consider the endocrine system first.

Hormones, the chemical messengers of the endocrine system, are produced by glandular tissues or specialized organs known as endocrine glands. After they are released, hormones circulate throughout the body until they reach the so-called **target cells** of tissues or organs prepared to respond to the hormones. The hormones bind to receptor proteins in the cell membranes of target cells, which generate an appropriate signal indicating the time has come for a physiological response of some kind. Such a response might involve a digestive process, a reproductive state, preparation for a state of emergency, a change in the degree of function of the circulatory or excretory systems, or changes in growth or metabolism. The endocrine system may also interact with the nervous system. In general, though, hormones induce changes in the physiological state of an organism where immediate changes are not necessary, prolonged changes are possible, and many different cells or organs can be affected at the same time. Moreover, the hormones only affect target cells, which have specific receptor proteins for the hormones in their cell membranes. The receptor proteins are produced by a selective activation of genes within the target cells.

Hormones also mediate the development of an animal through its larval stages and metamorphosis. This is particularly true of invertebrates, which tend to have more diverse larval life histories than fishes. In recent years, functions of the endocrine system have come to have a significant impact on fishery production, because fishery scientists are now learning ways to manipulate endocrine systems in order to increase harvests. Aquacultural practices, for example, have been greatly enhanced by injecting pituitary gland extracts into various fishes to bring them into a reproductive state. These extracts, taken from the pituitary gland of a sacrificed fish, contain hormones which enter the circulatory system of the broodstock fish and bring about the desired changes in its reproductive system. Among invertebrates, a very simple procedure is now used to artificially bring broodstock shrimp into spawning condition. The procedure involves removal of one of the female's eyestalks, which also serves to remove a gland that keeps gonadal development in check. Without reproductive inhibition by the gland, the ovaries mature and the shrimp can be artificially inseminated to yield the offspring needed to stock shrimp ponds.

Desired hormones can also be introduced by **genetic engineering**. New genes can be administered to an animal so that when the new gene is activated a desired effect is achieved. If a gene is transferred from a human to a fish, for example, this yields a **transgenic** fish. The first transgenic fish produced by scientists received a gene that codes for the human growth hormone, and significant increases in growth rates may be possible through such genetic engineering of aquatic animals. Genetic engineering is also useful for producing sterile fish and shellfish for use in aquaculture projects where escape and interbreeding of captive organisms with wild species is undesirable.

Nervous systems allow physiological and behavioral responses in a matter of milliseconds (thousandths of a second). Instead of circulating among a large number of possible target sites, messages conducted by the nervous system reach a very limited number of cells and travel at signal conduction rates of hundreds of feet per second. As stated earlier, signals that traverse the cells of the nervous system are called nerve impulses, and the specialized cells that conduct these impulses are neurons. Neurons have very long cellular processes called **axons,** which act in a manner analogous to electrical wires—they conduct an impulse from one part of the body to another. An example of a neuron is shown in Figure 6.8.

The nerve impulse itself is the result of a sudden movement of electric charges across the cell membrane at the surface of the axon. The electric charges are carried by positive **ions**, primarily **sodium (Na+)** and **potassium (K+)**, as they rush through hydrophilic pores within certain proteins in the membrane (Figure 6.9). These pores are called **sodium channels** and **potassium channels**, respectively. Ordinarily there is a high concentration of sodium ions outside of the axon (but little potassium), and a high concentration of potassium ions inside the axon (but little sodium). When an impulse arrives at a site on the axon, it causes a shift in electric charges that results in a sudden opening of the sodium channel (when the axon is at rest, the sodium channels are biochemically "capped"). Opening the sodium channel allows a sudden influx of sodium ions across the axon membrane from outside to inside, creating an excess of positive charges on the inside of the axon. Sodium ions flow through the open sodium channel at a rate of about a million ions per second. This inward movement of ions is reflected by a reversal of the electrical charge at the membrane called **depolarization**, and the resulting electrical potential is measured in millivolts by electronic devices (Figure 6.10).

After the sodium rushes in, potassium rushes out through potassium channels, to compensate for the excess of positive charges inside the axon. The nerve impulse is propagated down the axon by triggering the same sequence of events at each successive point along the length of the axon. After an impulse passes, the original distribution of sodium ions (higher on the outside of the axon) and potassium ions (higher on the inside of the axon) is re-established by active transport of the ions across the membrane at the expense of ATP.

When the impulse reaches the end of the neuron, a chemical messenger called a **neurotransmitter** is released, and the neurotransmitter diffuses across a narrow space (called a **synapse**) between the end of the axon and the next nerve cell. Protein receptors in the membrane of the next

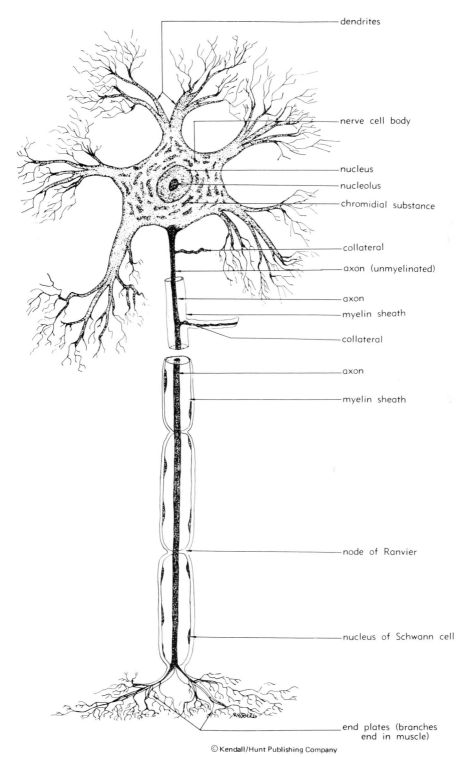

dendrites

nerve cell body

nucleus

nucleolus

chromidial substance

collateral

axon (unmyelinated)

axon

myelin sheath

collateral

axon

myelin sheath

node of Ranvier

nucleus of Schwann cell

end plates (branches end in muscle)

Figure 6.8. Diagram of neuron that relays electrical impulses to muscle tissue. (Courtesy of Kendall/Hunt.)

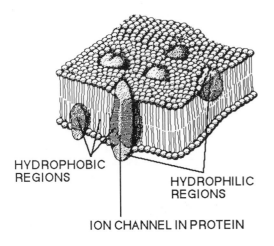

HYDROPHOBIC
REGIONS

HYDROPHILIC
REGIONS

ION CHANNEL IN PROTEIN

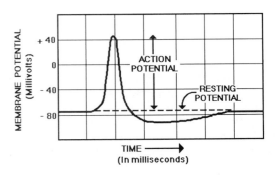

Figure 6.9. A greatly enlarged section of a cell membrane showing hydrophobic (''water fearing''), hydrophilic (''water loving''), and ion channel regions. The ion channels are located within membrane proteins.

Figure 6.10. A record of the electrical changes at a nerve cell membrane associated with a nerve impulse (as revealed by an oscilloscope).

cell bind to the neurotransmitter and initiate a nerve impulse that continues the signal down its axon. In this fashion, the propagation of a signal through an animal's nervous system involves conduction by both electrical and chemical means (electrical along the neuron and chemical between different neurons), while signals of the endocrine system are conducted only chemically.

It is interesting that many neurotoxins, including those produced by toxic fishes and toxic dinoflagellates, bring about their deleterious effects by physically blocking the ion channels in nerve cells. This results in an interruption of nerve impulses at the point where the channels are blocked, because the flux of ions through the channels is prevented at that point. Many anaesthetics that are used in medical practices also act on the nervous system in this way. We will learn more about the specific functions of the nervous system in the following chapter on senses and behavior.

7
Senses and Behavior

Frederick G. Johnson

Introduction

Knowledge of the senses and behavior of fishes and shellfishes can be important for a number of reasons. In capture fisheries, for example, fishermen need to predict when and where to find harvestable fish or shellfish. In addition, they need to know as much as possible about how fish react to the fishing gear or bait. Aquaculturists need to know even more about their animals than fishermen of wild stocks. For culture fisheries, knowledge of feeding and reproductive behaviors is critical. Moreover, any territorial or aggressive behaviors (such as cannibalism) need to be considered before placing organisms together in high densities.

In the foregoing chapter, we focused on the physiological processes that keep living things alive. In this chapter, we examine some of the many ways in which physiological processes allow fishes and other aquatic organisms to interact with their environment and with each other. Some of the same physiological systems that allow internal communication can be modified for the purpose of receiving information about the external environment. These systems form **sensors** and receive information in the form of a wide variety of **stimuli**, including chemicals, light, vibrations, touch, gravity, magnetism, temperature and pressure (sensors are also energy **transducers**). The sensors are linked to the nervous system so they can relay their information quickly and accurately. The sensors are also linked, via the nervous system, to **effectors**, which enable an organism to respond to the sensory input in ways that are appropriate to its survival, often on very short notice. The effectors include muscles and other organs. Sensors that receive stimuli originating outside the organism are grouped under the category called **exteroceptors**, while those that perceive internal stimuli are called **proprioceptors** and **interoceptors**. For most of this chapter, we will examine exteroceptors. A typical function of proprioceptors is to provide information about body position, while interoceptors provide information about the condition of the circulatory system.

An animal's overt actions constitute its behavior. Some of these actions may be responses to external (**exogenous**) stimuli, while other actions may be initiated by wholly internal (**endogenous**) cues. An animal will often have specific behavioral responses that are "preprogrammed" to follow the reception of a certain exogenous stimulus. A sensory stimulus that evokes such a behavioral response is called a **releaser**, because detection of the stimulus essentially "releases" the behavioral response.

In a sensor/effector system, the sensor relays its information to neurons that connect the sensor to the central nervous system—these neurons are called **afferent neurons**. Reception of the stimulus at a sensory cell generates a nerve impulse that propagates along the afferent neurons. The sensory information, now translated into nerve impulses, is evaluated and integrated in the central nervous system. If a behavioral response is to be initiated, this signal (in the form of nerve impulses) continues from the central nervous system to the effectors along **efferent neurons**. Efferent neurons that activate muscles are called **motor neurons**. Figure 7.1 shows an example of a sensor/effector system that begins with an exogenous stimulus (dissolved amino acids) and ends with the release of a behavioral response (search behavior).

In the rest of this chapter, we will consider the major senses by which aquatic organisms obtain information about their environment and a few examples of their typical patterns of behavior.

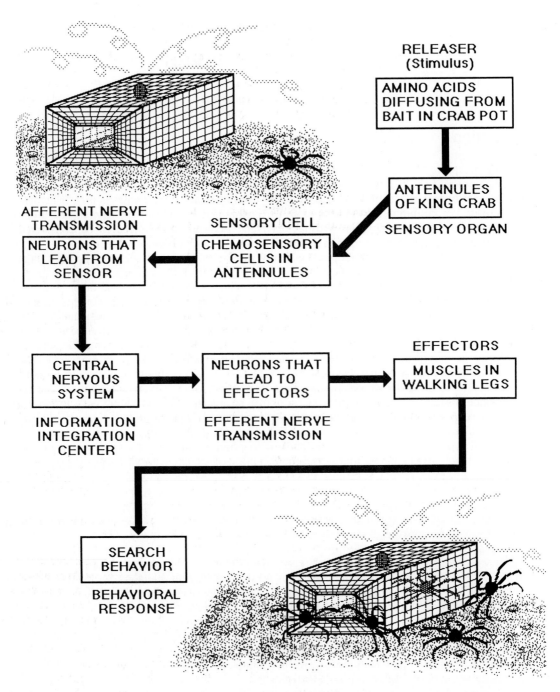

Figure 7.1. The sequence of sensory and neurophysiological events that allow a king crab to locate bait within a king crab pot.

Senses

Sensory cells and organs respond to a wide variety of stimuli by receiving them and initiating a signal that propagates along the afferent neurons. Besides the wide variety of stimuli that are detected, there is also an enormous range in the strength of a given stimulus that will generate a physiological response. For example, eyes are able to function over a wide range of light intensities and chemoreceptors are able to detect changes in the strength of a chemical stimulus over a wide range of concentrations. The general rule for biological sensors is that they are extremely sensitive to low levels of a stimulus but become progressively less sensitive as the stimulus level increases. As stimulus strength increases exponentially, receptor signals that register stimulus strength increase only linearly. This is known as **Weber's Law**, and it allows sensors to operate effectively over a much greater range of stimulus magnitudes than they would if the relationship between stimulus intensity and signal intensity was strictly linear.

Another general feature of sensors is that their output diminishes if the stimulus they are detecting remains steady and unchanging. This is known as **adaptation**, or habituation (some senses, such as sight and hearing, do not adjust in this way). If you walk into a room that holds the scent of perfume, for example, you notice the scent upon entering but your awareness of the scent soon fades. Only if you leave the room and then re-enter will you detect the scent again. Sensors are subject to both excitatory and inhibitory influences. Some sensory organs are able to convey a sense of directionality and distance. Our eyes, for example, can turn and focus on an object to give information on its direction and distance. Also, simultaneous comparisons by multiple sensors at different locations on the body (such as our ears or the nostrils of hammerhead sharks) can provide locational information about the source of a stimulus. Finally, as an animal moves about, the stimulus intensity may be compared over time (as long as adaptation is not a factor) to detect the source of a stimulus.

Sensitivity to Chemicals

The ability of an organism to detect chemicals is called **chemoreception**. Chemoreception is both the most primitive and the most ubiquitous sense among living things, and nearly all cells have chemosensory abilities. Chemosensory acuity is conferred by receptor proteins within the cell membranes, which bind in a specific fashion to the chemical agents that are detected. Even prokaryotic cells are capable of detecting chemosensory cues and responding to them. Flagellated bacteria, for example, can change the direction of rotation of their flagella to move either toward or away from a chemical stimulus. This simple form of chemosensory behavior is called **chemotaxis**.

A more complex organism such as a fish, mammal or other animal will use a host of chemosensory systems to discriminate among a range of chemical cues, both endogenous and exogenous (Chapter 6). We have already discussed some of the endogenous chemical cues—the hormones and neurotransmitters—which allow cells and organs within an organism to communicate with each other. Categories of exogenous chemical cues include the following.

- **Pheromones**—these are chemical messages that are released by an individual of a given species in order to elicit a response by another individual of the same species. The response may be behavioral or physiological in nature. An example would be a sex pheromone released by a female for the purpose of attracting a conspecific male (a male of the same species).
- **Allomones**—these chemical cues operate between individuals of different species. For example, an allomone emanating from a prey species might be detected by a predator and allow the predator to locate the prey.
- **Gamones**—these are released by gametes, or sex cells. For example, eggs that are released into open water for external fertilization may release a gamone that attracts the sperm to the eggs and thereby enhances fertilization.

Chemosensory perception may operate either upon physical contact with the source of a chemical stimulus (**contact chemoreception**) or over some distance between the stimulus source and its receptor (**distance chemoreception**). The sense of taste, or **gustation**, is an example of contact chemoreception, while the sense of smell, or **olfaction**, is an example of distance chemoreception. Gustation is typically less sensitive than olfaction. Olfaction tends to be an extremely acute sense—very few molecules of a chemical stimulus may be sufficient to initiate a response. An extreme example of olfactory sensitivity involves the male silk moth. If only a single molecule of sex pheromone released by a conspecific female binds to a receptor in the male moth's antenna, the male can initiate a search by flying upwind and can locate the female from a considerable distance. Among aquatic organisms, chemoreceptive behaviors are often linked with responses to water currents. An attractive chemical stimulus might, for example, cause a fish to swim against the current to locate the source of the stimulus.

In aquatic environments, chemical cues must travel through water by dissolving, and a chemical stimulus must therefore have aqueous solubility to be perceived at a distance. The chemoreceptors themselves may either be localized in certain organs, diffusely located over the body surface, or both. Catfishes have chemosensory barbels around their mouth and also have chemoreceptors distributed all over their bodies in the skin. Crustaceans have appendages near their eyes, called antennules, that they "flick" back and forth through the water to sense chemicals. Crustaceans also detect chemical substances with sensory hairs on the tips of their walking legs.

The primary distance chemoreceptors of salmonids are the nostrils, or **nares**, which are located on each side of the snout. The nares are oriented so that a one-way flow of water passes over chemosensory cells within each naris as the fish swims. Young salmon learn the chemical characteristics of their home stream by a process called **imprinting**, and are then able to find their home stream by olfactory cues when they return from the sea to spawn. If salmon nares are experimentally blocked with wax plugs, the fish are unable to find their home stream. The concentrations of chemical stimuli necessary to activate some of the chemosensory behaviors outlined above may be as little as one part per billion of the chemical stimulant in water!

Sensitivity to Light

Organs that detect light are sensitive to both the quantity and quality of radiant light energy. Light quantity refers to **light intensity**, while light quality primarily concerns the range of wavelengths that constitute the **visible spectrum**. (Some animals can also detect the plane of light polarity, which provides information about the position of the sun.) The range of visible light wavelengths is about 300 to 800 nanometers (nm, or billionths of a meter). The spectrum of light wavelengths that are visible to humans is 380 to 750 nm. The shorter wavelengths are at the blue/violet end of the spectrum and the longer wavelengths are at the red end. The energy in the light is inversely proportional to its wavelength. That is why ultraviolet light (which is not visible to humans but is visible to some other animals) can penetrate skin and cause damage to tissues, while longer wavelengths are less destructive.

In aquatic environments, the red wavelengths in sunlight are absorbed in the upper layers of the water column. (The exact depth to which red light extends depends on water **turbidity**, the degree to which light-absorbing particles are suspended in the water column.) Since marine environments tend to be deeper than freshwater environments, marine fishes tend to have eyes that are insensitive to red light. Freshwater fishes are sensitive to red light and often have red markings on their skin as visual signals to other fishes. Marine fishes that are colored red are actually camouflaged because their red colors are not seen by other fishes.

There are four basic kinds of light sensors. The first involves nonlocalized sensors in the bodies of simple organisms that respond to the presence or absence of light. These responses are called **phototropisms** if the animal or plant is sessile and **phototaxes** if it can move. Little is known about how these light sensors work, but they probably yield little information beyond light intensity and perhaps light direction. The other three types of light sensors are localized organs called ocelli, compound eyes and camera eyes (Figure 7.2).

Ocelli are simple light detectors which are sometimes called eyespots among some organisms. Each ocellus is bounded by a cup of darkly pigmented cells with sensory **retinal cells** on its inner surface and a lens-like layer of translucent or clear material at its external surface. Ocelli are probably not able to transmit a **visual image**, but do give information on light intensity and direction. Ocelli are common among zooplanktonic organisms and allow them to orient to light and detect sudden shadows.

Compound eyes occur in arthropods, although many arthropods have ocelli in addition to a pair of compound eyes. Compound eyes are constructed of hundreds of individual units called **ommatidia**, which are packed together and radiate outward from the middle of the eye to its surface, where they look like facets of a gemstone. Each ommatidium detects light emanating from a specific direction. In the aggregate, sensory input from the ommatidia probably does not allow the integration of a detailed visual image. However, any movements in the visual field would be signaled as an obvious "flickering" of light from one ommatidium to another, and this enhances detection of movements in their immediate environment. Compound eyes are very poor at detecting details at a distance.

Camera eyes form a **visual image**. Fishes, cephalopods and mammals have camera eyes, as do amphibians, reptiles and birds. Among all of these, even the invertebrate cephalopods, the structure of camera eyes is similar. There is an outer cornea, an adjustable iris, an adjustable lens and an inner lining of light-detecting tissue called the **retina**. The light-detectors within the retina are called **rods**, which function in low-intensity light and provide "night vision," and the **cones**, which function in high-intensity light and provide color vision. As a general rule, animals with acute vision tend to be predators, and animals with large eyes tend to be active at night. (Animals that are active at night are **nocturnal**, while animals active during daylight are **diurnal**.)

Light sensors act as transducers to change the quanta of energy in radiant light into a signal to the **optic nerve**. Regardless of the nature of the original stimulus, once signals from a light sensor or any other sensor are relayed to an afferent nerve, they are propagated through the nervous system in the same manner—as electrical nerve impulses. Only the format or timing of the impulses changes to reflect differences in the stimulus; the physiological characteristics of the impulses do not change. The mechanism of transduction between the incoming light and the nervous system is uncertain. We know that critical molecules, the **retinal pigments**, are involved. They may work like a fluorescent light (discussed in Chapter 6), only in reverse order. A given pigment is excited by light of a specific wavelength; it absorbs some of that energy (possibly by some of its electrons moving to a higher energy level) and then gives up that energy (as its electrons decay to their former energy level) to the signal propagation system. The eyes of most fishes are sensitive enough to provide them with more visual information than is usually available. That is because water clarity and the scattering of light in water limit visual acuity more than the structure and physiology of the eyes themselves.

Before we leave the topic of light detection, we should briefly consider biological light production. The production of light by organisms is called **bioluminescence** and is a very common ability among deep-sea animals. Sometimes the bioluminescence results from biochemical reactions produced by the animals themselves, while in other cases the light is produced by symbiotic bacteria within special organs. Bioluminescence has many functions, some of which are obvious and others mysterious. Most deep-sea fishes have bioluminescent organs called **photophores**. Each species has a particular pattern of photophores, and this allows them to recognize each other in habitats that lack sunlight. Many midwater fishes have photophores on their ventral sides only and are darkly colored on top. These fishes detect the amount of light coming down from the surface and match it with the output of their photophores, so they will not appear as a dark silhouette to a predator lurking below. Since their dorsal surface is dark, they also cannot be seen from above.

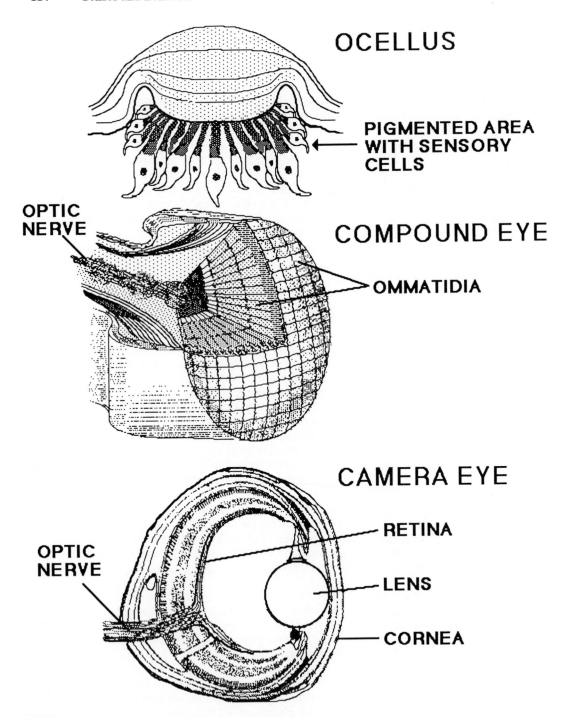

OCELLUS

PIGMENTED AREA
WITH SENSORY
CELLS

OPTIC
NERVE

COMPOUND EYE

OMMATIDIA

CAMERA EYE

RETINA

OPTIC
NERVE

LENS

CORNEA

Figure 7.2. Three types of light detecting organs, or eyes.

Other deepwater fishes have large photophores located above their jaws as lures for unsuspecting prey. Perhaps one of the most remarkable adaptations of a predatory fish involves the use of a photophore that is covered by a red filter and used as a beacon, or "spotlight," by the fish. This photophore is used to detect another fish as prey, and since most deep-sea fishes cannot see red light, the prey species is not aware that it is being "spotted." The predatory fish with the red photophore has a specially adapted retina that is red-sensitive, and it takes advantage of this system to consume other fishes.

Sensitivity to Gravity

We will consider two gravity sensors, one simple, one elaborate. The first is simple and the second is elaborate. Both allow an organism to orient itself relative to the surface of the water using gravity, where other cues, such as a visual detection of the surface, are absent. Both planktonic and nektonic organisms need to maintain an upright position in the water column, and they rely on information from gravity sensors to do so.

A simple gravity-sensing organ is called a **statocyst**, shown in Figure 7.3. Statocysts are the typical gravity sensors of zooplankton (except for larval fishes). Each statocyst is a capsule lined with sensory **hair cells** on its inside. Within the statocyst is an object called a **statolith** ("lith" means rock) that has a density greater than that of water. As the animal topples or loses its position relative to gravity, the statolith impinges upon different hair cells. The hair cells, in turn, then signal the nervous system that orientation has changed.

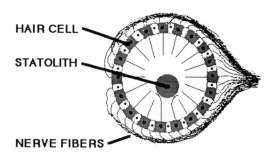

Figure 7.3. The statocyst, an organ that detects an animal's orientation with respect to gravity.

A more elaborate system that utilizes some of the same principles as statocysts appears in the **semicircular canals** of fishes (Figure 7.4). This is an organ of the **labyrinth system** (or acousticolateralis system) of fishes, and it is homologous with the inner ear of humans. The labyrinth system, however, facilitates detection of a number of stimuli in addition to gravity. There are three semicircular canals, each oriented in a different plane. They are filled with fluid and contain numerous tiny particles that are denser than the fluid. The otoliths are located near the semicircular canals. There are three pairs of otoliths; the two smaller pairs relate to the semicircular canals. **Neuromasts**, the sensory units of the labyrinth system, respond to the motion of both the small otoliths and the fluid in the semicircular canals. The large pair of otoliths are the ones used for ageing.

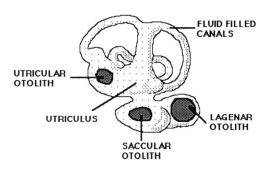

Figure 7.4. The labyrinth system of fishes, showing the three semicircular canals and otoliths.

If the fish is stationary, the dense particles settle on neuromasts and indicate body orientation with respect to gravity. If the fish moves, the fluid and the particles in the semicircular canals shift to different regions, depending on the plane of motion and the acceleration, and stimulate neuromasts in those regions. This sensitivity allows the fish to also detect changes in direction and rate of body motion. The labyrinth system thereby provides a feedback system for the effectiveness of muscular swimming movements and is important to the maintenance of muscle tone. The labyrinth system also serves an additional function—the detection of high frequency vibrations (sound).

Sensitivity to Mechanical Vibrations

Vibrations are pulses of mechanical deflection that propagate through a medium such as air or water, and the frequency of a vibration refers to the number of pulses that are propagated per second. What we know as sounds result from the high-frequency compression of air molecules that translate some of their mechanical force to our eardrums. In water, high-frequency vibrations travel even more effectively than they do in air, since the water molecules are packed more closely together than molecules are in air. As mentioned above, high-frequency vibrations in water are detected by the labyrinth system among fishes.

We humans hear high-frequency vibrations as high-pitched sounds, and as the frequency diminishes, the sounds have a lower pitch. When the frequency decreases to a point where the sounds are no longer audible to us, we can still feel the vibrations with our fingers if we touch material connected to the vibrational source. (The sense of touch is called the **tactile** sense, and most animals have tactile sensitivity.)

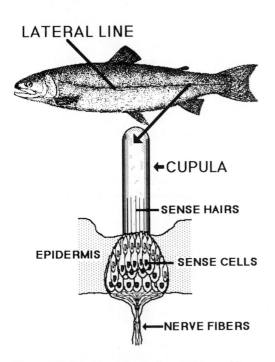

LATERAL LINE

←CUPULA

SENSE HAIRS

EPIDERMIS

SENSE CELLS

NERVE FIBERS

Figure 7.5. The lateral line system of fishes, with an enlargement of the sensory structures located in pits along the lateral line.

As it turns out, fishes have a more acute sensitivity to low-frequency vibrations than we do, and it provides them with a sense of what we might call "distant touch." This sense is provided by the **lateral line system** of fishes, which consists of a series of sensory pits usually located near the middle of the body along the right and left sides (Figure 7.5). Each sensory unit of the lateral line is a neuromast organ and has sensory hairs within a capsule known as a **cupula.** When waves of water movement pass over the lateral line system, each cupula is bent slightly. This action in turn deflects the hairs of the sensory cells of the lateral line, causing them to dispatch an appropriate signal to the nervous system. The lateral line system provides fishes with a host of abilities. Among these are the ability to avoid bumping into objects while swimming in darkness, the ability to maintain position with a school of fishes (and change swimming direction almost instantaneously as the school changes its direction of travel), and the ability to sense the approach of a predator or the abnormal swimming movements of a disabled prey species. All of these abilities accrue from detecting the low-frequency vibrations carried by water between solid objects and the lateral line sensors. Invertebrate animals such as crustaceans have hairlike exteroceptors, called **sensilla,** that are sensitive to vibrations and water movements. The abilities of the invertebrates to perceive this type of sensory information are not nearly so acute as those of the fishes, however.

Another fish organ involved with vibrations—the **swimbladder**—is a gas-filled chamber that lies within the body cavity of bony fishes, just ventral to the kidney (shown in Figure 5.36). An accessory function of this organ among some fishes is to act as a resonating chamber for the enhancement of sound detection, through its anatomical connections with the labyrinth system. For certain fishes, such as drums and croakers (family Sciaenidae), the swimbladder can produce sounds in concert with the actions of specialized muscles that are attached to the swimbladder. The primary function of the swimbladder, however, is to control the buoyancy of the bony fishes. Cartilaginous fishes lack swimbladders and must swim continuously or sink.

Miscellaneous Sensitivities

Aquatic organisms have a number of sensory modalities in addition to those considered above, but in many cases we know little about how they work. For example, it is known that most aquatic animals are sensitive to temperature, and motile animals often distribute themselves on the basis of water temperature gradients. Little is known, however, of the nature of the temperature sensors themselves and the means by which changes in temperature are detected. Another very common sense among aquatic organisms that scientists know relatively little about concerns sensitivity to **hydrostatic pressure**. Hydrostatic pressure is a function of depth within the water column. Many aquatic organisms are quite sensitive to changes in hydrostatic pressure, but in most cases, how they detect their depth remains a mystery. Somewhat less mysterious is the ability of marine mammals to produce sounds and detect objects by **sonar**, or the return of the reflected sound waves. Dolphins have the ability to locate unseen objects by sonar.

Perhaps one of the more interesting senses is the sense of time. We know that most organisms are aware of the passage of time. When we sequester an organism from all cues that might give it an idea of what time it is, such as photoperiod (light and dark cycles), temperature cycles, tidal cycles and the like, the organism continues to exhibit a daily activity pattern that corresponds approximately to a 24-hour cycle. Scientists call this endogenous sense of timing a **circadian rhythm** (circadian means "about a day," so this term means that organisms have an approximate, but not exact, sense of day length). Both animals and plants have this ability and often make use of it to establish the onset of daily activity patterns, as well as migratory, reproductive and other timed behaviors. We do not know exactly where the **biological clocks** that confer this sense are to be found within organisms, nor do we know how they work. We do know that the biological clocks are **temperature compensated** and have a Q_{10} of 1.0. That means the clocks do not speed up or slow down as temperature changes (which is unusual for biological processes, most of which are **temperature dependent**). We also know that if we replace an organism's normal water, H_2O, with D_2O (deuterium oxide, or **heavy water**—deuterium is a heavy isotope of hydrogen and has an extra neutron in the atomic nucleus), the biological clock of that organism slows down somewhat. Scientists therefore suspect that the workings of the clock may be a biochemical process.

Another sense that is attracting modern research attention concerns the ability of many aquatic animals to detect the orientation of the earth's **magnetic field**. In essence, this ability provides an animal with an internal compass, which is extremely useful during migrations for navigating when other directional stimuli are lacking. Recent studies seem to indicate that tiny iron-containing cellular inclusions might be involved in sensitivity to magnetic fields. Whatever the mechanism may be, the earth's magnetic field is relatively weak, and organisms that are able to orient their movements to it must have a very acute magnetic sense.

The last sensory mode we will consider concerns **electric fields**. As we learned in Chapter 6, the physiological processes of animals involve changes in electrical potential across cell membranes, and these electrical differences occur also at body surfaces as an animal proceeds with its normal patterns of activity and physiological functions. Some fishes have adaptations that allow them to take advantage of the electric fields that their prey produce as a result of these activities. Sharks, for example, have specialized organs in their head area called **ampullae of Lorenzini**, which are sensitive to weak electrical fields (they also detect magnetic fields). Sharks can locate a flatfish buried beneath the sand by detecting the electric field produced by contractions of the flatfish's respiratory muscles. **Electric fishes** (including electric eels and electric rays) produce their own electric fields by emitting pulses that project over short distances (a meter or less). The electric organs that generate the pulses are modified muscle tissues. Objects within the field cause deflections which are detected by modified neuromast organs, even in conditions of total darkness or low water clarity (high turbidity). Moreover, some electric fishes can produce strong pulses or bursts of electricity that can stun their prey or other organisms that disturb them.

Behavior

In connection with our discussions of the senses of aquatic organisms (Chapter 6) and of feeding strategies (Chapter 5), we encountered examples of animal behavior. In this section we will discuss a few more of the myriad ways in which animals exhibit predictable patterns of behavior.

In the study of behavior—**ethology**—we can organize categories of behavioral responses into a series that progresses from simple to more specialized and elaborate processes, as follows: tropisms, taxes, reflexes, instinctive responses, learned responses and reasoned responses. **Tropisms** are limited movements, usually accomplished by differential growth, towards or away from the source of a stimulus, by a sessile organism. The tips of a plant growing towards a light source is an example of a tropism (**positive phototropism**, in this case). **Taxes** are movements of motile organisms towards or away from the source of a stimulus. A zooplanktonic animal swimming away from a light source is an example of **negative phototaxis**. **Reflexes** are responses to stimuli that occur automatically in an animal, meaning that they occur without the integration of sensory information by the central nervous system. However, reflexes do involve the linkage of sensory organs with neurons and **ganglia,** or the **spinal cord**, but not with the brain. The knee-jerk response of humans is an example of a reflex that involves a neuronal pathway extending only as far as the spinal cord (the afferent pathway extends from proprioceptors in the tendons of the knee to the spinal cord, and the efferent pathway extends from the spinal cord to the leg muscles). **Instinctive responses** are behaviors that are genetically preprogrammed actions that organisms do without learning, although these behaviors may be modified by learning from experience. A birdsong is an example of instinctive behavior since a young bird can recite the songs specific to its species without previously hearing it from other birds. After it has heard the songs of older birds, however, the young bird will refine and perfect its own song, thus adding a learned component to its innate instinctive behavior. **Learned responses** are those based totally upon experience. A child learns not to touch a hot stove even though it is not born with that instinct. **Reasoned responses** allow an animal to respond in an appropriate fashion to a stimulus even if no prior exposure to that stimulus (no prior experience) is involved. Marine mammals seem capable of reasoned responses, as are humans. Fishes and some invertebrates are capable of learned responses. Instinctive, learned and reasoned responses all involve integration of information by the central nervous system.

The behavioral responses of fishes and invertebrates to various stimuli can have very important consequences to fisheries. We noted in the introduction to this chapter how this information may be useful to capture fisheries and aquaculture. Being able to predict behavioral responses can also be of critical importance to protecting aquatic organisms from harm that might result from unnatural modifications of their environment. When we build dams or other structures in aquatic habitats, we need to know how fishery resource organisms will respond to these new conditions. It may be necessary, for example, to guide fishes around structures that might harm them or impede their migrations. Screens, fish ladders and other guiding devices are often designed for this purpose, based on an understanding of the behavioral responses of the animals we wish to protect. In the following sections we take a look at some common patterns of behavior shown by aquatic animals.

Migratory Behavior

Migratory behavior concerns the purposeful and predictable movement of an animal through its environment. Migrations can involve large distances (thousands of miles) and long time periods (years) or short time spans and distances, depending upon the animal in question. In addition, both horizontal and vertical axes of movement may be involved in aquatic environments.

An example of vertical migrations that take place daily are **diel migrations**. These are movements of organisms up and down in the water column (shallower and deeper) in a daily cycle. Many zooplanktonic organisms display diel migrations, and they do so in such density that they are sometimes detectable by sonar. These animals are known collectively as the **scattering layer**. Organisms of the scattering layer move up in the water column when sunlight declines and move deeper when the sun is high. The scattering layer consists primarily of crustaceans and small fishes, and their migrations span distances of a few hundred feet up and down each day. Circadian rhythms and photoperiod are probably the most important sensory cues that mediate this behavior.

We are more familiar with horizontal migrations, some of which are spectacular. Many of the fishes, such as salmons and tunas, migrate thousands of miles in their lifetime (some marine mammals, turtles and birds also have spectacular migratory habits). Usually these long-distance migratory behaviors involve a partitioning of habitats for feeding and reproductive purposes (discussed further in Chapter 12). For example, it is best to grow in a habitat where there is an abundance of large food items, but it is best to reproduce in a habitat where there is an abundance of small food items for the offspring and a lack of predators. Most commercial fisheries involve the migratory phase of a target species, since the animals tend to be most concentrated and easiest to capture during migrations.

Animals that migrate between freshwater and saltwater environments are **diadromous**. There are two patterns of diadromy—anadromy and catadromy. **Anadromous** fishes such as salmons (family Salmonidae), sturgeons (family Acipenseridae), and smelts (family Osmeridae) spawn in freshwater habitats and migrate to seawater to feed and grow. They later return again to the same freshwater habitat where they were born, and this is known as **homing behavior**. **Catadromous** fishes such as American eels and European eels (family Anguillidae) do just the opposite—they spawn in the sea and migrate to freshwater habitats to feed and grow. Considerable physiological demands are placed upon diadromous fishes in terms of osmoregulation (Chapter 6), but the presence of food and suitable reproductive conditions, along with the absence of predators in certain areas, presumably outweighs the physiological difficulties. Anadromy is prevalent in temperate areas, where ecological productivity in the ocean is high, but in the tropics catadromy is more common because tropical marine productivity is relatively low compared with the productivity of tropical freshwater habitats. The primary sensory modes involved in horizontal migratory behaviors include geomagnetic sensitivity, orientation by the position of the sun and olfaction.

Associative Behaviors

Animals associate with each other for many reasons. We have already encountered examples of **interspecific** associations (associations between different species) in the three forms of symbiosis—mutualism, parasitism and commensalism (Chapter 4). The major purpose of interspecific associative behavior in aquatic ecosystems is predation. **Intraspecific** associations involve only members of the same species, and we will consider a few examples of these.

Reproductive behaviors are an obviously necessary form of intraspecific association. Many aquatic animals come together in swarms or dense aggregations of individuals to reproduce. Others pair off or form more limited assemblages. Elaborate behavioral rituals, including sexual displays, aggression and territorialism, may accompany reproductive actions among both vertebrate and invertebrate species. For those animals with relatively few offspring (low fecundity), **brooding behavior** is a common way to increase the chances of offspring survival. Many types of sensory stimuli are important in mediating reproductive behaviors among animals. Chemosensory cues are most important, along with visual cues and displays, tactile cues and, for some species, auditory signalling. Even displays of bioluminescence may be involved for those animals that pair off in darkness. Animals with reproductive behaviors that include mate selection, territoriality, sexual display and intraspecific aggression also tend to have **sexual dimorphism** of mature adults, meaning that the males and females will look different from each other.

Schooling behavior is a form of associative behavior that we tend to identify with the fishes. In general, it is small fishes that form schools, and it seems that schooling behavior is an adaptation to avoid or minimize predation. It may be easier and more effective for a predator to concentrate on attacking an isolated fish than trying to pick fish out of the confusion presented by a tightly packed and moving school. Moreover, fishes are often foraging while they are in schools, and it is inefficient to feed and watch for predators at the same time. In a school, the chances are high that at least one fish will notice an approaching predator, and fish in a school can therefore devote their primary attention to foraging. Their lateral line organs and their eyes provide the primary senses that enable fishes to form schools. It is perhaps ironic that schooling behavior, an adaptation to reduce vulnerability to predation, also leads to increased vulnerability to capture by fishing vessels. Modern fishery technology makes it possible for whole schools of fishes to be located, surrounded by purse seines or other gears, and captured (these techniques are discussed in Chapter 15). The capacity to overfish schooling species of fishes makes it particularly important to manage these fisheries carefully.

Many invertebrates also form dense aggregations, although they are not called schools. Juvenile king crab, for example, pile on top of each other at times, forming large mounds on the sea bottom. It is common for squids, invertebrate larvae and other zooplankton to form swarms in the water column. Spiny lobsters have an interesting behavior called queueing, where they line up along the bottom in long lines and travel together as a unit (this is probably related to migratory behavior). Another common pattern of invertebrate associative behavior concerns **habitat selection**, in which young animals preparing to settle on the bottom after leading a planktonic larval existence often do so only if they detect the presence of conspecific adults. Invertebrate settling behavior is often triggered by chemical cues emanating from the adults, or their shells.

Before we leave the topic of associative behavior, we should mention one of the most elaborate forms of interspecific ritualized behavior known, the **cleaning stations** of tropical coral reefs. These are specific localities in a coral reef that support a number of small fishes, juveniles of larger fish species and shrimps, all of which serve as cleaners of parasites from larger fishes. The symbiotic relationship between the cleaners and the fishes that are cleaned is mutualism, since both parties benefit from the association (parasitism of the larger fishes, of course, is also involved). Ritualistic displays by both the cleaner species and the larger fishes are involved in cleaning behavior. Some cleaner species even seem to signal human divers to pause at cleaning stations so that they too can be inspected for parasites. In larger cleaning stations, a dozen or more larger fishes may form a queue, waiting in line for their chance to be freed of parasites on their skin or in their gill chambers. The displays that mediate cleaning behavior appear to be mostly visual in nature and serve to indicate a lack of aggressive intent, for the larger fishes certainly could attack and consume the cleaners.

Defensive Behavior

Since predation is the major factor that determines the natural abundance of animals, defenses against predation are also of major importance in aquatic ecosystems. Defensive behaviors include color changes for the purpose of **camouflage** (accomplished through the actions of chromatophores), **escape behaviors**, erection of **spines** or activation of **venomous defensive structures**, becoming firmly lodged within **crevices**, ejection of **ink** (cephalopods), **burial** within sand or other soft substrates and a host of others. The puffer fish and the porcupine fish have a unique form of defensive behavior in that they suddenly inflate themselves by taking in water if they are threatened or swallowed by a predator. The surprised predator then spits the puffer or porcupine fish out immediately, or avoids it because it has become too large to swallow. Animals that have noxious defensive capabilities often are brightly colored, and this serves to remind an attacker that the result of an attack will be unpleasant for the predator. (For **warning coloration** to work as an adaptive strategy, the predator must have good eyes, color vision, the ability to learn from experience and a good memory.)

Another very important defensive behavior concerns **avoidance**. Avoidance behavior is particularly advantageous when it is chemically mediated. If avoidance is solely a function of visual cues, the avoidance response may be too late because predators usually move faster than their prey and can see better as well. For this reason, many species are able to detect their predators at a distance by olfaction and then are able to initiate an appropriate avoidance response in advance of being seen.

Since aquatic animals live in a different medium than we do, there is still much for us to learn about their behavior patterns. As our knowledge of these animals increases, we can often put that knowledge to good use to protect fishery resources and to make our fishery harvesting methods more efficient as well.

8
Age and Growth

Donald R. Gunderson and Katherine W. Myers

How big can a fish grow and how fast? How old is it? Most of us can remember asking these questions as children (Fig. 8.1). Fisheries scientists have developed a wide variety of techniques to answer these questions. But this is more than a trivial pursuit of interesting facts about fish. Collection and analysis of age and growth data can be very time-consuming and expensive. In fact, some scientists devote entire careers to this work, and fisheries management agencies spend major portions of their annual budgets to collect and analyze data on age and growth. The study of age and growth is very important. In this chapter, we consider some of the reasons for these studies and take a look at how they are carried out.

Growth is a universal biological process for all living things. Growth occurs in cells, tissues and organ systems; in the organism as a whole; in **populations,** or groups of inter-breeding organisms; and in **biological communities,** or groups of interacting plants and animals. Growth in individuals is simply the process of becoming bigger by taking in food. The energy obtained from food is used for the growth of new tissues. Growth of individuals is the basis of **production,** or total tissue elaboration, of populations. Ultimately, the study of fish growth is important because our very survival depends on the production of our food resources. We need to determine **age,** the length of life to a specified time, to estimate

Figure 8.1. Two young boys, ages 12 and 11, with a quillback rockfish, *Sebastes maliger,* and a lingcod, *Ophiodon elongatus,* caught at Possession Bar off Whidbey Island, Washington. (Courtesy of Ken Chew.)

growth. Information on such factors as rate of growth, longevity, age of sexual maturity, age at migration, the length of time it takes for fish to reach catchable size and catch by **age group** (a group of animals of the same age) provides us with important clues to the structure of fish populations and the type of management required to maintain the necessary levels of production.

We can characterize the wide variety of fishes, shellfishes and marine mammals that contribute to aquatic production by their age and growth. Species such as anchovies (Engraulidae), herring (Clupeidae), sand lance (Ammodytidae) and shrimp mature relatively early, have small maximum sizes, put a high proportion of available energy resources into reproduction and are short-lived. California anchovy (*Engraulis mordax*), for example, mature at 1 year of age, reach a maximum size of about 18 cm (7 in), and few live more than about 6 years. They spawn 5 to 24 times annually (depending on age), and the total number of eggs produced represents about 65% of the body weight of a mature female. Other species such as Pacific dogfish (*Squalus acanthias*), halibut (genus *Hippoglossus*), and blue whales (*Balaenoptera musculus*) are relatively old at maturity, reach a large maximum size, put a relatively small proportion of available energy resources into reproduction and are long-lived. Pacific dogfish mature at 23 years, reach a maximum size of about 112 cm (44 in), and commonly live about 55 years. Dogfish give birth to live young after one of the longest pregnancies of any vertebrate (almost 2 years). In contrast to the anchovy, a dogfish ovary with fully developed eggs weighs only 8% of the body weight of a mature female.

To study growth, we first need measurements of size, usually length or weight. Depending on the situation and the condition of the fish, we can use different measurements to determine its length (Figure 8.2). **Standard length** (from the tip of nose or snout to the end of the hypural plate or base of the caudal fin) is often used by taxonomists. Biologists working in the field and fishermen typically use **fork length** (the distance from the tip of the snout to the fork of the tail, **TSFT**) or **total length** (the distance from the tip of the snout to the tip of the tail). The snouts of spawning Pacific salmon (*Oncorhynchus* spp.), particularly the males, often become elongated and deformed; the length of these fishes is measured from the middle of the eye to the fork of the tail (**MEFT**).

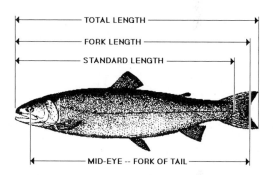

Figure 8.2. Types of measurements used to determine fish length.

We can measure weight directly, for example, by placing the fish on a balance or scale, or indirectly from length. For most species, growth in weight (w) increases as an exponent of length (l). We can use the **length-weight relationship** (w = al^b, where a and b are constants) to calculate weight from length. The value of the exponent, b, is usually near 3 because growth in length takes place in a single dimension, while volumetric (and hence weight) growth takes place in three dimensions. When b = 3, growth is **isometric** (that is, growth is similar in all dimensions). Another form of this relationship, **condition factor** (K = w/l^3), is an index of 'fatness' or general 'well-being.' The value of K is smaller for fish that are long and thin, and larger for fish that are short and thick. Indices of condition have been formulated for other types of animals, as well. For example, an index of condition for oysters is calculated by dividing the dry weight of the meat by the volume of the shell cavity and multiplying by 100. Condition indices are useful for determining when to harvest a stock. Oysters, for example, have a low condition index after spawning and are less marketable than when their condition index is high.

Traditional methods for studying fish growth include **direct observation, tagging and marking** and **back-calculation.** To directly observe growth, we can place fish of known size in an enclosure and measure them at intervals of time. We can also release tagged or marked fish of known size and measure them again when they are recovered. To back-calculate lengths at earlier ages, we determine the relationship between the length of a fish and scale size (Figure 8.3), and we infer that the growth of the scale and of the fish as a whole was proportional to each other over the life of the fish.

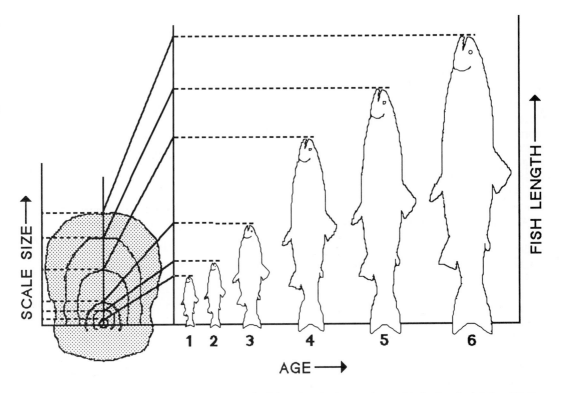

Figure 8.3. The relationship between the length of a fish and scale size can be used to back-calculate lengths at earlier ages.

Traditional methods of studying fish growth do not provide an index of growth rate at the time the fish is examined. Consequently, in the 1950s scientists began to study the relationship between nucleic acids, which play a fundamental role in the synthesis of proteins, and growth rates of fish (as described in Chapter 6). The total quantity of **deoxyribonucleic acid** (DNA) in the nucleus of normal cells is constant within a given species. **Ribonucleic acid** (RNA), which is involved in the synthesis of new protein, is present in variable quantities in the nucleus and cytoplasm of the cell. Fast-growing tissues have larger quantities of RNA than slow-growing tissues. Thus, the ratio of RNA to DNA (RNA per unit DNA) is considered a more accurate index of growth than RNA concentration alone, because the ratio is not affected by differences in the number of cells analyzed. The uptake or incorporation of radioactive **amino acids**, the building blocks of proteins, provides another index of the growth rate at the exact moment the fish is examined.

No universal law of growth has been discovered, but we often use mathematics to try and make generalizations about growth. Most of these mathematical models are based on the idea that an animal's growth rate is a function of its body size. The growth in weight of individuals as well as populations tends to follow a **sigmoid** or s-shaped curve (Figure 8.4). Growth is rapid prior to sexual maturation, and the growth curve is concave upward. Later, as the maximum size (W∞) or **asymptote** is approached, growth slows, and the curve bends and is concave downward. We can use the **von Bertalanffy** growth model to characterize the growth in length of most fishes and molluscs over their lifetime (Figure 8.5). The relationship shown in the figure bears significant consequences for a fishery, since fishing should be concentrated on individuals that have had a chance to pass through their most rapid growth phase.

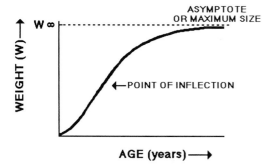

Figure 8.4. The growth in weight of individuals as well as populations tends to follow a sigmoid or s-shaped curve.

Growth of crustaceans is a special case, since species such as shrimp, crabs and lobsters must all molt (shed the confines of their rigid exoskeleton) before they can increase in size. Molting frequency varies with size and age, however, so individual growth follows a series of discrete stair-step trajectories (Figure 8.6). But the **stepwise** growth of crustaceans is not really that different from the growth of molluscs and vertebrates if considered on the basis of tissue elaboration because most of the sudden increase in size is due to addition of water.

Mathematical models help us to understand the general pattern of growth, but they do little to explain the true complexity and underlying mechanisms. Growth is extremely variable, changing with age, sex, migration, season, activity, amount of living space, quality and quantity of food-supply, temperature, and many other physiological and environmental factors. Our understanding of how physiological and environmental changes affect growth can be increased by studying **bioenergetics**—the partitioning of the energy

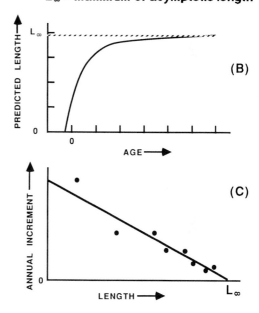

(A)

von Bertalanffy Growth Model:

$$\frac{\Delta l}{\Delta t} = K (L_\infty - l)$$

$\frac{\Delta l}{\Delta t}$ = growth increment
t = time
l = length
K = growth coefficient
L_∞ = maximum or asymptotic length

(B)

(C)

Figure 8.5. (a) The von Bertalanffy growth model can be used to characterize the growth in length of most fishes over their lifetime. (b) A graphical representation of the von Bertalanffy growth model in terms of length. (c) Plot of length increment against initial length. The growth coefficient (k) can be estimated from the slope $(1-e^{-k})$ of this line.

obtained from food among growth, maintenance and reproduction. This relationship may be generally expressed in terms of energy equivalents as $\Delta W/\Delta t = R - T$, where ΔW is growth, R is the total energy obtained from food rations, and T is the total energy used for maintenance, all during a unit of time Δt. The left side of the equation is a rate function or the growth rate. When R = T, the growth rate is zero (that is, no growth occurs). If R is greater than T, the growth rate is positive.

Partitioning of energy varies with the developmental stage of the organism. **Larvae**, immature forms which differ in structure from the adult, use energy for rapid growth and ultimately change to the adult form. **Juveniles**, which are of adult form but sexually immature, use energy primarily for growth, and secondarily for maintenance. **Adults** or sexually mature individuals, use much of their

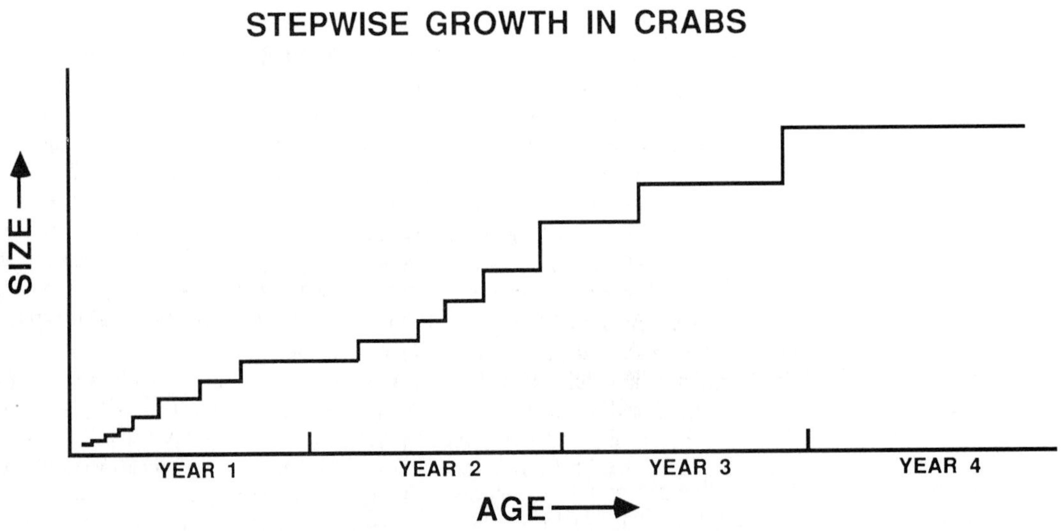

STEPWISE GROWTH IN CRABS

Figure 8.6. Crustaceans must molt (shed their rigid exoskeleton) before they can increase in size. Molting frequency varies with size and age, so individual growth follows a series of discrete stair-step trajectories.

energy for reproduction (development of eggs and sperm, spawning migrations, care of young, etc.) and replenish depleted stores of energy after spawning.

As aquatic organisms develop from eggs to larvae to adults, they pass through several stages or **stanzas**. These growth stanzas, separated by physiological or ecological transitions, can be dramatically different from each other, as in the case of larval crabs, which must undergo a complete change or **metamorphosis** to reach their adult form (Figure 8.7).

In the case of fishes and shellfishes, there is a high degree of plasticity in growth, and it can vary widely from one environment to another. Wild Atlantic cod (*Gadus morhua*) captured as juveniles and reared in captivity under favorable feeding conditions mature up to 5 years earlier and grow 2 to 3 times faster than their wild siblings. Laboratory-reared brown trout (*Salmo trutta*) can vary twofold in weight at comparable ages, depending on feeding conditions. This plasticity in growth has been widely exploited by aquaculturists, who may feed fish at high rates to produce such fish as 1-year-old "Donaldson" rainbow trout (*Oncorhynchus mykiss,* formerly *S. gairdneri*) that weigh over 1 kg and 2 1/2-year-old Atlantic salmon (*S. salar*) weighing 5 kg.

In 1891, C.G.J. Petersen of Denmark introduced **length-frequency** analysis, which is the simplest method of age determination. Age is interpreted directly from length, making this method rapid and inexpensive. Modes or a series of peaks that represent different age groups are formed by the data (Figure 8.8). The technique works well for distinguishing age groups during the fast-growing phase, but the modes of older fish may overlap. If no permanent growth record is available from the hard parts (for example, many tropical fishes, crustaceans that molt), length-frequency analysis may provide the only possible means of age determination. A wide variety of statistical and analytical techniques have evolved to aid in distinguishing individual **year-classes** or **cohorts** (that is, the fish spawned or hatched in a given year) by length-frequency analysis.

Growth of aquatic life forms in length or size is seasonal in northern and temperate climates. Most growth occurs during the late spring and summer months, when food is abundant and temperatures are high; growth is minimal during the winter. This growth pattern is recorded in hard-parts

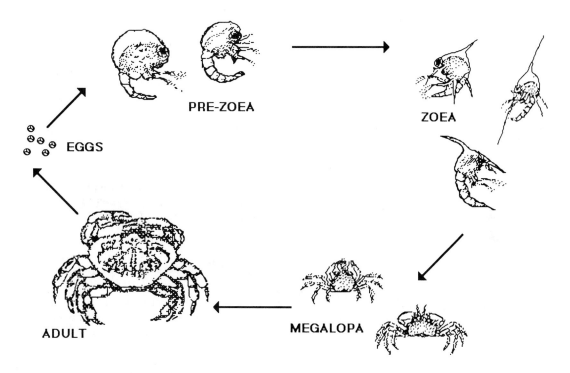

Figure 8.7. The stages of larvae and metamorphosis of Brachyuran crabs, which include the common *Cancer,* the rock crabs, kelp crabs and shore crabs. The free-swimming pre-zoea hatches from the egg and soon undergoes metamorphosis to the zoea. Upon final molt, the young megalops resembles a miniature adult crab.

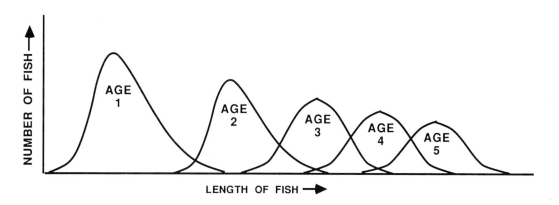

Figure 8.8. Length-frequency distribution. The data form a series of peaks that represent different age groups. The technique works well for distinguishing age groups during the fast-growing phase, but the distributions of the older age groups may overlap. Information on survival rates may also be derived from length-frequency analyses.

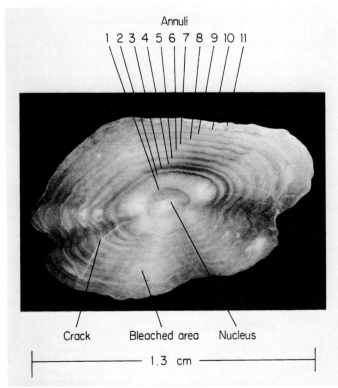

Annuli

1 2 3 4 5 6 7 8 9 10 11

Crack   Bleached area   Nucleus

├─────────── 1.3 cm ───────────┤

Figure 8.9. An otolith or ear bone from a Pacific halibut (*Hippoglossus stenolepis*), showing yearly growth rings. (Courtesy of Phillip Neal.)

such as scales, **otoliths** (bones in the inner ear; Figure 8.9), vertebrae, and fin rays of fish, shells of clams and other molluscs; and teeth or ear plugs of marine mammals. These structures often have obvious annual rings or **annuli**. Each annulus is formed during a winter growth period, when growth is slow and the rings, much like the rings of a tree, are closely spaced. This makes each annulus look like a dark band between the periods of more rapid growth. In tropical areas, reduced seasonality in temperature and feeding conditions results in growth being more continuous, and annual marks on hard parts cannot generally be used to determine age.

Scales were first used to age carp (*Cyprinus carpio*) in the late 1800s, and the method was soon applied to the North Sea fishes, Pacific salmon in Canada and other species in the United States. Most of the methods used today were developed by the late 1920s. Advances since then have been largely technological: examples include plastic impressions of scales, microfische readers, and computers with digitizers or video frame-grabbers to collect and store age and measurement data (Figure 8.10).

For fast-growing species such as salmon, the annual growth record is usually clear from an examination of scales alone, and we can easily distinguish the slow-growing years spent in freshwater from those spent in the ocean (Figure 8.11). This makes it possible to routinely age salmon on the basis of years spent in freshwater and salt water. Two common methods of age designation in salmon are the Gilbert-Rich and European formulas. In the **Gilbert-Rich** formula, a regular whole number is used to designate the **total age** or the year of the fish when the scale was collected, and a subscript number is used to show the year of the fish when it left freshwater. An age 6_2 fish, for example, is in its sixth year and left freshwater in its second year. The scale of this fish has one freshwater annulus and four ocean annuli, and is designated as an age 1.4 fish by the **European** formula. The number preceding the decimal point is the number of freshwater annuli, and the number following the decimal point is the number of ocean annuli, according to the European formula.

When possible, we rely on scales to age fish because scales are easy to collect, prepare and read, and also because we can collect scales without killing or seriously injuring the fish. However, our age determinations from scales may not be as reliable as those from otoliths and other bones. Scales may not form until well after hatching, so the record of early growth is missing. Calcium and other minerals may be resorbed from scales, causing annuli or entire growth zones to be obliterated.

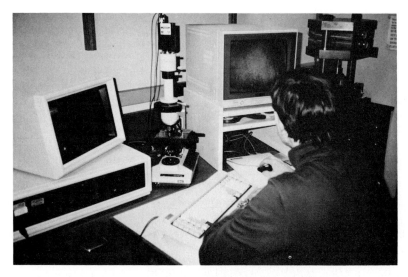

Figure 8.10. A fishery scientist measuring coho salmon (*Oncorhynchus kisutch*) scales on a BioSonics OPRS video digitizing system.

Scales can be lost through injury, and replacement or **regenerated** scales lack information on growth prior to scale loss. In addition, some fishes (such as catfish) do not have scales, and shark scales have no annuli.

For slow-growing and long-lived species, scale growth is almost nonexistent once the fish begins to approach maximum size, and we must use otoliths, fin rays or other hard parts to determine age. In the case of Pacific ocean perch (*Sebastes alutus*), for example, scale and body growth fall off rapidly after sexual maturity (about age 10 for females). Beyond age 15, scales are no longer adequate for age determination, yet cross-sections of the otoliths show that these fish live to be 90-years-old. The process of counting annual rings for long-lived species is far from straightforward, and a surprising number of interpretations of the growth record of a given fish by different scientists are possible.

Otoliths continue to grow as the fish ages, but the growth becomes **allometric** (that is, the shape changes) because deposition occurs predominantly on the inner surfaces. For age determination of long-lived species, the otolith is prepared by breaking it through the thickest portion and burning this section. Broken and burned sections have very distinct dark and light rings.

Giorgio Pannella's discovery of **daily growth rings** in fish otoliths, which was published in 1971, has been considered the most important recent advance in age determination. We can use daily growth rings to find the position of the first annulus, to age species with short life spans or larval and juvenile fish, and to validate early ages. Daily growth increments are composed of an incremental zone (aggregates of needle-like crystals of calcium carbonate) and a discontinuous zone (formed between incremental zones), which is predominantly organic matrix (Figure 8.12). These rings can sometimes be seen on an irregular basis after maturity, but are usually only observable in immature fish.

The general applicability of the scale method or any other age determination method should never be assumed. Agreement between readers of growth rings is a minimum criterion for reproducibility of results, but does not guarantee accuracy. Age determination comparisons have frequently been carried out for test collections of scales and otoliths, and substantial differences can exist between scientists in different laboratories even though the individual readers within each

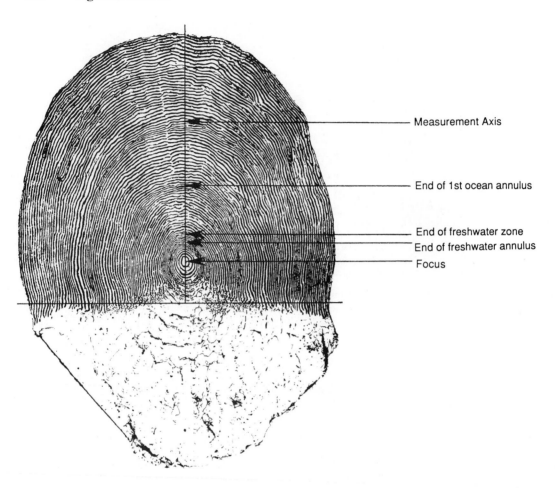

Measurement Axis

End of 1st ocean annulus

End of freshwater zone
End of freshwater annulus
Focus

Figure 8.11. An adult chinook salmon (*Oncorhynchus tshawytscha*) scale from the Kamchatka River, U.S.S.R. This is an age 1.4 fish, which spent one winter in freshwater and four winters in the ocean before returning to freshwater to spawn.

laboratory agree well with one another. **Age validation**, therefore, is a critical part of any age determination study.

The best validation studies come from examining known-age fish. A comparison of scales and fin ray sections from muskellunge (*Esox masquinongy*) planted in Wisconsin lakes, for example, showed that either scales or fin sections were accurate for ages 1 through 9 years; for fish between ages 9 and 11 years, the fin section method was more accurate than scales; for fish older than 11 years, neither method was accurate.

Tag recapture studies are also helpful in that they provide observations of individual growth that are independent of scale, otolith or fin section readings. But this is only useful for that portion of the life history when growth is rapid (and age determination easiest). The utility of tag-recapture studies for age validation can be improved greatly by inducing a mark on the otolith, scale or fin ray (for example, by injection with tetracycline) at the time of tagging. Tetracycline, an antibiotic, becomes incorporated into the growing margins of hard parts and can be detected by examination under ultraviolet light later, when the hard parts are examined. Counts of the number of annual rings laid down beyond the mark can then be compared with the number of days or years at liberty. The decay rate of radioactive isotopes such as Pb^{210} can also be used to verify the true age of an otolith or scale for long-lived species.

Despite problems in identifying the best structure to use in age determination, in defining repeatable criteria for annulus identification, and in validating the results that are obtained, this work is central to fisheries science. Age determination provides the "clock" necessary for quantifying growth, mortality and recruitment processes. Without it, most stock assessment and fisheries management would be ineffectual, and it is important that advances that improve the accuracy of age determination continue to be made.

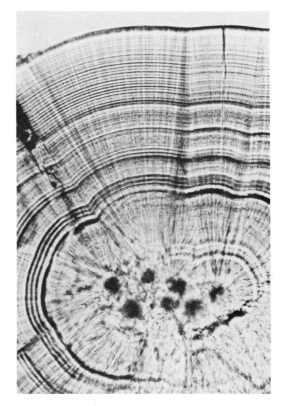

Figure 8.12. Daily growth rings in a juvenile chum salmon (*Oncorhynchus keta*) otolith. The dark spots at the center of the otolith are the primordial core, in which the first crystals of CaCO3 form. (Courtesy of Eric Volk.)

Additional Reading

Bagenal, T.B., ed. 1974. The proceedings of an international symposium on the ageing of fish. Unwin Brothers Limited, Gresham Press, Old Woking, Surrey, England.

Chilton, D.E., and R.J. Beamish. 1982. Age determination methods for fishes studied by the groundfish program at the Pacific Biological Station. Canadian Special Publication of Fisheries and Aquatic Science 60.

Royce, W.F. 1984. Introduction to the Practice of Fishery Science. Academic Press, Inc., N.Y.

Summerfelt, R.C., and G.E. Hall, editors. 1987. Age and growth of fish. Iowa State Univ. Press, Ames.

Weatherley, A.H., and H.S. Gill. 1987. The Biology of Fish Growth. Academic Press, Inc., N.Y..

9

Reproduction and Early Life History of Fishes

Bruce S. Miller

Introduction

This chapter summarizes basic information on the early life history of fishes. In general, there are four stages in the life history of a fish: egg, larva, juvenile and adult (Figure 9.1). (The life history of shellfish species also tends to involve these four stages. In addition, many invertebrates, the crustaceans in particular, have multiple larval stages where the larvae at each successive stage assume a distinctly different form.) The egg (embryonic) stage ends with hatching, the larval stage begins with hatching and ends with metamorphosis into a juvenile, the juvenile stage begins when the larva takes on adult characteristics, and the adult stage begins when the juvenile becomes sexually mature. The term "early life history" refers to the egg and larval stages and includes such topics as **modes of reproduction** or reproductive strategies, **spawning**, **fecundity** (the number of eggs produced), and **egg** and **larval development**. Where we choose to begin and end our descriptions of life histories is somewhat arbitrary in the sense that we are actually describing **life cycles**. The important factors to consider are those that affect the viability and growth of an animal during each stage in its life, including the ecological conditions in the various habitats it occupies.

Considerable research into fisheries biology has focused on the early life history of fishes because of its importance to understanding recruitment (discussed in Chapter 12) and protection/enhancement aspects in wild stocks. It is also important for successfully rearing fish for food (aquaculture), for testing water quality (toxicity bioassays), and simply for identifying undescribed or poorly described eggs and larvae (taxonomy).

Modes of Reproduction

The role of reproduction in fishes, as in all organisms, is to perpetuate the species. Each fish species has evolved in response to a unique set of selective pressures with the result that species often differ in their life history strategies, including mode of reproduction. All fishes reproduce sexually, but the range of sexuality in fishes is large (they stand out among vertebrates in this regard), including various forms of hermaphroditism, unisexuality, bisexuality and various combinations; however, the majority of fishes reproduce **bisexually,** and this large group is the focus of this chapter. Fertilization may be either **internal** (involving copulation) or **external** (involving the release of gametes), and this fact, along with the "placental" relationship (the degree to which the embryo develops in close association with maternal tissue), allows fishes to be classified reproductively as **viviparous, ovoviviparous** or **oviparous.**

Viviparous

Viviparous fishes have internal fertilization and are characterized by the embryo developing in close contact with the nourishing maternal tissue—no egg membrane covers the embryo. The embryos are retained within the mother until they are morphologically well advanced into the juvenile stage (there is no free-living larval stage). Fecundity in such fishes is low (less than 100 per

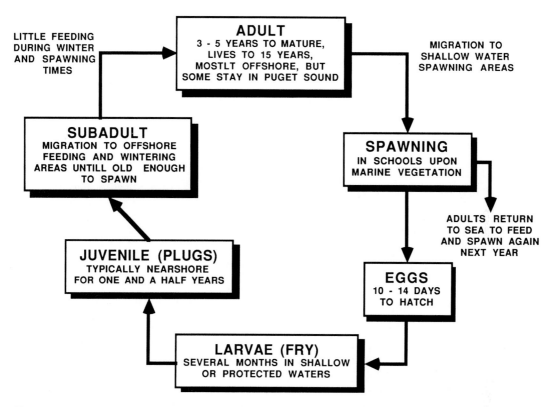

Figure 9.1. An example (Puget Sound herring) of how the four general stages of life history (egg, larva, juvenile and adult) are integrated in the life cycle of a fish.

spawning), but parental care of eggs and larvae is maximized since the young remain within the mother until they are juveniles, at which time they are released and parental care ends. Viviparous fishes are rare, but surfperches (family Embiotocidae) constitute a common and important recreational family of fishes on the U.S. west coast. A specific example is the shiner perch (Figure 9.2). It breeds from April to July, at which time the sperm is stored in a compartment of the ovary for 5 or 6 months until November or December, when the sperm penetrates the ovary, fertilizing the eggs and beginning gestation. Five to 35 young are born, tail first, usually in June or July of the following year. (Mammals, including humans, also reproduce in a viviparous manner.)

Ovoviviparous

Ovoviviparous fishes also have internal fertilization, and the mother may or may not nourish the embryos, but the embryos and the maternal tissue are separated by egg membranes. Typically, the eggs hatch internally and are released as early larvae, such as in the economically important rockfishes (genus *Sebastes*). Dogfish sharks and the coelacanth, also ovoviviparous, release their young as juveniles after a gestation period of a year or more. Rockfishes (Figure 9.3) generally mate in the winter when the male is "ripe" (sperm are ready to be ejected), but the female is not (eggs are not ready to be fertilized); instead, the female stores the sperm until early spring when fertiliza-

Figure 9.2. Shiner perch (*Cymatogaster aggregata*), family Embiotocidae, an example of a species with a viviparous mode of reproduction. (Photo courtesy of Robert Donnelly.)

Figure 9.3. Copper rockfish (*Sebastes caurinus*), family Scorpaenidae, an example of a species with an ovoviviparous mode of reproduction.

tion takes place. The eggs develop within the mother and in late spring, at hatching, the early larvae are immediately extruded to the marine environment where they drift on their own; thus the time from mating to larval extrusion is on the order of 4 to 8 months depending on the species and environmental factors (such as water temperature). Fecundity may be higher than in viviparous fishes (each female rockfish produces thousands of larvae), but many sharks (such as the dogfish) produce less than 20 young every 2 years. Parental care is at a maximum for eggs since they are retained internally, but larval rockfishes have no parental care.

Oviparous

Internal fertilization is actually rare among bony fishes although, besides the live-bearing fishes, there are also a few bony fishes with internal fertilization that lay eggs. However, eggs of the vast majority of bony fishes are released before fertilization—these are the oviparous (''egg laying'') fishes. In oviparous fishes, fertilization is external, eggs develop entirely external to the female, an egg membrane is present and the embryonic stages are nourished entirely by yolk. Among the oviparous fishes, there are both so-called demersal spawners and pelagic spawners—the eggs they extrude are either **demersal** (on the bottom) or **pelagic** (floating above the bottom, often at the surface).

Demersal Spawners

Almost all freshwater fishes are demersal spawners, with eggs that are attached to the substrate or are loosely in contact with the bottom (which reflects the higher specific gravity of fish eggs, which are mostly protein, compared to freshwater). Demersal eggs are also characteristic of many marine fishes that inhabit nearshore areas, where the salinity (and specific gravity) may be lower, where freezing is more likely and where turbulence and unpredictability are often the rule owing to river and stream runoff, storm effects and water currents. Parental care varies from none (e.g., herring) where the eggs are deposited and fertilized on a substrate (e.g., seaweed) and then abandoned by both parents; to guarding a nest of eggs by a single parent (usually the male, e.g., freshwater basses); to the extreme case of parental care displayed by the mouth brooding of eggs by some freshwater cichlids, and freshwater and marine catfishes. Fecundity may be low or moderately high among demersal spawners (usually hundreds to thousands), but is generally related to the amount of parental care; that is, the more parental care involved, the lower the fecundity. Although there are both demersal and pelagic eggs, most larvae that hatch from either demersal or pelagic eggs are pelagic and float freely or swim weakly for days or weeks.

Pelagic Spawners

As opposed to freshwater fishes, most marine fishes are pelagic spawners. Of over 12,000 marine fish species, three-fourths are pelagic spawners. In general, males and females swim close together and the eggs and sperm are broadcast into the water; after fertilization, the eggs float individually in the water column, often near the surface. Fecundity in pelagic spawners is high, usually in the hundreds of thousands or even millions (greater than 300 million eggs for a large ocean sunfish!). No parental care occurs among pelagic spawners.

Demersal and pelagic eggs have some distinctive characteristics that can be summarized as follows:

Egg Characteristic	Demersal Eggs	Pelagic Eggs
Specific gravity	greater than water	less than water or similar
"Shell" of egg	thicker	thinner
Appearance	opaque or colored	transparent
Amount of yolk	large	small
Period of development	longer (weeks)	shorter (days)
Parental care	common	none
No. of eggs produced	hundreds to thousands	thousands to millions
Egg dispersal	low	high
Larval behavior	swim/feed at once	float/yolk sac provides early nourishment

Among the oviparous fishes, there is the special case of the cartilaginous fishes, such as some sharks, ratfish and skates (Figure 9.4), which produce egg cases. These all expel their eggs in a chitinous case that rests on or is attached to the substrate, where the eggs develop into juveniles of the species—that is, the young are far advanced before leaving the protection of the case to live on their own. They receive nourishment inside the egg case from a large yolk sac attached to each embryo. Once the egg case slits open, seawater is continuously circulated by the beating of the tail (at least in some species).

Mating associations in both demersal and pelagic spawners are not related to the type of egg produced, but rather to whether the fishes are schoolers or non-schoolers. Schooling species in which the sexes are notably similar often form large mating aggregations (e.g., flatfish, herring; Figure 9.5). Territorial species that display **sexual dimorphism** (e.g., size, shape and color between the males and females differ) may generally be expected to form breeding pairs (e.g. greenlings, salmon; Figure 9.6).

Spawning

Spawning usually refers to the act of the female releasing eggs into the environment or, more rarely, the release of larvae or even juveniles. Three major factors that affect readiness to spawn are the nutritional state of the mother, physiological (hormonal) factors in both the male and female, and ecological factors.

Figure 9.4. Egg case and juvenile big skate (*Raja binoculata*).

Figure 9.5. Mating aggregation of Pacific herring (*Clupea harengus pallasi*).

Figure 9.6. Sexual dimorphism in a breeding pair of pink salmon (*Oncorhynchus gorbuscha*).

Factors Influencing Spawning

Diet is clearly important to spawning success. For example, laboratory studies have shown that the ovary draws from maternal tissue those amino acids that are found in high quantities in eggs; if the mother's diet has been deficient in those needed amino acids, egg production and subsequent spawning may be unsuccessful. Indeed, it has been found that some natural populations of fishes may spawn only every other year if environmental conditions, particularly those affecting food supply, are poor.

Hormones govern migration and timing of reproduction, associated morphological changes and mobilization of energy reserves; they also often evoke intricate courtship behaviors. The **pituitary gland** is the major endocrine gland involved with physiological readiness for spawning. It produces the hormone gonadotropin, which controls gametogenesis—the production of gametes (eggs and sperm) by the gonads; the pituitary also controls steroid production by the gonads, which in turn control yolk formation in eggs and spawning. This latter effect is often used to the advantage of fish farming. In the culture of carp and in the production of caviar (eggs) from sturgeon, spawning is induced by injecting an extract from the pituitary gland of the same or another species.

Numerous ecological factors may affect spawning, either directly or indirectly, including temperature, light, tides, latitude, depth, substrate, salinity, photoperiod (day length) and exposure to the atmosphere (emersion). Often, ecological factors are associated with timing, so that food availability is optimal for the larvae. In high (temperate) latitudes, spawning is often associated with light and temperature, which dictate seasonal pulses of primary production that assure larval survival. In low latitudes, where there is little variation in day length, temperature, food production and other factors may be most important to spawning (e.g., timing with the monsoons and competition for spawning sites, living space and food). An extreme example of an external ecological factor (moon/tide cycle) being closely associated with spawning is the case of the grunion spawning on California beaches. Grunion are adapted to spawning on the beach every 2 weeks in the spring during a new or full moon, just after the highest high tide, which allows the eggs to develop undisturbed by the surf for 10 days to 1 month, at which time the surf reaches the eggs and they hatch.

Number of Spawnings by Fish

One important question that confronts fisheries scientists involves how often individuals of a given species spawn. This is often a surprisingly perplexing problem. For some species, such as the Pacific salmon that spawn once in their lifetime and die, it is very simple, whether they are spawning on the Pacific coast or in the Great Lakes basin. Other species are much more complicated in this regard. For example, American shad spawn once in their lifetime in Florida, but one-half to three-fourths of American shad in New Brunswick, Canada spawn more than once in their lifetime. In general, spawning in temperate waters occurs in the winter and spring, duration of the spawning period is short (3 to 4 months), and spawning frequency of a species is once a year. In low latitudes (tropical waters), spawning is in the spring, summer or is continuous; spawning duration is long (5 to 6 months or more), and spawning frequency of the species is several times per year (or even nearly continuously). Although a spawning period may be short, and frequency of spawning may be only once per year, individual fish may take several days (herring) or weeks (plaice, a flatfish) to release all of their eggs. Another confounding factor is that, in general, older fishes usually spawn first and younger fishes later in a given season, which means that a prolonged spawning period for a species may not equate to a prolonged period for an individual fish. Histological techniques are particularly useful for determining the number of times a species spawns in a year; by this method, sections of gonadal tissue are sampled from a population of maturing individuals and examined by scientists for egg maturation stages.

Maturity Stages

A record of the state of maturity is often useful for determining the proportion of the stock of fish that is reproductively mature (i.e., near spawning), for determining the sequence of events in the reproductive cycle of a particular stock of fish and for determining the size or age at first spawning. This is most accurately done by classifying the ovaries histologically, at the cellular level; however, by using gross anatomical criteria, much of this work can be done on the fishing boat or in the fish market by simply looking at the gonads. Various scales of comparison have been established for this purpose, some of which are internationally accepted for certain fish species. While maturation changes are continuous rather than instantaneous, in general the gonad stages for oviparous fishes are as follows: The **immature** and **resting** stages usually cover the major part of the year and are characterized by gonads that are very small. The **mature** stage is characterized by larger gonads, eggs are visually distinguishable in the ovaries, and the testes change from being nearly transparent to translucent. In the **ripe** stage, the eggs have become nearly transparent and the testes are usually milky-white (if upon slight pressure to the gonad area of a live fish, eggs or sperm flow outward freely, the fish is referred to as "ripe and running"). Fishes are referred to as **spawned out** (or "spent") if the ovaries are soft, bloody and deflated with only a few remaining unshed eggs, and the testes are deflated with only a few residual sperm. Finally, fishes are considered to be in the **recovery** stage when signs of inflammation are gone, gonads are small and eggs are not yet visible to the naked eye—unshed eggs and sperm will be resorbed during this phase.

Fecundity

A common definition of **fecundity** is the number of eggs to be spawned by a female in a spawning season. Fecundity information can be used to estimate the number of eggs (reproductive potential) that a population of fish can produce; this information helps fishery scientists to estimate the population sizes needed to maintain the fishery on a sustained basis. Fecundity information is also often used in conjunction with egg surveys to back-calculate the size of the population that spawned the eggs.

In general, fecundity predictably varies with the mode of reproduction. The lowest fecundity is usually in viviparous and ovoviviparous fishes (often less than 100), intermediate fecundity in demersal spawners (100s and low 1,000s), and the highest fecundity in the pelagic spawners (high 1,000s and 1,000,000s). Thus, the fecundity of the dogfish is around 8 or 9, the fecundity of most salmonids is in the range of 1,000 to 5,000, and many flatfish and cod produce around 1,000,000 eggs.

Estimates of Fecundity

Although the most accurate way to determine fecundity is to count all the eggs, usually this is not practical and some form of subsampling is used. A typical example is the volumetric technique where graduated cylinders are use to measure the water displaced by a subsample of eggs and by the entire sample of eggs spawned; in addition, the number of eggs in the subsample are counted. Fecundity is then found by calculating F as follows:

$$F/n = V/v \text{ or } F = n \, V/v$$

where F = fecundity,
n = number of eggs in subsample,
V = volume of water displaced by the entire sample of eggs, and
v = volume of water displaced by the subsample of eggs.

Other fecundity assessment techniques are usually variations on the volumetric technique.

Fecundity Relationships

The relationship between fecundity and the size and age of fishes has long been recognized, although it is not always easy to sort out the specific contribution of length, weight or age to the relationship. Generally speaking, the increase in fecundity with body length can be described as follows:

$$F = aL^b$$

where F = fecundity,
L = length, and
a and b = constants for a given species.

The fecundity/length relationship is useful and, once established, is often used to compare populations of fishes, often in a graphic form (Figure 9.7).

Development of Eggs and Larvae

It is not the purpose of this section to discuss embryo development (embryology) or larval development in detail, but rather to just introduce these complex subjects. Although the basic terminology of the morphology and development of fish eggs and larvae has not been formally standardized, there is generally close agreement (Figure 9.8). The course of embryonic and larval development, or the progression of life stages of an organism, is called its **ontogeny.**

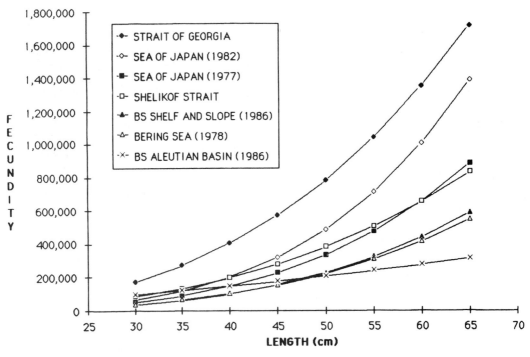

Figure 9.7. An example of a fecundity-length comparison for walleye pollock (*Theragra chalcogramma*) from several geographical areas in the northern Pacific Ocean. (From Miller et al. 1986, Final Report FRI-UW-8608, Univ. Washington, Fish. Res. Inst., Seattle, WA.)

Egg (Embryo) Development

Identifying fish eggs can be very difficult because morphologically they are often very similar. Most fish eggs are spherical, although some fishes have elliptical (e.g., anchovies) or oval (e.g., gobies) eggs. Marine pelagic eggs are often about 1 mm in diameter, while marine demersal and freshwater eggs are often much larger (e.g., salmon eggs may be 5 mm or more). Often the **envelope** (egg case or shell) is sculptured in a characteristic way and is often much thicker in demersal than in pelagic eggs. Oil globules are found in the yolk of some species, and the size and number of these globules may help in identifying the eggs of some species. Pigments in the yolk or embryo may also help to identify eggs of certain species.

After fertilization, cytoplasm concentrates at one pole (called the **animal pole**) of the egg, where incomplete or **meroblastic** cleavage (cell division) takes place. This leads to the formation of a **blastoderm** or **blastodermal cap** (Figure 9.8). The blastodermal cap hollows out, spreads over the yolk surface during the **gastrula** stage and leads to the poorly differentiated **early embryo** stage. During subsequent stages, the embryo grows in size, curves around the yolk and differentiates eyes, auditory organs, olfactory organs, a neural chord, heart and muscle masses (**myomeres**). (Differentiation refers to the progressive formation of specific body structures.) The heart begins to beat when the embryo curves halfway or more around the yolk. Finally, the tail becomes free of the yolk, the embryo grows in length, the organ systems become more developed, and the embryo begins to move within the envelope.

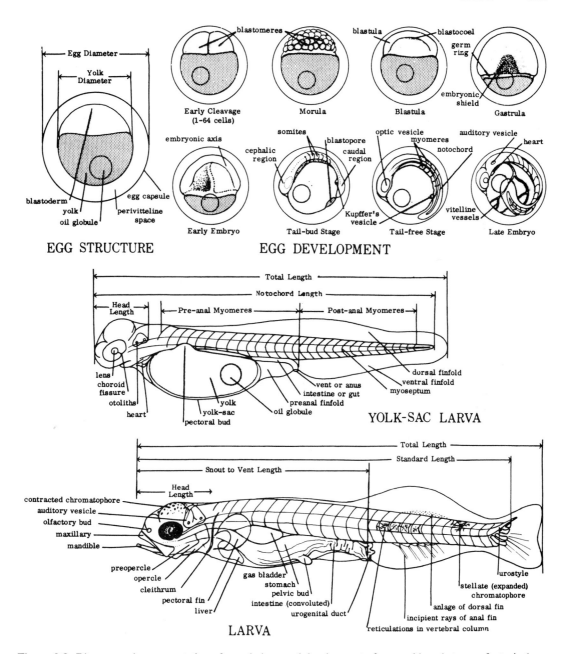

Figure 9.8. Diagrammatic representation of morphology and development of egg and larval stages of a typical teleost. (From FWS/OBS-78/12, Fish & Wildlife Service, U.S. Dep. Interior.)

Hatching

Once the embryo has reached an advanced stage prior to hatching, enzymes produced by the embryo digest proteins in the egg envelope and soften the envelope. The envelope is essentially digested from the inside out, with the embryo using any breakdown products that are useful as nutrients. Because the embryo eventually needs more oxygen than can be obtained by diffusion from the water through the envelope, the oxygen level inside the egg drops. A low oxygen level stimulates the embryo to wiggle, which helps it to escape the egg by rupturing the envelope. In large eggs where the larva is well developed and active, it emerges tail first; in small eggs where the larva is small, less developed and less active, it emerges head first. Larvae that are unable to rupture the envelope will die from oxygen deprivation.

The morphological stage reached at time of hatching differs considerably from species to species, even in closely related groups. However, in general, embryos of pelagic eggs hatch at an earlier stage of development than those of demersal eggs; in warmer waters, larvae of the same species hatch with much less yolk than those in colder waters.

Larval Development

The larval phase of most bony fishes is divided into two parts: the **yolk sac larva** (**alevin** in salmonids) and the **larva** (**fry** in salmonids) stages (Figure 9.8). Nutrition at the beginning comes entirely from the maternal yolk reserves. The major difference from the advanced egg phase is in the better direct supply of oxygen and increased excretion of metabolites. The end of the yolk sac stage is characterized by development of the jaws, improvement of vision and the gradual buildup of swimming and feeding behavior. The duration of larval life of bony fishes varies from a few days in some tropical fishes to several years in European eels.

Although a fish larva has a number of characteristic vertebrate features like a notochord and camera eyes along with a trunk with muscle segments divided into myomeres, it lacks the following features which are found in juvenile fishes after larval metamorphosis: internal gills, functioning kidneys, bones, scales, most fins and extensive pigmentation.

The eye is particularly important to the newly hatched larva, especially for feeding. Although the eye may be of little use or non-functioning at time of hatching, differentiation of the eye and the optic nerve is an extremely rapid process, although the image seen by larval fishes is probably fairly coarse. In general, the larval retina contains only cones, which means that most larvae are only light adapted (rods, and the ability of the retina to adapt to both light and dark conditions, usually only appear at metamorphosis to the juvenile stage). The visual parameters important for feeding are **acuity**, **distance of perception** and **field of vision**. However, the larval eye is so small, and the focal length of the lens so short, that even the dense packing of the small cones cannot provide the high acuity that is found in older fishes.

As the larva grows, its visual ability greatly increases, the mouth structure and other skeletal features develop more fully and its musculature fills out, until it eventually changes from its larval, planktonic form into a juvenile form with most of the morphological characteristics of an adult. Concomitantly, these features allow the fish to become a more effective feeder and to better avoid being preyed upon, which allows increased survival after the juvenile stage is reached (although the actual end of the larval stage, and subsequent metamorphosis to the juvenile stage, may be a particular period of high mortality for fishes).

10

Marine Environments

Robert R. Stickney

Introduction

Approximately 70% of our planet is covered by salt water. While in a few places the ocean is more than 30,000 feet (10,000 m) deep, the average depth is about 10,000 feet (3,300 m). The marine environment has been the focus of scientific investigation for many decades, but it is only within the present century that we have really begun to understand the complex processes that exist in the sea. Those who study the ocean are known as **oceanographers.** Their counterparts in freshwater are called **limnologists** (discussed in Chapter 12). Oceanography is generally divided into four subdisciplines: geology, physics, chemistry and biology.

Fisheries science might logically fit within the discipline of biological oceanography, but since fisheries involves both marine and freshwater species, it is generally considered to be an independent science. Scientists who are interested in marine fishes should be familiar with the science of oceanography since the distribution, abundance and growth of fishes are dependent upon the physical and chemical processes that are found in the ocean and on the assemblage of living things that provide the fishes with food. Geological processes are also important, particularly nearshore where sediments often exert a strong influence on the types of fishes present and on the food resources available to those fishes.

The marine environment can be classified on the basis of depth (Figure 10.1). The deepest region in the ocean is known as the **hadal zone**, where depths exceed 18,000 feet (6,000 m). The ocean **trenches** are the deepest places in the ocean, with the Marianas trench in the Pacific Ocean reaching depths of over 30,000 feet (10,000 m). Between 12,000 feet (4,000 m) and 18,000 feet (6,000 m) is the **abyssal zone**. From 600 feet (200 m) to 12,000 feet (4,000 m) lie the **continental rises**. All water 600 feet (200 m) or more in depth is said to lie over the **oceanic province**. The **neritic province**, on the other hand, overlies the continental shelves and includes all seawater less than 600 feet deep (200 m). Included are the continental shelf zone (low tide mark on the beach to 600 feet [200 m]) and the **intertidal zone** (region between the high and low tide lines), which is also known as the **littoral zone**.

Most commercially harvested fishes and shellfishes are taken over the **continental shelves** of the world; that is, over regions where the water is 600 feet (200 m) or less in depth. Open ocean fisheries do exist, but they tend to be focused on species that live within the upper portion of the water column since bottom fishing in deep water with trawls is often infeasible from an economic standpoint. Mid-water trawling, drift gill netting and trolling are among the means by which commercial species are captured in the open ocean.

Many commercially valuable fish species utilize **estuaries** during some portion of their lives. Estuaries are coastal regions where rivers enter the sea. They serve as a transition zone between the freshwater in rivers and the salt water of the ocean. The commercial shrimp of the southeastern Atlantic and Gulf of Mexico regions of the United States utilize estuaries as nursery areas. Many commercially valuable species like flounders, striped bass (*Morone saxatilis*) and other fishes also spend portions of their lives in estuaries. **Anadromous** fishes—those such as salmon (*Onco-*

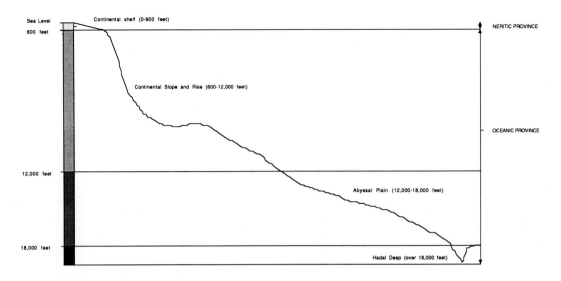

Figure 10.1. Classification of marine environments by depth. (Diagrams in Chapter 10 courtesy of the author.)

rhynchus spp. and *Salmo salar*), which spawn in freshwater but grow from juveniles to adults in the sea—pass through estuaries, as do **catadromous** species such as eels (e.g., *Anguilla rostrata*), which spawn in the ocean and spend much of their lives in freshwater.

Physical Oceanography

Physical oceanographers study such things as tides, waves, currents, temperature profiles and the interactions among these and other physical processes in the ocean. Various physical phenomena affect fisheries. For example, currents control the drifting of fish larvae and their food. Prolonged changes in temperature outside the normal range may allow fishes to distribute more broadly than normal or may decrease the area in which a species is distributed. Tides and currents not only influence the movement of fishes, they also impact the ability of fishermen to capture certain species. A few of the more important physical processes (the physical variables that can affect commercially or recreationally important fishes and shellfishes) are described in the following sections.

Tides

Tides are defined as the rhythmic rise and fall of the surface of the sea relative to some stationary reference point. The tidal range (difference in height between the highest and lowest tides) varies considerably from one location to another. In North America alone, for example, the tides range from a few centimeters along portions of the Gulf of Mexico to over 45 feet (15 m) in the Bay of Fundy, Newfoundland. Maximum tides of nearly 9 feet (3 m) are common along the Georgia coast, while those along the coast of Washington may reach 15 feet (5 m).

The tide is actually a standing wave caused by the gravitational pull of celestial bodies on the ocean. The greatest gravitational forces are exerted by the sun and the moon, but various planets and stars are also involved. If one assumes that the earth is completely covered by water, then the tidal wave would appear as a bulge on each side of the planet. The two bulges move around the surface of the planet as the earth rotates. When the sun and moon are on the same side of the earth or directly opposite one another, the height of the bulge is increased because of the additive gravitational effects of the sun and moon (Figure 10.2A and 10.2B). When the sun and moon are at right angles relative to the earth, they have a tendency to cancel one another out (Figure 10.2C). Thus, the location of the sun and moon relative to the earth impacts the tide range. When the tidal wave bulge is greatest, the highest high and lowest low tides of the cycle occur and we have what are known as **spring tides**. It takes the moon one lunar month (28 days) to circle the earth, and the sun and moon are aligned as in Figure 10.2A and 10.2B twice during that cycle. There are two spring tides each month. Note that the word "spring" has nothing to do with the season of the year. When the moon and sun are at right angles to the earth (Figure 10.2C shows one of the possibilities), we have what are known as **neap tides**; that is, the lowest high tides and highest low tides of the lunar month can be found under those conditions. Neap tides occur one week before and one week after spring tides.

If you look at a tide table, you might find that the actual timing of spring tide extremes is somewhat out of phase with the timing of the new and full moon. This results from the fact that the earth is not entirely covered with water but has continents interrupting its surface that create barriers to the tidal bulge that encircles the globe. The differences in tidal amplitude (difference between high and low tide elevation) from one region to another result from acceleration due to the rotation of the earth and friction, which slows the momentum of the bulge when it enters relatively shallow water. Those two forces tend to either work against or reinforce each other, depending upon where you are.

Since there are two tidal bulges or tidal waves present on opposite sides of the globe and the earth revolves once every 24 hours, there are typically two high and two low tides daily. In some areas of the world, there may be one high tide and one low tide daily. Where there are two high and two low tides each day, and the highs and lows are approxmiately equal in magnitude, this is known as a semidiurnal tide pattern. Areas having two highs and two lows of unequal magnitude have a mixed tide pattern. This occurs along the Pacific coast of North America and in many other regions of the world.

As a rising tide moves shoreward, it pushes large volumes of water into estuaries and river mouths. That tide generates currents. In some regions it is possible to float for considerable distances up a river on the incoming, or **flood** tide, and then to drift downriver with the outgoing or **ebb** tide. Fishes and shellfishes may use tidal currents to help transport themselves in and around estuaries. On the other hand, sedentary animals, such as oysters, are provided with food when algae and other microorganisms are transported to them by tidal currents.

As mentioned above, the tidal bulge can be called a **tidal wave**. We have all heard about so-called tidal waves that may reach heights of many meters and have been known to inundate coastal communities. Those waves, more properly known as **tsunamis** (the "t" is silent), are generated by submarine volcanic and seismic activity. They are not related to the daily gravitational tide.

Waves

Waves of various heights are generated when wind blows over the surface of the water. A wave is composed of a top, or **crest**, and a bottom, or **trough**. Groups of waves, the crests of which are moving by a given point, are known as **wave trains**. The distance between crests or troughs of two successive waves is known as the **wave length**, and the vertical distance between the crest and trough of a wave is known as the **wave height** (Figure 10.3). Waves break when the depth of the water is ~1/7th of the wave length. Whitecaps are waves that have their tops blown off by the wind.

The **surf zone**, where waves break on beaches or against rocks, is a generally hostile environment. While only the most resilient organisms can survive in a region of pounding surf, limpets, barnacles, clams and a variety of other animals and plants are able to thrive in the surf zone. Waves also impact man's activities both along the coast and in open water. To retard coastal erosion, which often happens when waves strike unprotected beaches, various types of structures have been

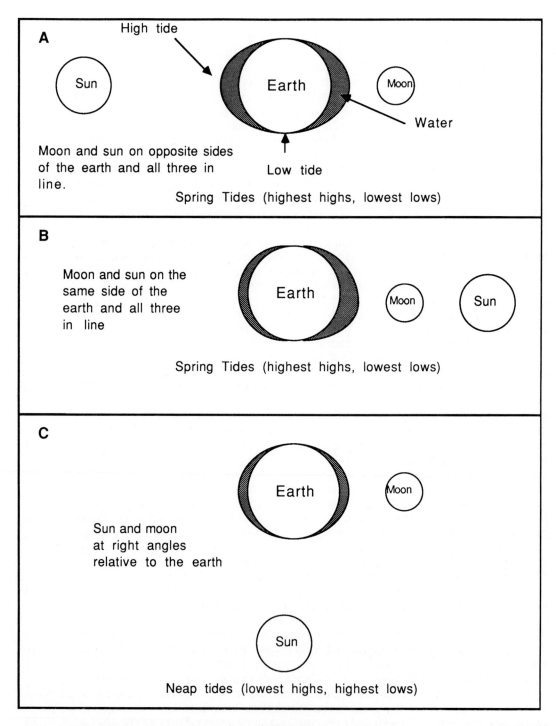

Figure 10.2. Relative positions of earth, sun, and moon during spring and neap tides.

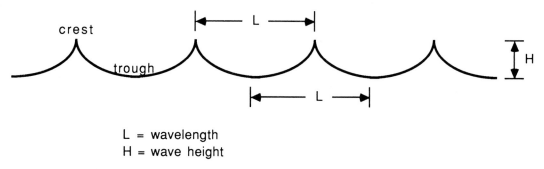

L = wavelength
H = wave height

Figure 10.3. Schematic representation of a wave train showing various parts of the waves.

designed, such as breakwaters, to prevent waves from undermining houses or commercial establishments. Waves have also spawned such recreational activities as surfing. In the open water, waves can lead to seasickness, which is inconvenient and sometimes disabling, or can damage and even sink vessels, which can be fatal.

When waves strike a beach at an angle, they can create a phenomenon known as **longshore currents** (Figure 10.4), which can carry sediments along with them, leading to erosion in some areas and deposition in others. The movement of sediments by longshore currents is known as **longshore drift**. The faster the water moves, the larger the sediment particles that it can carry in suspension. When two longshore currents meet from opposite directions, an offshore, or **rip current** can form (Figure 10.4). Swimmers caught in such currents (commonly known as rip tides, but not related to the celestial tides) can be carried seaward. Attempts to swim toward shore are often futile because the current moves faster than the swimmer. In order to survive a rip current, swim laterally across the rip current until you are out of its grip and then swim shoreward.

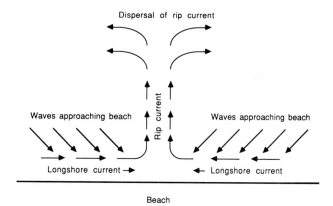

Figure 10.4. Schematic representation depicting the formation of longshore currents and rip currents.

Currents

Currents can be generated in coastal waters by winds, the outflow of freshwater into estuaries, the ebb and flow of the tide, and wave-generated longshore water movement. In the open ocean, there are a number of very large currents generated by the rotation of the earth and supported by prevailing winds. Most of us are familiar with the **Gulf Stream**, which carries warm water northward along the east coast of the United States from Florida to New England, after which the current flows eastward to Europe and then southward in the eastern North Atlantic Ocean.

The Gulf Stream's warm water tempers what would otherwise be a very cold climate in Great Britain. A similar current in the North Pacific Ocean carries warm water northward along the east coast of Asia, eastward from the Soviet Union and down the Aleutian Island chain of Alaska. The current continues southward along the west coast of Canada and the United States. That produces the relatively mild, or maritime climate of North America's west coast.

In the southern hemisphere, similar large currents exist. In the northern hemisphere, the rotation of the large ocean currents is clockwise; however, in the southern hemisphere they rotate counterclockwise. These rotations are a result of what is known as the **Coriolis effect**. Along the equator, rotation of the earth causes the formation of an equatorial current that flows from east to west. When that current hits a continental land mass (such as the east coast of South America), a northern flowing current is formed in the northern hemisphere and a southern flowing companion current forms in the southern hemisphere. The northern flowing current along the east coast of the United States is called the Gulf Stream as we have seen. After reaching Europe, the current (now known by other names) can be traced southward along the west coasts of Europe and Africa until it reaches the equator where the cycle is repeated. Circular flows of this type are called **gyres**.

Fishes and invertebrates can be carried thousands of kilometers by the major ocean currents. Florida lobster larvae, for example, have been found in the Gulf Stream far removed from their native habitat. They may eventually be carried into water that is too cold to support them. Both major and minor ocean currents distribute the larvae of various commercially important species such as Dungeness crabs (*Cancer magister*) along the west coast of the United States and king crabs (*Paralithodes* spp.) in Alaska.

Temperature

Surface temperatures in the ocean range from below 32°F (0°C) to over 86°F (30°C). Surface temperatures in the Arctic and Antarctic regions remain cold throughout the year, while those in the tropics are constantly warm. In temperate regions, surface water temperatures vary seasonally. As water temperature changes in temperate climates, fishes and shellfishes may migrate in order to remain within the temperature range that is optimum for them. Sedentary species such as clams and oysters are unable to migrate and must be able to tolerate extremes in temperature and other environmental variables.

As one goes deeper in the ocean, water temperature decreases. This is true at virtually all latitudes. Wind leads to a considerable amount of mixing in the upper layer of the ocean, so temperature changes relatively little above a depth of a few hundred feet. Below that, temperature changes rapidly with increasing depth in a zone known as the **thermocline**. In the thermocline, temperature can change as much as 2 or 3°F for every three feet of increased depth (the standard definition is that there is a 1°C temperature drop for every meter of increased depth). Below the thermocline, which is generally many feet thick, the temperature changes very slowly with increasing depth. Yet, bottom water temperatures in the deep ocean are typically below 32°F. Such temperatures are not unusual because the high salt content of the ocean prevents water from freezing. Fishes living at temperatures below 32°F often have chemicals in their tissues that act as antifreeze and keep the water in their bodies from freezing.

Since the metabolic rate of organisms is affected by temperature, one can expect very slow metabolism in animals inhabiting the polar seas and the deep ocean throughout the world. Such species generally exhibit slow growth, but may live longer than animals that inhabit warmer waters.

In the deep ocean, shallow equatorial water and the polar seas, there is little or no seasonal change in temperature. Animals living in such areas may be exposed to nearly constant temperature throughout their lives. In regions of the ocean where seasonal changes in temperature are found, animals living near the surface and those that migrate vertically through the water column may be exposed to a broad range of temperatures. Seasonal temperature changes are often the stimulus for the onset of reproductive activity in animals living in temperate regions. Many species of fishes and invertebrates spawn in the spring as the water is warming toward its summer maximum. Others spawn in the fall as the water is cooling, or in the winter when it has reached its lowest temperature of the year. In areas where there is little temperature change over the year, other stimuli may trigger spawning (for example, changing daylength or **photoperiod**). Some species living in stable environments spawn year-round.

Salinity

The total quantity of dissolved chemical constituents in a sample of water is called the **salinity** of that sample. Because water often contains undissolved organic particles (either living or dead) and sediment particles, salinity cannot be accurately measured by merely evaporating a water sample and weighing the residue. Instead, formulas for calculating salinity have been developed. One of the most widely used formulas is:

$$\text{Salinity} = 0.030 + 1.805 \times \text{chlorinity}$$

Chlorinity is a measure of the halogen ions (chlorine, bromine and iodine) in the water and can be determined by a chemical technique called titration.

Relative to pure water, salt water has higher **density** (weight per unit volume). Salinity can, therefore, be measured by measuring the density of a water sample with a device known as a **densitometer**. Also, water density will affect the path a light beam takes when it passes through a sample (refraction of light), so salinity can be determined by measuring the **refractive index** of a water sample with an instrument called a **refractometer**. Salinity can also be measured by **freezing point depression**. As the salinity of water increases, the temperature at which it freezes goes down. The measurement is made by comparing the temperature at which a sample freezes with a chart that shows the corresponding salinity. Finally, waters of different salinity will conduct electricity at different rates, so salinity can be related to the **conductivity** of a water sample. **Conductivity meters** are available for making that determination. All of these types of measurements are currently being employed by scientists and others who need to know the salinity of water.

Salinity is reported in parts per thousand (ppt or °/oo); a ppt is 0.1%. While water contains virtually every element known to man, some of the most important elements contributing to salinity are sodium, chlorine (as chloride ion), potassium, calcium and magnesium. Sulfur exists in the form of sulphate, and carbon is often present as carbonate and bicarbonate.

Freshwater contains varying amounts of the above elements and many others. By definition, a sample is considered to be freshwater when the salinity is less than 0.5 ppt. Seawater ranges from 0.5 to 35 ppt, though it can reach higher levels in such places as warm tide pools and lagoons where evaporation rates are high. When water evaporates, the salts in it are concentrated and salinity increases. A human begins to detect salt in water when the salinity reaches about 2 ppt.

Marine animals that are able to tolerate a broad range of salinity are referred to as **euryhaline**, and those that tolerate only a very narrow salinity range are called **stenohaline**. Many fishes that live only in freshwater or only inhabit the open ocean during their lives are stenohaline, while species that live in estuaries or intertidal zones, along with catadromous and anadromous species are euryhaline.

Whether a species of fish is stenohaline or euryhaline, most maintain solute levels within their tissues around 10 to 12 ppt. The skin of a fish is a **semipermeable membrane**. That is, the membrane allows water and certain chemicals to pass through it, while not allowing other chemicals to pass. When water of low salinity is separated from water of a higher salinity by a semipermeable membrane, there will be movement of water through the membrane from the low salt concentration toward the high salt concentration. This movement, known as **osmosis**, will continue until the salt concentration is equal on both sides of the membrane. Unless a fish is living in an environment where the salinity is exactly the same in its tissues and in the surrounding water, there will be a continuous flow of water through its skin.

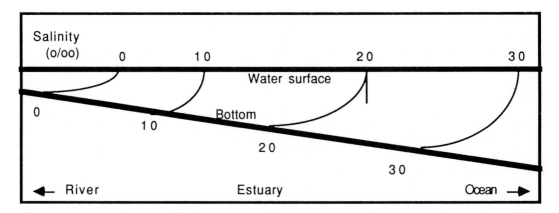

Figure 10.5. Cross-section of a hypothetical river and estuary showing the difference in salinity from surface to bottom when a salt wedge is present.

Freshwater fishes have internal salinities that are higher than those of the surrounding water. Thus, the water outside the fish is moving through the skin and into the fish in an attempt to dilute the salt concentration in the blood and cells. To compensate for this, freshwater fishes excrete large volumes of dilute urine. In that way, they are able to rid their bodies of excess water while retaining their salts. Freshwater fishes maintain their internal salt levels by obtaining the required elements from their food. They do not drink water because there is excess water entering their bodies through the skin.

Most saltwater fishes face the opposite problem from that described for freshwater fishes. The body of a marine fish contains a lower level of salt than that present in the ocean, unless the fish is in an estuary at a location where both the animal's tissues and the surrounding water are of equal salinity. Thus, the water in the fish migrates outward through the skin into the surrounding water. To maintain the required amount of water in the tissues, saltwater fishes must drink large amounts of water. When they drink, they accumulate excess salts that are eliminated by active transport and chloride cells. Thus, saltwater fishes produce small volumes of urine that contains high levels of salt.

Maintenance of water and salt balance in fishes is known as **osmoregulation**. Euryhaline species are efficient osmoregulators, while stenohaline ones often have limited ability to osmoregulate and, thus, may die if the salinity changes appreciably.

In freshwater streams and lakes, as well as in the open ocean, salinity is generally constant from surface to bottom. In estuaries, where fresh and salt waters mix, there may be distinct differences in salinity from surface to bottom. Currents, the shape of the basin, the amounts of freshwater and salt water that are involved in the mixing process and other factors impact what a salinity profile will look like in an estuary. In general, salinity increases with depth in an estuary since freshwater and low salinity saltwater are less dense than high salinity water and essentially float on top of it. Thus, the incoming water from a river will tend to flow out over the surface of the salt water, though the two waters also mix and produce intermediate salinities. That process is responsible for the creation of what are known as **salt wedges** (Figure 10.5). Salt wedges can move many miles up a river, carrying salt water well inland. The bottom water of the Mississippi River, for example, may contain salt as far inland as New Orleans, well over 100 km upstream from the Gulf of Mexico!

Chemical Oceanography

Chemical oceanographers are interested in all of the chemicals that can be found in sea water and how those chemicals influence geological and biological processes. The interests of chemical and physical oceanographers overlap in the areas of temperature and salinity. Temperature, while a physical parameter, is a controller of the rate of chemical reactions, so it is important to chemical oceanography. Salinity, while a product of the chemicals dissolved in water, is important to physical oceanographers because of density relationships and other effects.

Early chemical oceanographers determined that while different water samples might have different salinities, the relative amounts of the major chemicals in all marine water samples are the same. This is known as the **Law of Constant Proportions.** Thus, if there are two molecules of chemical x present for each molecule of chemical y in a water sample at a salinity of 35 ppt, then the ratio of chemical x to y will continue to be 2:1 if the salinity is reduced to 5 ppt or changed to any other salinity.

The Law of Constant Proportions does not apply to chemicals that are affected by biological activity. Among those affected by the activities of plants and animals are compounds containing **carbon, oxygen, phosphorus** and various forms of **nitrogen** (ammonia, nitrite, nitrate).

Carbon dioxide is removed from water during **photosynthesis,** a process by which plants are able to transform inorganic carbon into sugars that are converted biochemically into various other chemicals required to support life. A simple formula for photosynthesis is as follows:

$$\text{Carbon dioxide} + \text{water} \xrightarrow{\text{light}} \text{simple sugars} + \text{oxygen}$$

Written as a chemical formula, the equation looks like this:

$$CO_2 + H_2O \xrightarrow{\text{light}} CH_2O + O_2$$

Carbon dioxide is dissolved into seawater from the atmosphere, where it is present at low levels and is also introduced as a product of **respiration.** When an organism respires, it burns oxygen and produces carbon dioxide, almost the exact opposite reaction of photosynthesis. As shown above, photosynthesis occurs in the presence of light, whereas respiration occurs both in the light and in the dark.

Photosynthesis is restricted to water depths where the light level equals at least 1% of the light intensity that strikes the water surface. The 1% light level—known as the **compensation depth**—is the depth at which photosynthetic production of oxygen and respiratory uptake of oxygen are equal. Below the compensation depth, respiration rate exceeds the rate of photosynthesis.

Approximately 20% of the earth's atmosphere consists of oxygen. Thus, in 1,000,000 parts of air, 200,000 of them would be comprised of oxygen molecules. Oxygen dissolves in water to a limited extent. Depending upon temperature and salinity, seawater may contain up to 8 or 9 parts per million (ppm) of dissolved oxygen, as compared with 200,000 ppm in the atmosphere. Marine animals, like their freshwater counterparts, are extremely efficient at removing dissolved oxygen from water through their gills. It is generally recognized that levels of dissolved oxygen at or above 5 ppm are sufficient to promote aquatic animal growth and survival.

Oxygen is added to the upper layers of the ocean through photosynthesis and by dissolution from the atmosphere. The deep-water layers are replenished with oxygen by diffusion and the natural mixing of seawater, though there is often a region that has the lowest concentration of oxygen within the water column known as the **oxygen minimum layer.** The amount of life in the deep waters of the open sea is not generally sufficient to deplete oxygen supplies. Oxygen depletion problems are most likely to happen in warm shallow water when there are high rates of metabolism.

Biological **productivity,** or the rate at which new tissue is produced by living organisms, is influenced by the availability of various required chemicals, not the least of which are oxygen and carbon. A **limiting nutrient** is one that controls the amount of living material that a region in the ocean can support. Nitrogen and phosphorus have often been the limiting nutrients in aquatic ecosystems. **Amino acids,** the building blocks of protein, all contain nitrogen, so living organisms cannot survive without that element. Nitrogen is present in water as a dissolved gas (N^2), but it is not generally available to living organisms in that form. Plants absorb nitrogen in the form of nitrate (NO_3) or ammonia (NH_3) from the water. Ammonia enters the water as an excretory product from the gills of animals. Nitrate is formed from ammonia by bacteria and also enters the ocean from land runoff. Certain types of algae known as **blue-greens** are able to convert atmospheric nitrogen to nitrate, as are some types of bacteria. Plants are able to convert nitrate directly into amino acids. Animals, on the other hand, obtain their amino acids by digesting protein obtained from food composed of other plants or animals.

Phosphorus is found in skeletons of animals and is an important source of energy in certain types of biochemical reactions. Population explosions of living organisms often end when the available phosphorus supply is exhausted. Sources of phosphorus include decaying tissues of dead organisms, runoff from the land and rainfall.

Other elements critical to the maintenance of life in the ocean include manganese, sulfur, calcium, magnesium, copper, zinc, iodine, silicon and iron. For example, diatoms (a type of algae) require silicon for their skeletal structures. Molluscs such as oysters and clams have skeletons composed of calcium carbonate. The other elements mentioned have roles in various types of biochemical reactions.

In addition to the various interactions between chemicals dissolved in seawater and the plants and animals that inhabit the ocean, there are inorganic reactions that occur between the chemicals in water and both the atmosphere and the sediments.

Geological Oceanography

Plate Tectonics

Geological oceanographers are interested in such topics as the shape, structure and origin of ocean and estuarine basins, the origins and movement of sediments, undersea mining of petroleum and minerals, undersea volcanic activity and **continental drift** or **plate tectonics.**

Years ago, a theory was developed which hypothesized that the continents are floating on liquid rock, called **magma,** which allows slow movement of the land masses relative to one another. While the theory was heavily criticized for many years, evidence gradually accumulated in support of it until, by the mid-1970s, there was general acceptance of continental drift.

As presently explained, the entire earth's surface, including the crust immediately below the oceans, is made up of a series of large sections or plates that exist above the magma. Where the plates meet one another, there is a great deal of tension. In some instances, the plates are moving apart, while in others they are moving toward each other. Generally, two plates move away from each other along the ocean ridges that tend to be located midway between continents. These ridges are actually quite spectacular undersea mountain ranges. As the plates move apart, molten magma is brought to the surface of the crust, where it meets the water, cools and forms new rock that is added to the crust (Figure 10.6).

Volcanic activity is often associated with regions where two plates push against on another. Typically, ocean trenches can be found where one plate is trying to force itself under another. The crustal material that is forced down into the trench is destroyed and becomes magma. The upper plate is placed under a great deal of pressure that may result in the formation of volcanic mountains

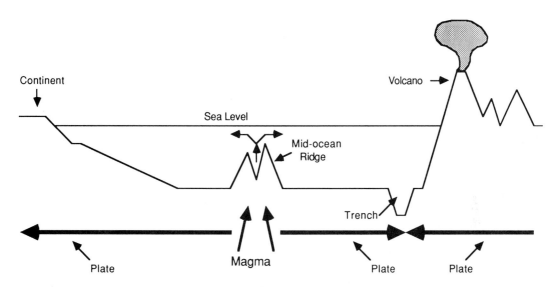

Figure 10.6. Diagram depicting the theory of plate tectonics. Magma flows up into the crust at the mid-ocean ridges pushing the plates apart. Crustal material is destroyed under the trenches, leading to volcanic activity in the mountains.

(Figure 10.6). **Tectonic activity** (earthquakes and the eruption of volcanos) in Alaska and along the coast of California results from two actively moving plates pushing against one another.

While the continents are moving as a result of plate tectonics, the rate is so slow that we need not fear that North America is going to crash into another continent anytime in the near future. The rate of movement is a few centimeters a year on average, so visible manifestations of the phenomenon require geological periods of time to produce. The volcanic activity that we witness around certain parts of the world is something that can be readily observed, though predicting exactly when sufficient pressure will build up to produce such tectonic activity remains difficult. Other manifestations of tectonic activity include tsunamis, which may be generated as a result of undersea earthquakes.

Sediments

We are all familiar with beaches composed of sand that results from the erosion of rock. When this process is carried on even further, silt and clay particles of microscopic sizes are produced. Some coastal areas do not have sand beaches at all, but consist of marshlands where the sediments are dominated by mud (mixtures of silt and clay). Examples are the extensive salt marshes of Georgia and South Carolina along the southeastern Atlantic coastline of the United States. Coastlines that have not been massively eroded may have rocky areas typical of northern California and Oregon.

Various coastal features are the direct result of geological activity. Embayments such as San Francisco Bay were created by tectonic activity. Many estuaries, on the other hand, have been formed by erosion over long periods of time. One particularly interesting type of coastal feature is the **fjord** (Figure 10.7). Fjords are common in parts of the British Isles, Norway and the west coast of North America, particularly in Washington, Alaska and British Columbia. Fjords were created by glacial activity that gouged out coastal embayments and, in some instances, inland lakes such as the lochs of Scotland. **Sills** (areas of shallower depth) at the mouths of fjords were produced when the

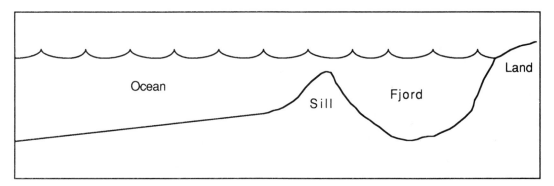

Figure 10.7. Cross-section through a fjord.

leading edge of the glacier met seawater and melted, with the result that the terrestrial material trapped in the ice was deposited.

Not all sediments in the marine environment are formed as a result of the breakdown of continental rocks. In the open ocean, the sediments are often comprised of the skeletons of organisms such as **diatoms** (algae) and **foraminifera** or **forams** (protozoans). The living diatoms and forams generally drift around in the water column as members of the **plankton** community, but when they die, their skeletons (called **exoskeletons** because the living tissue of the organism is housed within the skeletal material) settle to the bottom and accumulate. Countless billions of such skeletons have fallen over millions of years over vast areas of ocean bottom to depths of many meters. While alive, the creatures that contribute to these types of sediments are of interest to biological oceanographers, while the sediments themselves are of interest to geologists. It is possible to date sediments by the species of animals that contributed their exoskeletons in the formation of the deposits. It is also possible to get an indication of what the climate was like when the sediments were created. If the sediments are dominated at a given depth by a species known to have existed in tropical seas, but the region is presently associated with a temperate or polar climate, that is evidence of a major climatic change between the time the sediments were formed and the present.

The types of sediments present in a given location can also be used to evaluate the potential for the presence of oil, gas and minerals. Offshore production of oil and gas has been extensively developed in portions of the Gulf of Mexico, Alaska and off southern California. Increasing concern by environmentalists has slowed development of offshore production in other regions of the United States, and the issue continues to be highly controversial. Concerns associated with protecting the environment from possible oil spills are being balanced by the need for petroleum to provide energy. As gas and oil reserves become increasingly depleted, the pressure will grow toward developing known and expected offshore reserves.

Biological Oceanography

A large number of habitats are found in the marine environment, including: coral reefs, tidepools, abyssal plains, continental shelfs, saltmarshes, mudflats, kelp beds and fjords.

Each habitat type has its own rather unique group of plants and animals that interact with each other in various ways. One of the most important forms of interaction involves the flow of energy from one group to another. Energy is transferred when organisms of one level of the feeding hierarchy, known as a **trophic level,** feed upon those which occupy a lower level. Top level **carnivores** (meat-eating animals), such as sharks, killer whales and various others, feed upon smaller, or

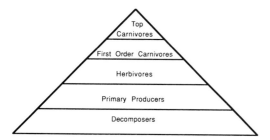

Figure 10.8. Pyramid of numbers of individual organisms within the marine environment (also applies to freshwater and terrestrial environments).

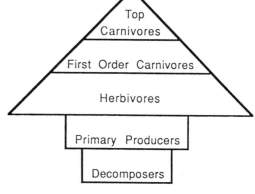

Figure 10.9. Relationship of standing crop biomass for organisms within the marine environment (also applies to freshwater and terrestrial environments.)

second-order carnivores and **herbivores** (plant-eating animals). The herbivores feed upon plants (the **primary producers**). Since individual size increases as one moves from the lower to the higher trophic levels, there must be larger numbers of herbivores than there are carnivores, or the food supply of the latter will become exhausted. Also, the transfer of energy is only about 10% efficient as it moves up the trophic ladder. The same relationship holds between each of the various trophic levels and has led to the concept of the **pyramid of numbers** illustrated in Figure 10.8.

At the base of the pyramid of numbers is the **decomposer** community which is made up primarily of bacteria and fungi. Those organisms are responsible for the decay of any dead organic matter present (dead plants and animals or their body parts, feces, dissolved organic materials). The decomposers are discussed in a bit more detail in Chapter 12. The decomposer community performs the same function in both the marine and freshwater environments.

If an oceanographer measures the weight of organisms present in a particular spot in the ocean at any given time, the result will be an estimate of **standing crop biomass.** Standing crop relates to the types of organisms present on an instantaneous basis, and biomass refers to the weight of those organisms. Typically, the weight of herbivores and carnivores follows the pattern shown in Figure 10.8, but whereas the numbers of individual decomposers and primary producer organisms almost always exceeds the numbers of animals, the weight of the two former groups may not. This is depicted in Figure 10.9.

The question might well be asked: How does a small weight of primary producers support a large weight of animals inhabiting the higher trophic levels? The answer is that the turnover rate of animal populations is much slower than it is for plants. For example, many species of fishes reproduce only once a year, so new individuals enter the population fairly infrequently. Some of the small invertebrates in the marine environment have short lifespans and may be replaced at intervals of a few weeks or months, though others live for many years and the turnover rate is slower. Single-celled algae, which are the most important primary producers in most marine environments, may have turnover rates of hours to days. Thus, their productivity is very high. Even though the standing crop biomass of primary producers at any given instant is low compared with that of the animal populations, new cells are being created as rapidly as they are being consumed.

Primary Producers

Marine plants range from single-celled algae (**phytoplankton**) to seaweeds (also types of algae) to rooted species (**tracheophytes**). Since plants grow by photosynthesis and, thus, depend on the presence of sufficient quantities of light, they are restricted to that part of the ocean known as the **photic zone,** where light is present. The photic zone extends from the surface of the water to the depth at which the light level is about 1% of the surface, or incident light level. Plants transform carbon dioxide and water into organic molecules through the process of photosynthesis which requires light energy. The plant tissues that result from the process of photosynthesis provide the food resource base upon which most other marine organisms depend. As noted earlier, photosynthesis results in the following process:

$$6CO_2 + 6H_2O \xrightarrow{\text{light}} C_6H_{12}O_6 + 6O_2$$

$$\text{carbon dioxide} + \text{water} \xrightarrow{\text{light}} \text{sugar (glucose)} + \text{oxygen}$$

Once carbohydrates are available to the plant, they can be converted into lipids, protein, or used as an energy source by the plant.

Phytoplankton

The phytoplankton community is comprised of plant species that are found in the water column. They are most commonly microscopic single-celled species, though there are species of algae that exist as groups of cells (Figure 10.10). Some seaweeds (macroalgae) are also phytoplanktonic. Phytoplankton drift with currents and have little or no motility.

The density of phytoplankton cells present in a given part of the ocean will depend upon turbidity (high turbidity limits the light penetration and thus reduces phytoplankton productivity), total amount of light present and the levels of nutrient available. The ultimate potential for growth of the community is typically reached when one or more nutrients become exhausted. Thus, phytoplankton growth is said to be nutrient-limited. The first limiting nutrient will be the one that disappears first, as discussed above.

Phytoplankton cells are nearly always present in measurable concentrations. When environmental conditions are appropriate, the population density may increase very rapidly from a few cells per liter of water to millions of cells per liter. Explosive growth of this type is called a **bloom.** The increase in cell concentration is slow at first but proceeds at an increased rate as doublings of the population occur through cell division. The period of slow increase is known as the lag phase. Thereafter, the bloom enters the logarithmic, or log phase of growth. At some point, a factor becomes limiting to further development and the population density will stabilize or, more likely, there will be a decline. These phases are shown diagrammatically in Figure 10.11. If the decline phase is precipitous, it is known as a phytoplankton **crash.**

Various parts of the ocean become conducive to the establishment of phytoplankton blooms at different times and under different circumstances. In coastal waters, periods of heavy rainfall produce nutrient-rich runoff water, which enters the marine environment and may stimulate phytoplankton blooms. In other areas, a current that moves upward (vertically) from deeper to shallower depths leads to a phenomenon known as **upwelling,** which may bring nutrient-rich deep water into the photic zone, where the nutrients can stimulate photoplankton growth. Upwelling can develop when wind blows across the surface of the water and induces a circulation pattern that brings up the deeper water (Figure 10.12). When the wind direction changes or the wind stops blowing, the upwelling may be discontinued and the source of nutrients is cut off, thereby limiting the growth of the phytoplankton and leading to termination of the log phase. The well known **El Niño** phenomenon, which develops along the coast of Peru and Chile, is associated with disruptions in the normal upwelling that develops in that region off South America. During periods of El Niño, fish production may be greatly suppressed.

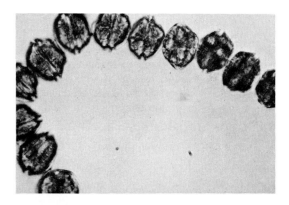

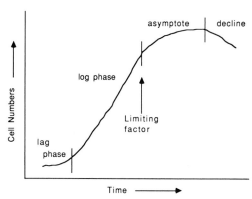

Figure 10.10. Photomicrograph of *Protogonyaulax catenella,* a phytoplanktonic species that occurs in chains. (Courtesy of Ken Chew.)

Figure 10.11. Diagram of phytoplankton growth curve showing the stages in bloom development and the point at which some environmental factor becomes limiting to further growth.

In some cases, an expanse of the ocean will have relatively small areas where phytoplankton growth is rapid, while adjacent areas show much lower rates of primary productivity (Figure 10.13). Such "patches" of primary productivity are commonly rich in animal life because herbivores are attracted to the phytoplankton bloom as a food supply. The presence of herbivores attracts carnivores, so the overall biomass present in patches of phytoplankton may be very high compared with adjacent areas. Patches may be from hundreds of square meters to thousands of hectares in area. It is often difficult to determine with precision what leads to a patchy distribution of phytoplankton, but the phenomenon is well-documented.

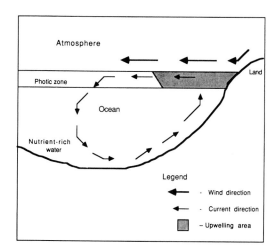

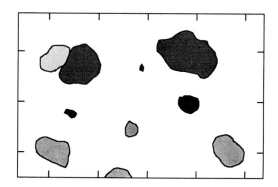

Figure 10.12 Diagram showing how wind-driven ocean currents can create upwelling, which brings nutrients to the surface, where they support plankton blooms.

Figure 10.13. Conceptual view of a portion of the open ocean showing an overall low density of phytoplankton (light shading), with patches of higher concentrations (darker shadings) existing in various locations.

Benthic Algae

In shallow waters, where sunlight is able to penetrate to the bottom of the water column, single-celled algae can often be found in abundance growing on the surface of the sediments or in association with other types of organisms. Some species of corals, for example, have algae living within their tissues that apparently help provide nourishment and energy to the coral animals. Other coralline algae secrete calcium carbonate and actually contribute to the skeletal structure of coral reefs.

Algal cells that colonize shallow water areas may be exposed to the atmosphere at low tide. Mats of blue-green algae are common, as are various types of green algae and diatoms. Intertidal algae species must be able to tolerate desiccation as well as extremes in temperature and other environmental factors.

Benthic algae provide a food supply for various types of animals that either live in conjunction with the sediments or inhabit the water column and browse along the surface of the sediments. In addition to being found attached to silt, sand, gravel and other natural materials, benthic algae can be found on the surface of virtually any type of available substrate, including boat hulls, pilings and other man-made objects.

The relative contribution of benthic algae to total primary production varies greatly. While that contribution is obviously small to nonexistent in the open ocean where most, if not all, substrates are located below the photic zone, in shallow embayments, along shorelines and in tide pools, the productivity of benthic algae may be significant.

Seaweeds

Seaweeds are large, sometimes incredibly large, algae. On the west coast of the United States, perhaps the best known of the seaweeds is the California giant kelp, *Macrocystis pyrifera*, a genus of brown algae which grows to depths of up to many tens of meters. Reports of this species reaching over 600 feet (200 m) in length have been made, though the longest reliably measured specimen was less than one-quarter of that length. Forests of kelp are maintained in place because each individual has a holdfast organ that anchors the alga to the sediments. Pieces of the main part of the plant sometimes break off and are set adrift. Kelp beds provide food and shelter for a variety of marine animals.

In the Atlantic Ocean and Gulf of Mexico, gulfweed or sargassum weed is found, sometimes in large concentrations. Gulfweed, genus *Sargassum,* is a somewhat unique brown alga. Whereas most seaweeds grow attached to substrates, Gulfweed floats at the surface of the sea. An area of the north central Atlantic Ocean is known as the Sargasso Sea because of the dominance of the seaweed.

Gulfweed supports an animal community, many species of which are not found in other habitats. The sargassumfish, *Histrio histrio,* for example, has developed colors and fins that mimic the seaweed and is found only in association with the plant.

Seaweeds have significant commercial value. Extracts from them are present in the products that each of us uses every day, including ice cream, toothpaste and medicinal products. The demand for seaweeds as human food and for the extracts that come from them is sufficiently high that a large aquaculture industry has been developed in many countries. Several hundred thousand persons are employed in the seaweed culture industry in Japan. A fledgling industry exists in the United States.

Rooted Plants

Rooted plants may exist below the water (**subtidal**), in the zone between the high and low tide (**intertidal**) or just above the normal elevation of high tide (**supratidal**). Subtidal rooted plants grow in shallow water since sunlight must be able to penetrate to the depths of the plants if photosynthesis is to occur. Such plants as eelgrass (*Zostera marina*), manateegrass (*Cymodocea* *manatorum*), widgeongrass (*Ruppia maritima*) and turtlegrass (*Thalassia testudinum*) grow on sandy sediments and provide habitat for a variety of associated animals. Of those four, eelgrass is the most cold adapted, ranging from North Carolina northward on the Atlantic coast and along the west coast into Alaska. Certain species of marine turtles eat rooted plants, but the grassbeds are not grazed by most of the species associated with them.

Saltmarshes are comprised of various species of intertidal and supratidal plants. On the southeastern Atlantic coast of the United States, from northern Florida into North Carolina, are extensive areas of saltmarshes dominated by smooth cordgrass, *Spartina alterniflora*. This grass, one variety of which can grow nearly 9 feet (3 m) tall, covers hundreds of thousands of hectares in South Carolina and Georgia (Figure 10.14). The grass grows lushly during much of the year, but in the winter the foliage dries up and breaks off. As the countless tons of vegetation are broken down by the decomposer community, nutrients are released and recycled. It has been hypothesized that in areas where extensive marshland exist, the food web is largely supported by the so-called **detritus food base** provided by the decaying plants.

A much smaller species of cordgrass, *S. patens,* is often found in conjunction with *S. alterniflora,* but is usually a minor constituent of the marshlands in the southern states, whereas in New England marshes, *S. patens* often dominates. Both species tolerate relatively high salinities. *Spartina patens* may be replaced in the upper estuary by another cordgrass, *S. cynosuroides.* Ultimately, strictly freshwater species of plants take over as salinity approaches zero. On the west coast of the United States, another species, *S. foliosa,* can be found. Various other species of intertidal plants exist, but they are often of secondary importance to the cordgrass, though there are exceptions, as with the needle rush, *Juncus roemerianus.* Extensive regions of needle rush can be found in conjunction with, but at slightly higher elevations than, the cordgrasses.

Mangrove swamps are found primarily in tropical regions. They are most commonly found between 25°N and 25°S latitudes. In the United States, mangrove swamps are largely restricted to south Florida (Figure 10.15) and Hawaii; however, isolated black mangroves (*Avicennia germinans*), which grow in regions that are covered with water only at high tide, are sometimes found along the coast of the Gulf of Mexico as far west as the Texas coast. The most seaward plants are red mangroves (*Rhizophora mangle*) which grow subtidally. Between those two are white mangroves (*Laguncularia racemosa*) and, sometimes, button mangroves (*Conocarpus erectus*).

Mangroves provide habitat for many species of economic importance, including shrimp, pompano, red drum, snook, seatrout and mullet. The leaves from the mangroves add nutrients to the area when they fall, and the roots and stems of the plants provide substrate for the growth of algae and a variety of associated animals. Turtlegrass is also commonly present in the immediate vicinity of mangroves.

Supratidal plants must be able to tolerate high levels of salt that may be present in the soil and also in the air because of salt spray. Some plant species are found predominately in association with sand dunes along beaches. The roots of those plants help stabilize the dunes. Numerous other salt-tolerant plants grow on beaches and adjacent upland areas, as well.

Zooplankton

Larvae, juveniles and, in many instances, adults of marine animal species representing most phyla can be found at one stage or another in their life cycles. This community of organisms, known collectively as **zooplankton,** is comprised of individuals unable to swim strongly enough to overcome currents.

Many zooplanktonic species can move vertically in the water column and may migrate from one depth to another depending upon environmental conditions. In the open ocean, **hydroacoustic profiles**—sound waves generated on a vessel that reflect off the bottom and off objects within the water column to determine the amount of water under the keel and to locate fishes and other objects (see the example in Figure 1.4, p. 8)—often show the formation of a strong area of reflection at rela-

Figure 10.14. Intertidal salt marshes in Georgia and South Carolina feature the cordgrass, *Spartina alterniflora,* which is commonly over at 2 meters near the low tide line.

Figure 10.15. A mangrove community, such as this one in Florida, provides habitat for a wide diversity of marine plants and animals.

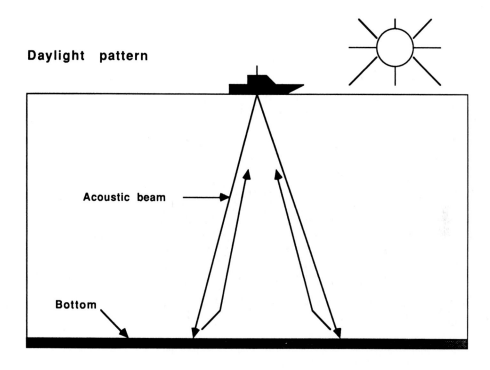

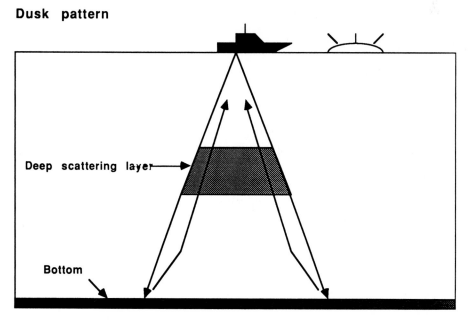

Figure 10.16. Comparison of hydroacoustic patterns during daylight and at dusk following formation of the deep scattering layer.

tively shallow water depths near sunset. This layer of reflection, known as the **deep scattering layer**, was a mystery for some time, but it was eventually determined that as the sun is setting, zooplankton and fishes begin moving upward in the water column. The zooplankton, which are **negatively phototaxic** (avoid light), remain below the compensation depth during the daylight hours, but move up at dusk into the phytoplankton-rich water that live in the photic zone. The phenomenon is shown diagrammatically in Figure 10.16.

Many species of zooplankton feed by filtering phytoplankton from the water. Such herbivores include protozoans, copepods, shrimp larvae and mollusc larvae, among others. Feeding upon the filter feeders will be a variety of carnivorous zooplanktonic species, including jellyfishes, ctenophores and larval fishes.

The zooplankton community may be relatively stable in terms of community structure or it may vary considerably with time as the larvae of animals that spawn during restricted periods of the year enter the zooplankton population and are dominant for brief or extended periods. Species that spend their entire life in the zooplankton are called **holoplankton,** while those which are planktonic for only a portion of their lives and then leave the plankton to join some other community of animals are called **meroplankton.** Pelagic copepods are an example of the first type, and most fishes, along with oysters, mussels, crabs (Figure 10.17) and corals, represent the latter category.

While fishes are not generally considered to be planktonic, except as larvae, the ocean sunfish, *Mola mola,* has been characterized as a zooplanktonic species. Reaching several hundred kilograms in weight and a few meters in length, this fish has extremely weak swimming ability and can often be found drifting with the current at the water surface.

Benthos

Animals that live on or within the sediments are called **benthic** and make up a community known as the **benthos.** In reality, there are many types of benthos communities. The animals inhabiting muddy sediments are quite different than those on sandy substrates. Also, the animals that populate coral reefs are distinct from those found in association with oyster beds.

Benthic animals that live on the surface of the sediments are known as **epifauna,** while those living within the substrate are called **infauna.** Some of the epifaunal organisms are immobile, including barnacles, mussels, oysters, corals and sponges. Others, such as crabs and lobsters, may move around with ease. Infaunal organisms are usually quite sedentary. Clams, geoducks, polychaetes (segmented worms) and ghost shrimp are examples.

Coral reefs (Figure 10.18) develop only in tropical regions. Dominated by a variety of different coral species, the reefs are homes for sponges, sea urchins, lobsters, crabs, polychaetes, shrimp and a host of other types of animals, plus benethic algae and, of course, the associated fish populations. Coral reefs are delicate environments. Not only is warm water of high salinity required for their development, but the water must be clear since corals and some associated organisms are sensitive to sedimentation.

Oyster beds are a type of benthic community that is dominated by one species, the oyster, but which also has associated with it a variety of often less conspicuous forms. Included are algae, seaweeds, polychaetes, crabs, barnacles, certain types of fishes, shrimps and others. Oyster beds are found within the photic zone since the oysters are filter feeders and depend on the presence of phytoplankton for a major portion of their food. Oysters feed by filtering small organic particles from the water.

The formation of an oyster bed depends, in part, on food, but is also dependent on an appropriate substrate. Oyster larvae, called **spat,** are planktonic. They remain in the plankton for a period of days to a few weeks, after which they find suitable substrates upon which to settle and attach. If a larva settles in an unsuitable area, such as on mud, it may be buried and will die. If it settles on a hard substrate, for example another oyster shell (either dead of alive), it may survive and contribute to the productivity of the oyster bed.

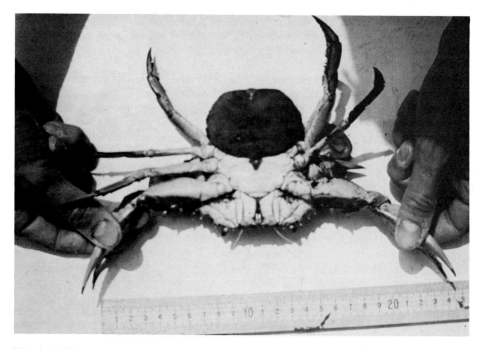

Figure 10.17. An adult female blue crab, *Callinectes sapidus,* carrying her eggs on the abdomen. Blue crab are common along the east coast of the United States and in the Gulf of Mexico.

Figure 10.18. Coral reefs are found in clear, warm marine waters. They are comprised of various species of corals, sponges and other attached invertebrates, and provide habitat for large numbers of motile invertebrates and fishes. (Courtesy of Fred Johnson.)

Muddy sediments typically support primarily soft-bodied benthic animals such as polychaetes, shrimps and tube worms. Sandy sediments, on the other hand, are often dominated by molluscs and crabs. Faunal assemblages do overlap from one sediment type to another, but there are often animals present that are indicative of a particular sediment type.

An interesting group of organisms can be found attached to hard substances such as rocks, pilings and bulkheads. Intertidally the community may be composed of limpets, oysters, mussels, barnacles and associated motile creatures such as small crabs and isopods. Subtidally, sea anemones and echinoderms may dominate, depending on the particular region of the world in which the habitat exists. When a wood substrate is present, either intertidally or subtidally, shipworms such as *Teredo* sp. may be present. Benthic species that colonize structures constructed by man are called **fouling organisms.** While some of them, such as barnacles, are only a nuisance, shipworms and a few other animals will bore into wood and can eventually destroy a piling or wooden boat hull.

Accumulations of fouling organisms on the hulls of boats have been a considerable problem for generations of mariners. Fouling organisms reduce the efficiency with which a boat moves through the water by increasing friction. Thus, the speed and, in the case of powerboats, fuel efficiency are sacrificed. Frequent scraping of boat hulls may be necessary in areas where heavy fouling is a problem. Antifouling paints that release toxic chemicals have been developed for use on boat hulls. The toxins kill the larvae of fouling organisms attached to the paint. The use of some of the most effective antifouling paints has recently been restricted because non-target organisms have been impacted and the environmental threat from the paint is considered to be significant.

Many fishes are found in association with the sediments and, in some cases, they are actually found within the sediments. Species of flatfishes, for example, bury slightly in sandy sediments with just their eyes visible. Moray eels (family Muraenidae) can typically be found hiding in crevasses and caves, while other eel varieties may burrow into the sediments. Sea robins (family Triglidae), toadfishes (family Batrachoididae), rockfishes (*Sebastes* spp.) and stingrays (family Dasyatidae) are among a host of fishes that are commonly found at the sediment surface or slightly above the sediment. Coral reefs are crowded with an array of fishes that remain in close proximity to the reef structure. Fishes are not usually considered to be members of the benthos community, buy many certainly have an affinity for sediments.

Nekton

Animals that have strong swimming ability are members of the **nekton** community. Most are fishes, but we should not overlook squids, nor organisms that may be intermittent visitors. The latter include swimming crabs, scallops, shrimps and others which spend most of their lives in association with the sediments but have the ability to enter the water column for various periods of time and can swim strongly enough to overcome currents.

As we have seen, there are species of fish that are most commonly found in association with the sediments and, in at least a few cases, may be associated with the water surface or are found in conjunction with vegetation that is floating at the surface. Most swim in the water column. They may occupy limited ranges or roam the world ocean. Ranging in length from a few millimeters to several meters, the thousands of species of modern fishes have become adapted to a wide variety of marine habitats.

Conclusion

The ocean is vast and has retained many of its secrets from mankind for millennia. Our understanding of the ocean realm has increased in quantum leaps over the past few decades. We now have a good picture of how the physical, chemical and geological features that exist impact biological productivity. We have also come to realize that the ocean is not the source of unlimited numbers of animals and plants that can be utilized as human food. In order to meet the growing demand for animal protein from the sea, we have expanded fishing activity into previously untapped areas and, perhaps more importantly, we are catching and processing previously ignored species at astounding rates.

As traditional seafood catches have stabilized or declined in the wake of steadily increasing world demand, other formerly ignored or underutilized species have come under intensive fishing pressure. Squid are one example. There has been interest in utilizing **krill** (small shrimp-like animals that are present in huge quantities in the polar seas) as human food.

Man has reached the point in terms of developing new fisheries where additional production increases will be difficult to achieve. Few extensive new fisheries can be envisioned. We face a situation where the total recovery of products from the capture fisheries of the world ocean will stabilize at somewhere near 100 million tons annually. That is, those fisheries will stabilize if proper management is employed. Without good management of our marine fishery resources, we could see annual harvest rates accrue at a lower level.

Additional Reading

McConnaughey, B.H. 1974. Introduction to marine biology. C.V. Mosby Company, St. Louis. 544 p.

Reid, G.K., and R.D. Wood. 1976. Ecology of inland waters and estuaries. D. van Nostrand Company, New York. 485 p.

Ross, D.A. 1970. Introduction to oceanography. Appleton-Century-Crofts, New York. 384 p.

Stickney, R.R. 1984. Estuarine ecology of the southeastern United States and Gulf of Mexico. Texas A&M University Press, College Station. 310 p.

Tait, R.V., and R.S. de Santo. 1975. Elements of marine ecology. Springer-Verlag, New York. 327 p.

11

Freshwater Environments

Robert R. Stickney

Introduction

While over 70% of the earth is covered by salt water, less than 2% is covered by freshwater lakes, ponds and rivers. Though small in volume and surface area relative to the world's oceans,* the freshwater environments of much of the world are in great demand by man as they are sources of water for recreational, domestic, agricultural and industrial uses. There is a great deal of interest in the study of natural freshwater environments and upon the impact that man has had, and continues to have, on those environments. **Limnologists** are biologists interested in the physical, chemical, geological and biological activities that occur in freshwater environments, thus making them the freshwater equivalent of oceanographers.

Approximately 3 ft (1 m) of water evaporates from the surface of the ocean annually and enters the atmosphere as water vapor. That water then returns to the surface of the earth as precipitation. About 91% of the precipitation that falls returns to the oceans, with the remainder falling on land and freshwater bodies. Some of the water falling onto the land percolates into the ground to recharge water tables. Some runs off the land to form creeks, streams and rivers. The circuit of evaporation, precipitation and runoff is known as the **hydrologic cycle**.

Freshwater bodies are divided into two major types, **lentic** and **lotic**. Standing water bodies, such as ponds, lakes and reservoirs, are lentic, while flowing waters, such as creeks and rivers are lotic. Channels for lotic waters are formed as water flows downslope from its source to a receiving water body that may contain either freshwater or salt water. Springs, snowmelt and rainwater runoff are sources of lotic waters. Lakes have typically been formed from the scouring out of depressions by glaciers and the subsequent filling of the depressions with water as the glaciers melted. Lakes can also be formed in various other ways. Rock that is soluble in water may eventually dissolve to provide a lake basin. **Oxbow lakes** (Figure 11.1) are formed when a meander in a river is cut off by sediment deposition. Ponds are small bodies of standing water that may be natural or man-made. Reservoirs are man-made lakes. They are typically formed when a dam is constructed on a river.

Many of the subjects discussed in Chapter 10 with respect to marine waters are also applicable to freshwater environments. Freshwater contains various chemicals dissolved within it. We have already seen that the salinity of freshwater is, by definition, less than 0.5 parts per thousand (ppt). Freshwater organisms require the same nutrients as their marine counterparts. The freshwater food web is based largely upon primary producers, and many of the same geological processes occur in both freshwater and salt water. Freshwater bodies are not large enough to have daily water level fluctuations caused by tides, but they do have currents, waves and other physical phenomena in common with the oceans.

---

*98% of the water on the earth's surface can be found in the oceans. Of the remainder, 0.02% is present as liquid freshwater and the remainder is ice.

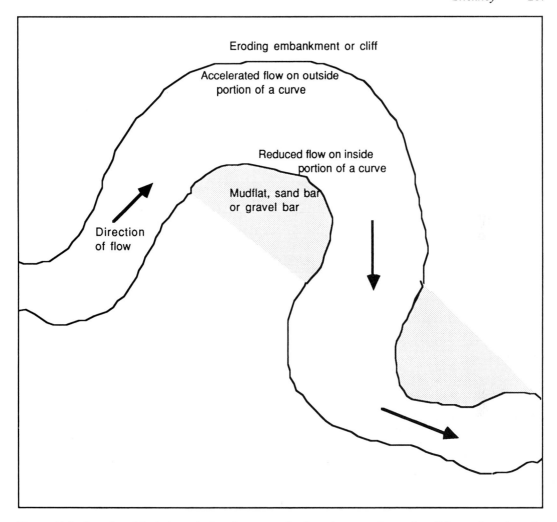

Figure 11.1. An oxbow lake is formed when the course of a river changes and a portion of the previous channel becomes isolated. Meanders in a stream lead to erosion on the outer aspect of a turn where water velocity increases, and deposition on the inside of the turn where velocity is reduced. (Diagrams in Chapter 11 courtesy of the author.)

Thermal Stratification of Lakes

Flowing water tends to be fairly well mixed, and the mixing process is aided by the fact that most streams are relatively shallow. Currents and wind allow the water column to be fairly consistent from surface to bottom with respect to temperature. Shallow lakes are also generally well mixed by wind except during the winter, when they may be covered with ice if the climate is sufficiently cold.

In temperate climates, lakes typically go through some interesting seasonal patterns with respect to temperature and water depth. In the winter, the surface of such lakes will be frozen with ice (32°F [0°C] or colder), which overlays liquid water that becomes slightly warmer with increasing depth. During winter, the maximum temperature found in ice-covered lakes is about 39°F (4°C) at the bottom. That is because freshwater has its highest density (most weight per unit volume) at 39°F (4°C) and the heaviest water will sink below water of lower density. Most liquids become more dense as they solidify. If water acted like other liquids and sank, the bottoms of the oceans and temperate lakes would be ice-covered. Life, as we know it, might not have been possible under those conditions.

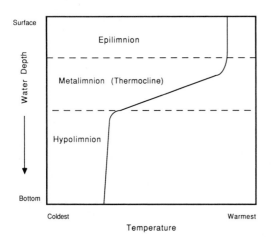

Figure 11.2. Temperature profile of a temperate lake during summer stratification showing the epilimnion and hypolimnion, where temperatures change slowly with depth, and the metalimnion, or thermocline, where temperature changes rapidly with depth.

As the ice melts in spring, wind will mix the water body and the temperature will become uniform from surface to bottom. Also, the water will begin to slowly increase in temperature. As the season progresses, the amount of heat entering the water will increase. Where there is little or no wind, the only way warm surface temperatures can be passed to the deeper water is through diffusion, which is a slow process. Warm water at the surface is less dense than the underlying water and tends to float on top in the absence of strong wind. The result is that an upper layer of warm, relatively uniform temperature water forms that has below it a zone of rapidly decreasing temperature. The upper layer is called the **epilimnion**. The area below it, in which the temperature drops about 1°C* for every meter in depth, is known as the **metalimnion**, or **thermocline**, which is similar to the oceanic thermocline (Chapter 10) but not nearly as deep. Below the thermocline is the **hypolimnion**, a layer in which temperature is, once again, nearly constant. The layering of a lake by temperature is known as **vertical stratification** (Figure 11.2).

Once stratified, a lake will often remain so until fall, when cooling water temperatures cause the lake to become **isothermal**; that is, the temperature will be again constant from surface to bottom and the wind will cause the water column to become mixed. The two seasons of the year when the water column is mixed are known as periods of spring and fall **overturn**. Water circulation patterns during the various seasons of the year for a temperate lake are depicted in Figure 11.3.

The epilimnion of a lake may be several yards (meters) thick (usually less than 10) or only a yard or two (1 to 2 m) deep. In some instances, a distinct difference in temperature can be felt from the legs to the upper body by a person standing chest deep in the water. That happens most commonly in shallow ponds. A hypolimnion does not usually occur in such ponds because the thermocline extends to the bottom.

While the atmosphere contains about 20% oxygen, the ability of oxygen to dissolve into water is limited. Normally, water contains no more than 10 parts per million (ppm) of oxygen. The amount of oxygen that can be dissolved in water under normal circumstances is affected by temperature, altitude and salinity. As any of the three increases, the oxygen-holding capacity of water is diminished.

In deep lakes, where a significant portion of the water volume is within the hypolimnion, problems with oxygen depletion may occur during summer because of the respiratory demands of

---

*The definition of epilimnion is based on a change in Celsius temperature of one degree for each meter of depth increase. Since the relationship between Celsius and Fahrenheit is not linear, a simple definition based on the English system of measurement cannot be presented.

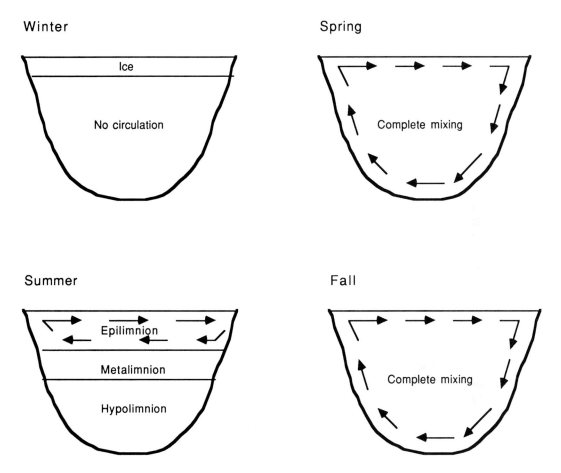

Figure 11.3. Cross-section of a temperate lake showing seasonal circulation patterns (arrows show mixed areas).

animals living within that zone. Oxygen is transferred to water from the atmosphere by diffusion at the surface and is produced photosynthetically by vegetation below the surface in the upper parts of the water column when light is available. When a thermocline is established in a lake, there is little or no physical mixing of the water in the epilimnion with that in the other strata, so there is little opportunity for oxygen to mix into the deep waters of a stratified lake.

Animals that inhabit the deep waters of the lake utilize a portion of the available oxygen in the hypolimnion. Bacterial decomposition of organic matter such as leaves, the bodies and body parts of animals, and feces that fall to the bottom of the lake and decompose also place a considerable demand on oxygen. This demand is known as **BOD** (biochemical oxygen demand). In some instances, the oxygen supply can reach critically low levels or even be eliminated (become **anaerobic**) before fall mixing occurs. In such instances, resident animals may be killed, particularly those that cannot migrate to shallow water, where sufficient dissolved oxygen may be available.

Oxygen problems can also occur in the winter. Ice cover on a lake reduces light penetration, thus reducing the volume of water that is available for primary production. Snow on top of the ice can further reduce light penetration. Ice cover also prevents the addition of atmospheric oxygen.

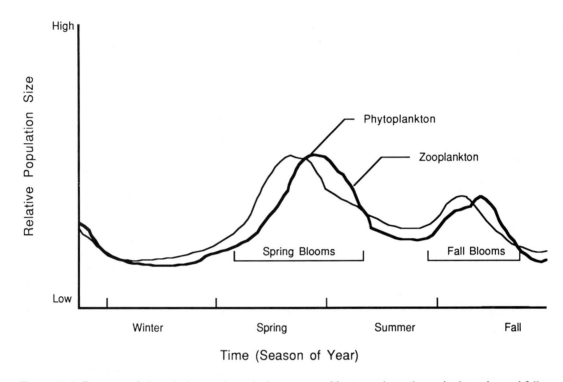

Figure 11.4. Response of phytoplankton and zooplankton communities to nutrient releases in the spring and fall.

The demands for oxygen are sometimes greater than the rate of replenishment from primary production, so there can be oxygen depletions leading to **winterkill**. Large numbers of fishes have died from that cause in northern lakes.

Nutrients released from decomposing organic matter in the deep waters of a lake during summer and under the ice during winter become available to plants in the spring and fall when mixing occurs. As the phytoplankton population increases, there is more food available to the zooplankton community, and zooplanktonic animals begin to reproduce at a rapid rate. The result is an eventual decline in phytoplankton abundance. When the food of zooplankton becomes limited because of overconsumption, zooplankton biomass begins to decline (Figure 11.4). Plankton blooms often occur when fish larvae are present because of the timing of fish reproductive cycles. Planktonic organisms can provide small fishes with ample food supplies. On the other hand, failure of a plankton bloom can deprive young fishes of food and lead to a subsequently poor year-class.

As indicated, thermal stratification does not occur in all lakes. Polar lakes and lakes located at high altitudes may be covered with ice throughout the year, so no mixing occurs. When such lakes have only a brief period in the year of ice-free conditions, the water may never warm above 4°C (39°F) and, when the water column mixes, it remains at that temperature or slightly warmer. It is soon covered with ice again and there is insufficient time and temperature to allow for stratification. Tropical lakes may remain mixed and at relatively constant temperature from surface to bottom throughout the year. Deep tropical lakes that remain constantly stratified may actually become devoid of oxygen.

Productivity in Lakes

Some lakes have very little life in them, while others are teeming with a variety of living organisms. Lakes often differ in productivity because of the amount of nutrients present. We have seen in Chapter 10 that various forms of nitrogen as well as phosphorus are critical for maintaining living organisms. Those nutrients enter a lake with runoff from the surrounding land and are found in streams that flow into the lake. Lakes surrounded by rocky ground and even many that occur in forests often have low levels of nutrient input. On the other hand, lakes that receive runoff from agricultural land or sewage input from a city often contain high levels of nutrients.

In the normal evolution of a temperate lake, there is a period after the lake is formed when productivity is very low. Take, for example, the Great Lakes. When such lakes were first filled, the land had been scraped clean by glaciers, so there was little in the way of organic matter to supply nutrients to the water. Forests eventually became established, but the amount of nutrients in the runoff from them was insignificant relative to the volume of the lakes. Yet, over thousands or tens of thousands of years, the nutrient content of the lakes has increased, but at a rate that would have been difficult to measure had there been limnologists around to do so. Nutrient levels were certainly high enough that by the time North America was settled by Europeans there were large numbers of fishes being supported. Over a relatively short period, at least in terms of the lifespan of a lake, cities sprang up and a great increase in the amount of nutrients entering the Great Lakes occurred. Sewage and an array of other organic materials that accompany the development of human communities were deposited in the lakes, so nutrient loading increased dramatically.

The changes that occurred in the Great Lakes could be and, in fact, have been measured by scientists. Changes that once took decades or centuries became measurable in periods of a few years. To the average observer, most of the Great Lakes seemed to outwardly change little as a result; however, the pressure on at least one of them, Lake Erie, was so great that significant changes were apparent to even the most casual observer. More intense, longer-lasting blooms of algae led to an increase in the amount of phytoplankton and rooted aquatic plants and increased populations of all kinds of often undesirable animals. This led to a concomitant loss of more desirable species. High levels of organic pollution led to depletions in dissolved oxygen, eliminating many sensitive, desirable animals and promoting undesirable ones that could withstand the new environment. Fish kills during the 1960s led some people to claim that Lake Erie was dying and that nothing could save it. The establishment of sewage treatment plants, removal of phosphates from laundry detergents and various other measures have turned the situation around in the case of Lake Erie, and the lake has been revitalized beyond what many thought would ever be possible. The Lake Erie experience stands as a warning, however, about what can happen when the natural aging process in a lake is accelerated by human activity.

Most natural lakes show characteristically low levels of production when they are initially formed. As a lake ages, nutrient levels increase and so does biological productivity. Sediments from incoming streams, as well as organic matter, cause the lakes to become more and more shallow.

As the process continues, a lake will eventually become so shallow because of sedimentation that it will support only rooted vegetation such as cattails and marsh grasses throughout what remains of its surface area. Ultimately, the lake will become a bog, and then dry land.

A young lake that has very low nutrient levels is called **oligotrophic.** When the nutrient levels increase, there is typically a change in the type of fauna, and the lake becomes **eutrophic.** The cold, clear lakes of the Rocky Mountains, and such places as Alaska, northern Minnesota and Wisconsin are generally oligotrophic. They have fish populations dominated by such groups as trout and salmon (family Salmonidae), northern pike (*Esox lucius*), muskellunge (*E. masquinongy*) and walleye (*Stizostedion vitreum vitreum*). Eutrophic lakes tend to be more turbid and are characterized by largemouth bass (*Micropterus salmoides*), crappie (*Pomoxis* spp.), bluegill (*Lepomis macrochirus*) and catfish (family Ictaluridae). That is not to say that some of the same species don't exist in both oligotrophic and eutrophic lakes. Lake aging (called **eutrophication**) is a continuous process, and there is no clear line that distinguishes an oligotrophic from a eutrophic lake. A lake that is choked with rooted vegetation and nearing the end of its life is called **dystrophic**.

The process of lake eutrophication is a natural one, and while we cannot typically observe it in a given water body because it happens very slowly, examples of lakes at various stages are not difficult to find. The process can be accelerated by the activities of man, as was the case of Lake Erie. Lake Washington, which borders Seattle, Washington, is another example of a lake in which the eutrophication process was accelerated by man's activities. When action was taken to curtail the inflow of nutrients from municipal wastes to Lake Washington, the process was reversed and the lake, as was the case in Lake Erie, reverted back to a condition that had existed many years before.

Man-induced eutrophication has been the result of allowing untreated or partially-treated sewage and other compounds that contain high levels of nitrogen and phosphorus to enter lakes. Phosphorus is often the limiting nutrient for plant growth in freshwater environments. When large amounts of phosphorus are available, phytoplankton blooms are encouraged, which can lead to reduced water clarity (increased turbidity from the millions of algae cells present per liter of water). Turbidity is easily monitored with a Secchi disc in lakes, just as it is in marine waters. Increased zooplankton production generally occurs, and the great increase in biological activity can lead to oxygen depletions, particularly if the plankton population crashes and is subjected to bacterial decomposition.

It was largely the nationwide eutrophication of our freshwater lakes that led to the passing of legislation mandating the establishment of secondary sewage treatment in U.S. cities. In primary sewage treatment, solids are settled from the water that passes through a treatment plant, but the dissolved nutrients are not significantly reduced. In secondary treatment, there is some nutrient removal by microorganisms, though the effluent water continues to have elevated levels of dissolved nitrogen and phosphorus compounds. To completely remove nutrients from water requires what is known as tertiary treatment, a process that is extremely expensive and one that is not in common use with respect to the treatment of sewage.

Acid Rain

Lakes in the United States, particularly in the northeastern region, and in various other parts of the world have become increasingly acidic in the past several years. This has largely been due to acids that enter the atmosphere and return to earth in rainfall. While the acid rain phenomenon was first described many decades ago, the relationship between acid precipitation and the disappearance of fishes in affected waters was first recognized in the late 1950s. Since then a large body of scientific evidence has been gathered with respect to the impact of acid rain on fishes and other aquatic organisms. To understand the problem, it is necessary to examine the way in which water is **buffered**. Buffer capacity is the ability of water to resist a change in prevailing acid or base concentration. The buffer system described below operates in salt water as well as freshwater, but because of the strong buffer capacity of the ocean, acid water has not generally been a problem.

Water may be acidic, neutral, or basic. Acids are compounds that release hydrogen ions (H^+) into solution, while bases release hydroxide ions (OH^-). The state of the water with respect to acidity is measured as **pH** (defined as the negative logarithm of the hydrogen ion concentration in the water). The pH scale ranges from 1 to 14, with 7 being neutral. Water with a pH below 7 is considered acidic, while levels above 7 are basic. A buffer system is established when a weak acid (e.g., acetic acid) is mixed with a weak base (e.g., sodium bicarbonate). Buffer systems resist pH changes through a series of chemical reactions. Since the pH scale is logarithmic, the difference in hydrogen ion concentration between whole pH units involves an order of magnitude. For example, water of pH 10 is ten times more basic than water of pH 9 and 100 times more basic than water of pH 8. On the acidic side (pH <7.0), the lower the number the more acidic the sample, so water at pH 5 is ten times more acidic than water at pH 6.

As indicated at the beginning of this section, freshwater lakes in some parts of the United States and Canada have become increasingly acidic in recent years. The problem is associated with the phenomenon of acid rain. The acidity is thought to be due to the production of acid-forming gases that are released when coal containing high levels of sulfur is burned for the production of electric energy in power plants or for other industrial or domestic purposes. When sulfur is released into the atmosphere and combines with water vapor, sulfuric acid is formed. There are other compounds in fossil fuel power plant gases that form nitric acid in the atmosphere. The acids fall to earth with rain and enter lakes directly and through runoff. As the acids accumulate, the buffering capacity of the lakes is overcome and the pH begins to fall. Critically low pH levels have been reached in some lakes, to the extent that the type of fishes present may be altered. In extreme cases, all fish life has been threatened or eliminated.

The acid rain problem is a biological one to be sure, but it also has political implications. The government of Canada has been negotiating with that of the United States to control gaseous emissions from power plants, which are often carried north over Canada where they eventually fall as acid rain. This issue has strained trade and other relationships between Canada and the United States.

Short-term remedial solutions have included adding lime and other basic compounds to affected waters to raise the pH. Such solutions are impractical on a large scale when either water bodies or forests are involved (hundreds of thousands of hectares cannot be treated economically). Ultimately, the problem can be solved only by cutting off the source of the acid. We need to strictly control the emissions of power plant exhaust gases and cease the burning of high-sulfur coal, except in instances where the acid-forming compounds are tightly controlled to keep them from entering the atmosphere. In fact, some strides have been made with respect to reducing the problem, but as regulations are imposed and the emission controls placed on power plants take effect, the acid rain problem continues. Reversal of the situation in heavily impacted waters may continue to require the use of chemicals to neutralize the acidity.

Characteristics of Streams

Creeks, small streams and major rivers all carry water downhill. Most streams flow throughout the year, but intermittent streams are common in arid areas where they actually dry up between the infrequent rainfalls. Lotic systems vary tremendously in size and the volume of water they carry. The largest river system in the world is the Amazon, accounting for one-fifth of the world's river discharge. Most streams have variable flow rates depending on the amount of runoff that occurs at any given time. The flow rates in streams typically increase during the spring, when snowmelt begins. In some regions of the world, relatively predictable rainy seasons occur, during which streams become swollen with runoff water.

The rate at which a stream flows depends upon the volume of water entering the stream channel, the shape of the channel itself and the degree of slope present. The steepest part of any river system generally occurs in the vicinity of the upper end or headwaters of the stream. The head of a stream is often located in hills or mountains where the river is narrow, the slope is steep and the volume of water is much lower than further downstream. As a stream moves toward the ocean or a terminal lake, more and more water enters from direct rainfall, land runoff and tributaries that join the main stream. At the same time, the degree of slope may decline and the channel broadens, so while there is a much larger volume being carried than at the headwaters, the lower portion of a stream may actually flow at a slower rate.

Many streams begin at high elevations where there is little sediment entering with runoff. As the stream moves out of hills or mountains and into more flat terrain, often characterized by agricultural land, the amount of silt and clay in the runoff waters increases, and the stream becomes increasingly turbid. Some streams are quite turbid at their headwaters, but those are exceptions. For example, glaciers are the sources of some streams. Glaciers contain heavy burdens of particulate matter. As a glacier melts, a stream may be created that contains high levels of what is known as "glacial flour," made up of the various types of particles trapped when the ice was formed. Such a stream will appear milky, while the typically turbid stream that receives runoff from agricultural land appears muddy.

The aging, or eutrophication process that was described for lakes is not duplicated in streams, though streams do not remain the same through long periods of time. Stream channels may change dramatically, moving from one place to another. Over geological time periods, canyons are created as water dissolves and erodes the earth's surface. The gradient of a stream will change as mountains and hills erode, thus the flow rate will change. Volumes of water carried by streams change both over the short term (as the result of intermittent rains and annual snowmelt) and over long time periods (as a result of climatic changes and other factors).

Streambeds are often made up of a series of relatively shallow areas where the water moves rapidly and deeper areas where flow rates can become quite slow. The former are known as **riffles** and the latter as **pools**. The two habitats are quite different and often support different types of flora and fauna.

Another feature of streams is that they rarely move in straight lines. Rather, they follow somewhat circuitous routes in flowing from the hills or mountains through the lowlands and into the ocean. They can often be seen to cross from one side of a valley to another and back again repeatedly. These side-to-side movements are known as **meanders**. As a stream turns, water on the outside of the turn moves faster than that on the inside of the turn (Figure 11.1), resulting in erosion of the outside turn. This may lead to the formation of a deeper channel, and it can mean the creation of a cliff or bluff. The velocity of the water at the inside of the turn may be sufficiently slow that suspended particles like gravel, sand and silt are deposited. Gravel bars, sand bars and mud flats are the result. Sufficient deposition can cause a river to seek a new channel. In such cases, oxbow lakes can be formed (Figure 11.1). The erosion occurring on the outside of a channel may break through into a region where the elevation is actually below that of the original stream channel, in which case the stream will begin taking a new route.

Stream channels are thus, quite dynamic. They may change considerably over time. Historically, this was not much of a problem. However, since man has begun constructing cities along rivers, there have been attempts to maintain river channels in place.

Valleys are actually the floodplains of the rivers they contain. When high volumes of water are present, a river will overflow its banks and the water will spread laterally across the expanse of the river valley. That is not much of a problem if there isn't a city, town, industrial complex, farm, or some other site of interest to mankind located in the valley. When there is something in the valley that people feel needs protection from floods, the battle is on! The result has been the development of various types of flood control structures such as dams and levees. A levee is basically a wall constructed parallel and on either side of a river to contain that river within a confined area if the water level rises high enough to overflow the natural channel. Hundreds of miles* of levees have been constructed along the Mississippi River and other rivers to control flooding.

Dams are often constructed for flood control. A reservoir will form behind each dam on a river. In the case of flood control reservoirs, the volume of water may be reduced periodically, particularly during periods of low inflow. Thus, when there is a great increase in runoff, the reservoir will fill, but the amount of water being released to the river below the dam will not increase significantly. If a dam and reservoir system are properly designed, they can protect downstream areas

*1 mile = 1.7 km

from flood damage. Dams can also be used to generate electrical power, and reservoirs can be used for various types of recreation.

One problem with reservoirs is that they accumulate sediments. It has been estimated that the typical reservoir will become filled with sediments within a few hundred years. At some point before that, the ability of the reservoir to hold the amount of floodwater for which it was designed will be surpassed and downstream flooding will occur despite the presence of the dam and reservoir.

Of significant importance is the fact that the communities of organisms that develop in reservoirs differ considerably from those that inhabited the stream prior to impoundment. In some cases, plant and animal species have become threatened or endangered because of this type of habitat alteration. It is only within the past 2 decades that the impact reservoir construction might have on indigenous flora and fauna has been considered either by those who are responsible for construction of the dams or the agencies that have the job of conserving natural resources.

Freshwater Communities

Decomposers

The **decomposer** community in freshwater environments is comprised of bacteria and fungi, organisms which colonize the surface of any type of dead organic matter present. Those substrates include such things as leaves and other plant parts that enter from the surrounding terrestrial environment, as well as a host of plants and animals and their parts that are available within the aquatic ecosystem.

Logs, tree branches and other types of woody debris may remain in the water largely intact for many years. On the other hand, such things as dead algal cells and the bodies of various types of small animals and fishes will be decomposed so that only skeletons remain after a few days or weeks. The bacteria and fungi that carry on the decomposition process may be consumed by various types of animals themselves; thus, the nutrients present in the dead organic matter are recycled. Nutrients may be liberated into the water as a part of the process, and those required chemicals help support the growth of the primary producer communities.

Primary Producers

The same basic types of communities that we have seen in the marine environment occur in freshwater, though the same species are rarely found in both types of environment. There are, for example, no species of coral inhabiting freshwater. While most people never see them, there are freshwater sponges, but they are relatively uncommon and have no commercial value. The freshwater environment, on the other hand, teems with insect life, while very few insects are associated with the marine environment. Similar comparisons can be made with respect to the plant communities in the two environments. Algae and tracheophytes occur in both, but there are few, if any, species in common.

Rooted Plants

A wide variety of rooted aquatic plants can be found in freshwater environments (Figure 11.5). In rivers, rooted plants (tracheophytes) tend to be restricted to pools where flow rates are sufficiently low to keep them from becoming uprooted. In all freshwater environments, plants grow only to depths where light can penetrate because when the seeds sprout in the sediments, there must be sufficient light for the plants to grow. Plants in very clear lakes will grow in much deeper water than is the case in turbid lakes.

Some rooted plants are entirely or almost entirely submerged. They may introduce oxygen into the water through photosynthesis. Others float on the surface of the water or grow in very shallow water with the bulk of the vegetation occurring above the waterline. In those instances, the oxygen produced through photosynthesis is largely transferred to the atmosphere.

Figure 11.5. Rooted plants of various kinds are common near the shorelines of streams, ponds and lakes.

Rooted aquatic macrophytes provide substrate for the development of attached algae and give shelter to various types of animals, including fishes. Many aquatic plants also have aesthetic value—most people are familiar with the beauty and diversity of water lilies, for example. But aquatic plants can cause problems as well. Weed-choked lakes make fishing difficult because lures become tangled and lost. They interfere with other recreational activities such as swimming, boating and water skiing. People have even been known to drown after becoming entangled in aquatic vegetation. Selection and maintenance of the "proper" level of aquatic vegetation in lakes has become a controversial issue in fishery management in recent years.

Aquatic plants can be controlled in various ways. Mechanical harvesters have been developed, but the cost of purchasing and operating them is quite high. Also, once harvested, the plants must be disposed of in some manner. Transporting them to a landfill is costly, and they are often not very useful as livestock foods. A second method of plant control is through the use of herbicides. Treatment is expensive and, as is also true of mechanical harvesting, frequent retreatment may be necessary because both methods tend to leave the roots intact and the plants will grow back. Dead plant material decays and exerts a high BOD. This can lead to oxygen depletion and resulting fish kills. The herbicides themselves may poison fishes and other aquatic animals directly, particularly if the chemicals are improperly used.

Another form of plant control depends on natural plant eaters (herbivores) to keep the unwanted vegetation in check. This is known as biological control. When a problem is identified, it may be possible to introduce an herbivorous animal that is normally absent from the system and have it eat the unwanted plants. One caution is that the introduced animal should not, itself, cause problems. For optimum conditions, it may be desirable to remove only a portion of the plant community. Thus, the biological control organism may also need to be under the control of the biologist. Creating the proper balance between the plants and herbivores is often difficult.

Biological plant control organisms include various types of invertebrates such as snails and some fishes. An African freshwater fish from the family Cichlidae, *Tilapia zillii*, has been introduced into irrigation canals in California to control aquatic vegetation. The fish die when the water temperature falls below about 50°F (10°C), so annual restocking is generally required to maintain the desirable level of control. Another, more controversial species, the grass carp (*Ctenopharyngodon idella*), has been used in some states, but is illegal in many others. Grass carp, native to the Amur River in China—hence, another common name, white amur—eat aquatic macrophytes to the exclusion of virtually everything else. By stocking them at the proper density, they will control many undesirable species of rooted plants.

In theory, grass carp will not reproduce in small streams, lakes or reservoirs. While the eggs might be dispersed into the water and fertilized, a set of highly specialized conditions is required for hatching. Those conditions do not generally occur in the United States. There have, however, been reports of grass carp (apparently produced from natural reproduction) in the Mississippi River and the Rio Grande. As a result, there is concern that grass carp might become established sufficiently to become a problem in the United States by consuming desirable as well as nuisance plant communities. This concern has led many states to ban grass carp introductions altogether or to only allow the stocking of sterile fish. Those fish can be produced by hybridizing grass carp with a distantly related species or by exposing fertilized eggs to certain physical and chemical treatments that cause the production of an extra set of chromosomes (triploidy). Triploid fishes are sterile.

Periphyton

The communities of algae that colonize substrates are known as **periphyton**. Many individual algae species are common to both the periphyton and phytoplankton communities. Thus, while rivers have phytoplankton present, individual cells that remain attached to the substrate ensure that all members of the species will not flow downstream and become lost to the upstream environment.

A rock or waterlogged piece of wood picked up from a streambed or the bottom of a lake will typically be covered not only by single-celled and filamentous algae, but will also have various types of animals living on it. Insect larvae, small worms, crustaceans and so forth can be found. The term that has been used to encompass both the plants and animals found growing in association with substrates is *Aufwuchs*. These communities can make significant contributions to overall primary and secondary productivity and have been the focus of hundreds of scientific investigations. Organisms will grow on top of one another, sometimes creating thick mats. *Aufwuchs* communities typically grow until light or nutrients become limiting to individuals inhabiting the lower layer of the community. Those underlying individuals will die and portions of the *Aufwuchs* community that include both living and dead organisms will slough off the substrate. Recolonization will occur and the process is repeated. Organisms that are still alive when a piece of the *Aufwuchs* mat is sloughed off may find a new substrate to become attached to, or in some cases, they may enter the plankton community.

Phytoplankton

As is true in marine environments, a large percentage of the primary production in most freshwater bodies is the result of phytoplankton growth. Phytoplankton provides a source of food for many species of aquatic animals. Thousands of species of mostly microscopic algae make up the freshwater phytoplankton communities. Many are single-celled algae, but others occur as filaments of various lengths. Each filament is comprised of a number of individual cells linked together or bound into a matrix in the manner of fruit chunks suspended in gelatin.

The growth of phytoplankton and other plant communities in water is supported by the presence of nutrients and light. We have already seen how spring and fall phytoplankton blooms in lakes occur in response to nutrients released from the sediments into the water column during overturn (Figure 11.4). Phytoplankton blooms crash when the nutrient supplies become exhausted, but they can also decline because of consumption by zooplankton and light restriction. An obvious instance of reduced sunlight is during periods of cloud cover. Also, if the turbidity of the water increases for some reason, the penetration of light will be restricted. Turbidity increases when silt and clay enter a stream, lake, or pond with rainfall runoffs.

When an intense phytoplankton bloom is present, there can be hundreds of thousands to millions of algal cells present in each quart (approximately one liter) of water. The presence of so many cells colors the water (typically green or brown) and adds to the turbidity level. Thus, as the phytoplankton community develops, it actually inhibits light penetration and reduces the amount of energy available for stimulating further algal growth. This is known as **self-shading**.

Reproduction in phytoplankton is asexual. The cells divide and new individuals are formed as a result. The time between cell divisions—as little as a few hours or considerably longer—depends upon various environmental conditions like light and temperature, as well as the levels of nutrients available. Cell division may virtually cease when conditions are not conducive to growth.

Zooplankton

A variety of animals comprise the zooplankton communities in freshwater environments. Among them are protozoans (single-celled animals), rotifers, crustaceans and the larvae of a variety of animal groups, including fishes. Any animal that lives suspended in the water, but at the mercy of currents because it has weak swimming ability, is a member of the zooplankton community. Larvae of clams, fishes and many other animals eventually begin their lives as zooplankton, but leave the community to take up residence as members of the nekton, *Aufwuchs*, or benthos, while other animals spend their entire lives as zooplanktonic organisms. Some of the more common animals that are zooplanktonic throughout their lives are cladocerans and copepods (both are in the class Crustacea).

Many zooplanktonic species feed by grazing on phytoplankton cells. It is that feeding strategy which accounts for at least part of the decline in the spring and fall zooplankton blooms (see Figure 11.4). When sufficient food is present, zooplankton populations can expand very rapidly because their reproductive rate increases, thereby allowing the population to grow until food becomes limiting. During periods when food supplies are extremely low, many zooplanktonic species produce resting-stage eggs that will not hatch until environmental conditions become conducive for their growth and survival. The eggs of zooplankton may not be able to sense the increased availability of food but respond, instead, to changing environmental conditions. For example, the spring bloom of phytoplankton is correlated with increasing water temperature.

Benthos

The benthos community of any freshwater environment is often dominated by the larvae of aquatic insects (Figure 11.6). Many groups of insects are terrestrial as adults (e.g., mosquitoes, mayflies, caddisflies, dragonflies), while others (e.g., backswimmers) are found in the water throughout most their lives. In many cases, the larvae of aquatic insects scarcely resemble the adults. Many larval insects are carnivorous, while the adults either do not feed (their only function being reproduction) or are able only to consume liquids (e.g., mosquitoes). Dragonfly larvae are highly carnivorous. They will attack larval fishes and have even been known to bite humans, yet the adults are innocuous to humans and fishes. Various types of worms, molluscs and crustaceans also inhabit the benthos communities of freshwaters.

Limnologists are sometimes able to obtain a good indication about conditions in an aquatic environment by examining the benthos. Certain organisms, called **indicator species**, only occur or

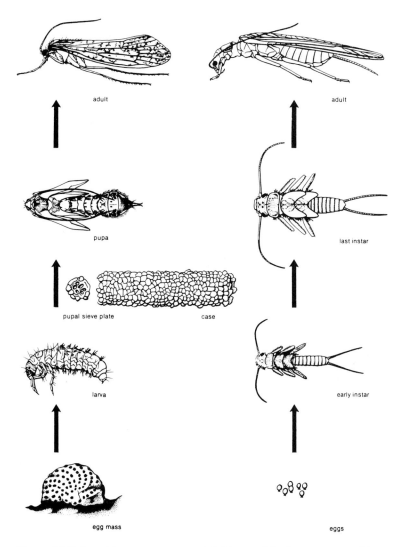

Figure 11.6. Insect larvae are common inhabitants of freshwater communities. (Merritt-Commins: *Aquatic Insects of North America.* Copyright © 1984 by Kendall/Hunt Publishing Company. Used by permission.)

become dominant under certain sets of conditions. For example, some types of aquatic insect larvae are highly tolerant of poor water quality and dominate the benthos when certain types of pollutants are present. Under normal conditions, an indicator species may be absent or contribute only a small fraction of the total number of benthic organisms per unit area. But if the environment becomes particularly suited to an indicator species, conditions promote a population explosion of that species, while other, normally successful species, are eliminated or their numbers are drastically reduced.

Benthic organisms contribute greatly to the diets of various fish species. Carp, sturgeon and catfish are examples of fishes that feed primarily on animals (and sometimes plants) found in association with the sediments. All three groups of fishes have sensory barbels located near their mouths that help them detect benthic food through the sense of smell.

While the numbers of benthic species per unit area may change relatively little on a temporal basis in marine environments, there are definite seasonal changes that occur in freshwater benthos communities, particularly in temperate regions. This is because many benthos communities are dominated by insect larvae. Those larvae are present in highest numbers following the breeding seasons of the particular species present. Most insects in temperate climates prevail during the warm months, so the aquatic larvae are present in large numbers prior to the appearance of the adults. In some instances, large outmigrations of adult insects will leave a pond, lake, or stream section almost simultaneously, causing a cloud of insects to rise above the water body, consequently depleting the food supply of animals that had previously been feeding on the plentiful larvae. In other cases, larvae metamorphose into adults and leave the aquatic environment over extended periods. Those losses are continuously augmented during the breeding season by the deposition of new eggs, which develop into larvae.

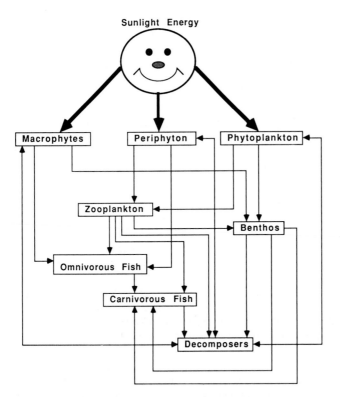

Figure 11.7. Energy flow patterns through a simplified freshwater food web.

Freshwater Food Web

The food web in freshwater environments closely parallels that found in the marine environment. The weight of organisms is dominated by the plant (primary producer) community, followed by the herbivores and, finally, the carnivores. The pyramid of numbers presented in Chapter 10 (Figure 10.8) and the diagram in that same chapter which shows the relationships between the trophic levels and standing crop biomass (Figure 10.9) both apply as readily to freshwater environments as they do to the oceans or estuaries. A simplified lake food web is presented in Figure 11.7. Clearly, the latter figure shows that the actual situation is not the straight-line relationship indicated in Figures 10.9 and 10.10. If we were to look at individual genera or species rather than higher taxonomic levels, the picture would become even more cluttered.

As energy moves through the food web, only a small fraction of it remains after each transfer. At best, only 2 or 3% of the light energy that strikes a plant becomes new organic matter. When a plant is eaten by an herbivore, only about 10% of the energy that was present in the plant forms new tissues in the animal. Thus, if there are 100 units of energy present in the phytoplankton, the zooplankton will be able to recover 10 units, the small predatory fishes 1 unit, and so on.

Riparian Zone Interactions

The **riparian zone** is defined as the land adjacent to the banks of a river or stream and the shoreline of a pond, lake or reservoir. Thus, riparian landowners are those who have land that borders any type of freshwater environment. The riparian zone has, in some instances, become a region of controversy. This has been particularly true with respect to logging in various parts of the United States. From the standpoint of a logging company, profits can be enhanced if all marketable trees can be removed from a particular site, often through the employment of the clear cutting technique where every tree is felled. This includes trees that occur in the riparian zone. Environmentalists argue that the riparian zone should be excluded from logging activities, not only to protect the aquatic environment, but also to provide habitat for the various types of wildlife species that use the riparian zone. Also, fishery resources are believed to be conserved when we maintain the riparian zone. Shade provided by shoreline trees may be important in helping maintain the constancy of the aquatic environment (for example, protection from daily temperature fluctuations that might occur if the water were exposed to direct sunlight). Also, timber harvest introduces logs and other debris, along with silt into the water, resulting in habitat alteration.

There is no doubt that logging right up to the bank of a stream can lead to an increase in turbidity due to soil entering the water. The same is true of plowing on agricultural land adjacent to ponds, lakes, streams and rivers. Many feel that a buffer zone should be left between the region of clearcutting, agricultural practice, or other human activity and the water. What is often unclear is how broad the buffer zone should be. Depending upon soil type, slope of the land and various other factors, including the types of wildlife in a given area, suitable buffer zone widths could vary considerably. There is growing interest in studying the problem in greater depth than has been done in the past and in establishing realistic criteria for use of the riparian zone.

Nutrient Cycling

We have already seen the importance of such nutrients as nitrogen and phosphorus in stimulating the growth of primary producers. Much of the nutrient load that is stored in the cells of plants eventually moves through the food web. In order for a body of water—for example, a lake—to produce a given amount of organic biomass each year, a steady supply of nutrients must be available, and those nutrients need to be recycled or replaced in order to be available to organisms at all levels of the food chain.

Chemical exchanges between the water and the sediments provide one mechanism by which nutrients are made available to aquatic organisms. If the level of a particular nutrient reaches a certain concentration in the water, the sediments will begin to absorb that element because of equilibrium reactions. If, on the other hand, organisms begin to remove the element from solution, the sediments may give it up to the water.

Some nutrients will leave the aquatic system each year. The migration of adult aquatic insects from a lake or stream takes with it large amounts of nutrients. Commercial and recreational fish harvests also remove nutrients from the system, sometimes in large quantities. In addition, terrestrial animals may feed on aquatic plants and animals.

Nutrients are added in runoff water and in rainfall. Bird and animal droppings contribute nutrients, as do falling leaves and other organic debris that enter the water. In a balanced system, nutrient input will equal that lost from the system on an annualized basis. Realistically, few freshwater systems will be in true nutrient balance. In lakes, for example, the trend is toward increasing total nutrient levels with time; thus, we see natural eutrophication taking place even in the absence of inputs under the influence of mankind.

Some systems receive only low levels of nutrient input, often resulting in oligotrophic lakes. Most freshwater systems in the United States are not presently nutrient limited. Instead, the problem has been an overabundance of nutrients due to the runoff of fertilizers, sewage, manufacturing wastes and other compounds produced by man's activities. Thus, the problem has often become one of nutrient excess. Plant growth has clogged waterways, excessive production of various types of organisms has led to oxygen depletions and consequent mass mortalities of fishes and invertebrates and, in some cases, previously pristine water bodies have become foul-smelling sewers.

In North America, water pollution control legislation and the clean-up campaign that resulted have been highly effective in many areas. The problem of excessive nutrient loading into our freshwaters has certainly not been eliminated, but in many cases it has been reduced and is under some level of control. That does not mean that people can become complacent about the quality of our inland waters. Legislation exists that should help us avoid large-scale degradation of our water in the future, but individual water bodies remain at risk. With a growing population that enjoys living in close proximity to water, the pressure to develop riparian zones for human activities continues to be extraordinary. Those pressures are particularly significant in that they carry with them economic implications much more far-reaching than the use of the land for agriculture or timber production. Economic development of riparian zones will continue, but it should be done in an environmentally sound manner. If it is not, the desirability of living and working near water could be turned into a liability.

Additional Reading

Frey, D.G. (ed.). 1966. Limnology in North America. University of Wisconsin Press, Madison. 734 p.

Johnson, R.E. (ed.). 1982. Acid Rain/Fisheries. Proceedings of an international symposium on acidic precipitation and fishery impacts in northeastern North America. American Fisheries Society, Bethesda, Maryland. 357 p.

Reid, G.K, and R.D. Wood. 1976. Ecology of Inland Waters and Estuaries. D. Van Nostrand Co., New York. 485 p.

Wetzel, R.G. 1975. Limnology. W.B. Saunders Company, Philadelphia. 743 p.

12

Fishery Ecology

Frieda B. Taub

Introduction

Ecology is the science of interrelationships between living organisms and their environment. Fishes are the major vertebrates of freshwater and marine environments. Although some amphibians, reptiles, birds and mammals also live in aquatic environments, they do not have the enormous variety and ecological impact of fishes.

Once fish become large enough to actively swim, they become members of the **nekton** community and actively control their place in time and space within the water column. Fishes are capable of vast migrations and need not drift with the currents, as do plankton. Shellfishes such as oysters and crabs are also important fisheries resources. Although we recognize that oysters and crabs are not fishes, the types of information needed to manage them (standing crop; birth, death and harvesting rates) are similar to those collected for fishes. To properly manage our fisheries, we need to understand the attributes of harvested species.

From an animal's point of view, obtaining food without being eaten is critical—it must be able to gather enough food to meet its energy requirements and reproduce the next generation, but not fall prey to another animal. To accomplish this, an aquatic animal may change its morphology and preferred habitat during its lifetime. Aquatic animals are subject to many environmental influences, including the availability of food, which is necessary for a population to demonstrate high growth and reproduction. The animal also must find favorable temperatures and oxygen conditions, while avoiding predators, diseases and parasites.

As we examine environments from small ponds through rivers, lakes and oceans, understanding the factors controlling fish production becomes progressively more difficult. In small ponds, food availability often controls fish production, but the factors that control the survival of young oceanic fishes and their ultimate productivity is one of the greatest challenges to the fisheries scientist.

Only certain fishes, shellfishes and algae have the characteristics necessary for harvesting by humans. They must be abundant and, at some time during their adult life, should be concentrated in a predictable area so they are accessible to fishing gear. A species' value as human food depends on its acceptability in our diet, nutritional quality and accessibility. Organisms that are highly valued by one society may be ignored or considered repugnant by another. Many species having no value as human food are fished for reduction to fish meal used in livestock and aquaculture feeds. These species constitute some of our largest fisheries in terms of catch weight.

Ecosystems

Ecology is best studied within the context of ecosystems, rather than considering only the species of interest. An **ecosystem** includes all of the interacting organisms and physical-chemical components. Within an ecosystem, the flow of chemical elements, such as carbon, oxygen, nitrogen and phosphorus, and the flow of energy from sunlight through photosynthesis into organic material and its utilization by respiration are linked. **Trophic** or feeding relationships, introduced in previous chapters, are also important in understanding ecosystem interactions. Each trophic level is limited in energy to the amount it can eat of the next lower level. Thus, to understand what limits fishery harvests, one must understand what limits the lower trophic levels.

As explained in Chapters 6 and 10, photosynthesis involves using the energy of sunlight to convert carbon dioxide and water into simple sugars and oxygen. Photosynthetic organisms are also termed **autotrophs**, since they make their own food from chemicals and sunlight. Plants convert simple sugars into complex carbohydrates, proteins and fats, and they use much of their photosynthetic production for their own respiration and growth.

Organisms that cannot perform photosynthesis, called **heterotrophs,** rely on other sources, including autotrophs, for their food. These include all animals, bacteria, fungi and protozoa that cannot convert sunlight into organic chemicals; thus, the amount of plants available for consumption limits heterotrophic production. Animals must consume enough energy (measured in calories) and nutrients to supply their needs for respiration, growth and reproduction, and losses of energy in waste products (urine and feces). Animals can make limited biochemical transformations among carbohydrates, proteins and fats, but require some **essential nutrients** in their diet.

The chemical elements within ecosystems cycle indefinitely, but some of the energy bound in organic chemicals by plants eventually converts to heat at each step in the process (Figure 12.1). This energy must be replaced by new organic material resulting from photosynthesis; this, in turn, is eaten by herbivores, which are eaten by predators (carnivores). In most aquatic communities, a substantial amount of the important elements and energy are in detritus, which is processed by decomposers. Some portion of the decomposer community serves as food for bottom animals, which, in turn, are eaten by predators.

Each **trophic level—primary producers** (green plants), **herbivores** (consumers of primary producers), **first order carnivores** (consumers of herbivores), **top carnivores** (consumers of other carnivores), **detritivores** (consumers of detritus)—consists of many species that compete for resources. Also, animals may feed on different levels as they mature, or may simultaneously feed on several levels. Thus, the "food chain" is better described as a **food web.**

Because the trophic levels are linked, changes anywhere in the ecosystem are likely to have impacts elsewhere. For example, if nutrients are increased by sewage inputs, the amount and types of algae present in the ecosystem may change. These changes in algal abundance and type may produce unsuitable foods for the types of **grazers** (**herbivores**) present, and this may cause algal biomass to accumulate or the type of grazers to change. These grazers, in turn, may not be suitable as food for the previously dominant predators. Thus, changes in the algal nutrients may change the entire community structure (Figure 12.2). Conversely, if the top predator is removed by overfishing or a winter fish kill (caused by lack of oxygen under the ice), such a loss can have dramatic effects on the lower trophic levels. Similarly, introducing exotic species to enhance sportfishing, allowing the escape of ornamental plants and applying pesticides can have unexpected impacts on other components of the ecosystem.

Primary producers generate the initial source of organic energy. Each successive trophic level has progressively less energy available. Thus, it is most efficient for humans to harvest and feed on plants, somewhat less efficient to harvest herbivores and least efficient to harvest top predators.

Types of Ecological Interactions

Predation

Predator-prey interactions are the most dramatic of the ecological interactions between organisms—by definition, the predator kills and consumes the prey. The vast majority of aquatic organisms die by being eaten; few die of old age. If an organism derives food from another without

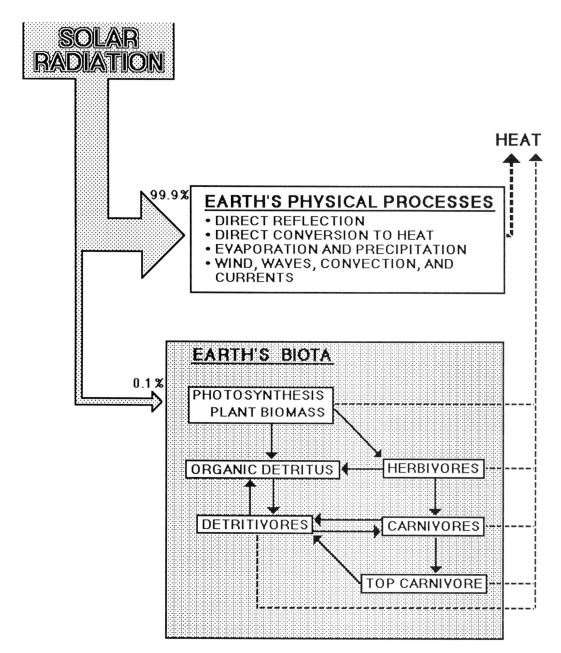

Figure 12.1. Solar radiation is the ultimate source of energy to fuel biological systems. Only a small part of that which strikes the earth is actually used for the production of organic matter. This figure shows how energy enters as light and leaves the earth's ecosystem as heat.

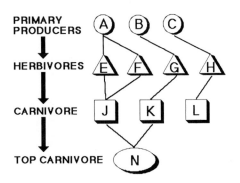

PRIMARY PRODUCERS

HERBIVORES

CARNIVORE

TOP CARNIVORE

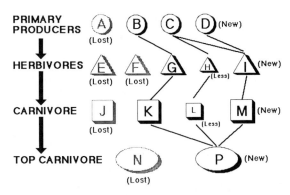

PRIMARY PRODUCERS

HERBIVORES

CARNIVORE

TOP CARNIVORE

Figure 12.2. Well-established food webs are present in stable aquatic environments as illustrated in the top portion of the figure. When a change occurs that leads to the loss of a species at any trophic level (bottom portion of the illustration), that change can have impacts throughout the food web.

killing it, we usually consider it a parasite. Predator-prey interactions are discussed in more detail below.

Competition

Of the ecological interactions, competition has received much study because it has been hypothesized to be a major controlling mechanism of species dominance. Most trophic levels have several species that compete for limited resources; the species most successful in capturing the resource will increase at the expense of the less successful. Scientists now recognize that factors other than open competition can structure communities; for example, competition may be prevented from becoming acute if a community is decimated by storms or other climatic events, or if predation on the populations of potential competitors is so severe that they never become resource limited. Although we now know that competition is not the only relationship that establishes communities structures, it is important in many situations.

Different organisms using the same resources do not necessarily compete if the resource is so abundant that its use by some species does not negatively impact others. Competition can be divided into **intraspecific competition** (between members of the same species) and **interspecific competition** (between different species). Competition can be most severe among organisms that use the same resources in the same way; thus, the most severe competition is likely to be intraspecific competition among members of the same age group because they will be competing for similar food or habitats.

Intraspecific competition is most apparent when high species densities result in reduced survivorship, growth and reproduction of that species. Often, the most desirable prey items will be depleted, and at least some are forced to feed on less desirable foods or in more dangerous (less protected) locations. During years when egg production is reduced, a greater proportion of young obtain enough food to grow rapidly. Intraspecific competition among young fishes is considered important in controlling the population of such fish species as sardines and herring; if the number of larvae hatched require more food than is available, many larvae starve and few from that year-class survive.

Intraspecific competition can also occur between age groups of a species when the young and adults use the same resources. This is often prevented by migration patterns in which the adults migrate to an area to deposit eggs, and subsequently migrate elsewhere to feed. The eggs often drift to a nursery area where suitable food is likely to be present at the appropriate season. When the fish mature, they return to the spawning area to release the eggs and sperm that initiate another generation. We have already learned about **anadromous** and **catadromous** fishes. Many marine fishes, such as tuna, herring, cod and plaice (a European flatfish) have extensive migrations within the ocean and are referred to as **oceanadromous**.

Interspecific competition occurs when one species interferes with anothers' use of a resource. Simple population models suggest that the better competitor should always eliminate the weaker competitor (**competitive exclusion**). This has been demonstrated by introducing exotic fish species, which displace native species. Often, however, two or more species can coexist if predation or other forces prevent either from becoming extremely abundant. Evidence suggests that fishes change their feeding areas or types of food when competing with or being preyed upon by other fishes. Thus, various mechanisms reduce direct interspecies competition and allow apparent competitors to coexist.

Symbiosis

Symbiosis, or living together, could include any interspecies relationships, but usually includes only (1) **mutualism** in which both species benefit, (2) **commensalism**, in which one species benefits without harming the other, and (3) **parasitism**, in which one species benefits at the expense of the other. Mutualistic behavior is shown by one species cleaning the external parasites and diseased tissues from the bodies of another species. The fish being cleaned gains by the removal of parasites, and the ''cleaner'' fish or shrimp gains, as food, the parasites, as well as bits of mucus and fin. Other mutualistic relations are shown between small gobies and burrowing shrimp; the fish benefits from having the burrow as a refuge, and the shrimp benefits from the predator warning behavior of the fish. Commensalism can be found in a species such as the remora that uses its sucker to attach to large sharks and hitches a ride to new feeding grounds.

Fishes can also be parasites. Lampreys, which attach temporarily to other species of fish and suck blood, are the best known. Pearlfish live in association with sea cucumbers and enter the gut through the anus; they often eat part of the gonads but do not kill the sea cucumber. Some small South American catfishes attach themselves to gills and feed on gill filaments and blood. Other fishes feed on scales of fish and may, while looking and behaving like cleaning fishes described above, take painful bites out of fins.

Life History Strategies

Aquatic organisms have a variety of life styles that aid in ensuring species survival in different kinds of environments. Some, like Pacific salmon (*Oncorhynchus* spp.), reproduce only at the end of their life span. This pattern, termed **semelparity**, may be most successful in environments that favor the survival of at least some offspring each year. Some organisms reproduce many times after they have reached sexual maturity—this pattern is termed **iteroparity**. Multiple reproductive events are better suited to highly variable environments, where reproduction may be successful only some years and not others. If an organism reproduces only at the end of its lifespan, it can expend all of its energy on reproduction, whereas if it survives to breed again, it must use some of its energy for body maintenance.

Another variable life history strategy is reproductive rate. Many aquatic organisms become sexually mature at a young age, produce numerous eggs and provide no parental care; the proportion of eggs that survive is usually low. However, if environmental conditions are favorable, the population can increase rapidly. These organisms are referred to as ''r'' selected. Other organisms become sexually mature at an older age, reproduce fewer eggs, may have extensive parental care (such as mouthbreeding fishes, which incubate and protect the eggs and young) and have higher proportions of their few eggs survive, even if the environment is not at its most favorable. Those species that have adaptations that allow them to maintain their populations during difficult times are said to be "K" selected. The terms come from growth rate equations that define "r" as the intrinsic rate of increase and "K" as the carrying capacity.

Another variable is related to the manner in which offspring are distributed through the environment. Many aquatic organisms, especially **sessile** ones (those that are attached to a substrate as adults, such as oysters and mussels) have planktonic larval stages. The eggs and larvae drift over a wide area, where they may find suitable surfaces on which to settle, far away from the parents and siblings, and thus not compete. This pattern of broadly distributing the young also aids in repopulating areas that were denuded by storms or predators.

Food and Feeding Habits of Fishes

Fishes occupy all trophic levels except the primary producer level. Various fishes eat plants (herbivores or **grazers**), animals (first order carnivores or predators), other carnivores (top carnivores or top predators), detritus (detritivores) and all of the above (omnivores). Most species of fishes are at least moderately specialized to feed on a certain type of food, and many are highly selective of the prey they will eat at specific times and locations. Some fishes feed on very specialized resources, such as fish scales or clam siphons. **Food habits** refer to the types of food consumed by a fish, while **feeding habits** refers to the behavior a fish uses in obtaining its food.

The type of food a fish eats is often determined by its body form, mouth and digestive system. As the morphology may change during the life of an animal, its food habits may also change. This is most dramatic in flatfishes, which begin life as normal-shaped fishes that feed on small zooplankton. As they grow, they metamorphose: one eye migrates toward the other, so both are on the same side of the head, and the eyeless "blind" side develops smaller scales and lacks color. These fishes assume their adult shape as "flatfish" and lie on or under the sediment, where they can catch their food while being protected from predation by their camouflage. Even fishes that do not undergo such dramatic changes in morphology often alter their food habits as they go through their life stages. Carnivores often have bodies that allow fast chases, teeth that grasp prey and digestive tracts that are short because animal food digests easily. Fishes that feed on plants and detritus often have longer, more complex guts that provide more surface area for processing foods that are less easily digested.

Herbivorous fishes may feed on phytoplankton (filtering it from the water), macroalgae, rooted vascular vegetation (in freshwater), or attached algae (by scraping or rasping coral or rocks and sucking the loosened materials into their mouths). Most herbivores must spend the major part of the day feeding and must handle considerable quantities of coral, rock or sand along with the plant material. Plant material is moderately high in energy and protein, but may require a long time to digest. Along with the plant material, most herbivorous fishes also consume small animals that supplement their diet.

Some very abundant fishes, such as mullet (*Mugil cephalus*), will eat live plant material if available, or detritus if live algae is not abundant. Algae supplies more energy and protein per mouthful than detritus, but is of limited supply. Detritus is an abundant resource, but of lower energy and protein quality. Shallow water environments such as ponds and estuaries often have abundant supplies of detritus, and detritus directly or indirectly (through invertebrate production) may support much of the fish production. The growth of bacteria, fungi and protozoa on detritus may be responsible for significant amounts of the digestible energy.

Predators (carnivores) feed on other animals and are generally able to satisfy their feeding requirements in less time than either herbivores or detritivores. Animal food is higher in calories and protein than plants or detritus. Many fishes feed on small animals, mostly invertebrates, when they are small and switch to feeding on fish when they are larger. Because energy has been lost as plant tissues are metabolized and incorporated into the tissues of animals, less energy is available to carnivores than to herbivores. Thus, carnivores produce less than herbivores, and top carnivores produce the least total biomass.

Omnivores, by definition, feed on a wide variety of foods, but at a given time and place, they may be highly selective in that their choice of prey may vary depending on what is available in different habitats. To some extent, most fishes are omnivores in that plants, animals and detritus are often mixed, especially on the sediment. Evidence indicates that herbivores and detritivores depend on occasional live or dead animals to meet their protein requirements.

Food Selectivity, the Niche and Optimal Foraging

Within a specific environment, many species of fish may live together. Their food habits often have relatively minor overlap. This partitioning of food occurs by two methods: (1) competition between species that use the same resources may eliminate the less successful of the competing species, while allowing non-competing fishes to co-exist, and (2) some fishes change their feeding behavior when other fishes are present. The result of food partitioning is to reduce competition. The term **niche** describes the ecological role of an organism in its community; it includes its requirements for feeding, temperature, etc. Some fishes are very specialized (have a narrow niche breadth), and this may allow ''species packing'' so that a complex environment may support many species of animals with reduced interspecific competition. Other species may be more generalized in their ability to use a variety of prey, but in a given habitat may feed selectively on prey not being used by a more dominant species (described in greater detail below).

Because fishes have the ability to shift their food habits, **stomach content analyses** are needed to determine what a particular population is consuming in the environment being studied. This is often done to define ecosystem food webs, to determine if competition for food is occurring and to establish if the food supply is adequate. When analyzing stomach contents, the researcher notes the number of empty stomachs, the degree of fullness in those with food and the types of food items present. By comparing the relative abundance of prey items in the stomach with the relative abundance of those items in the environment, the **selectivity** of the fish for that item can be determined. If the relative abundance in the stomach is the same as in the environment, the fish is feeding non-selectively. If the relative abundance is greater in the stomach than in the environment, the fish is selecting those prey items. By selecting different prey or different sizes of prey, fishes may reduce competition for a common resource.

Fishes often migrate to a dense supply of food. For example, during the daily cycle, many fishes migrate toward the surface to obtain an ample supply of the zooplankton that are feeding on phytoplankton. Annually, some fishes make extensive migrations to areas that have ample food supplies; when they deplete one area, they migrate to another. Many fishes have regular feeding, breeding and migration routes. As mentioned earlier, by leaving their eggs in one area and migrating to another for feeding, they minimize competion with, and cannibalization of, their young.

Feeding selectivity can also serve to optimize a diet so that a fish can get the greatest amount of food energy in return for the time and energy it spends in feeding activity. To feed, an organism must expend time and energy in searching, attacking and handling prey. If the prey has effective escape mechanisms, the effort invested in search and attack may be lost. Some prey items have higher food value than others. Large items and those rich in fats have higher energy value because the energy content is proportional to the prey's biomass and chemical composition. Fats provide more energy per-unit-weight than carbohydrates or protein.

There is a trade-off between the value of a food item and the time or energy to capture and handle it. A prey item may be preferred if it is easy to see and catch and can be handled quickly, resulting in the energy gain being greater than for another prey item, which may contain more energy, but is so hard to catch or takes so much time to process that the net yield is poorer. The **Optimal Foraging Theory** provides a mathematical method to describe the ''profitability'' of various prey items. According to the theory, a predator should ignore prey that are less profitable and attack the more profitable prey items if they are available. If a desirable prey item is no longer available, the fish may switch to another prey item that it previously ignored. The theory predicts that large prey will be preferred over smaller prey if the ease of capture and other factors are similar. Ample evidence from laboratory studies and from some field studies shows that actual feeding is closer to optimal foraging than to random feeding. This theory has helped explain why a potential prey item may be ignored during some seasons (when better prey is available), but will be actively eaten during another season (when it is the best prey available).

Predator-Prey Interactions

Feeding behavior is often modified to reduce vulnerability to other predators. The same traits that make a prey item desirable (easy to see, catch and handle, and high in energy) make it vulnerable to other predators. Many animals migrate to lighted surface waters to feed, where they are particularly vulnerable to predators. By limiting the time they spend in lighted waters, they limit both their food and their vulnerability. Predator avoidance strategies include being hard to see and catch, requiring high handling time, having bad taste or poison and obvious warning coloration (**aposematism**), or looking like such organisms (**mimicry**). Schooling behavior may also minimize encounters with or confuse a predator.

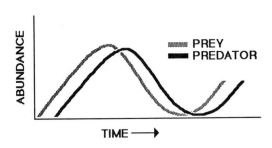

Figure 12.3. As prey populations increase, an increase in the abundance of predators follows. Eventually, the predators will reduce the prey population below that needed to sustain the predator, and both species decline and initiate another cycle.

As predators, fishes often control their prey abundance. Predator-prey cycles in aquatic environments are common. Such a cycle is usually described (Figure 12.3) as an increasing population of prey that is subsequently reduced as predators increase in abundance and overeat their food supply. If only one type of prey is available, the predators will starve and allow the prey to increase. Thus, a sequence of increasing and decreasing predator-prey cycles will occur. While single predator-prey models are easy to simulate mathematically, most real cases are more complex because in most environments the predator will switch to another type of prey or migrate to another area. Either of these strategies will prevent local extinction and allow the prey population to increase.

Fishes have numerous responses to food availability. If food is abundant, fishes will grow faster, feed more effectively, mature and reproduce at an earlier age, produce more eggs per mature female and thus have a faster growing population. Also, fishes tend to migrate to areas with an abundant food supply. If food is not abundant, their growth rate may slow, maturation can be delayed and the number of eggs per mature female may be reduced. They also tend to migrate away from areas in which they have depleted the available food. If they cannot migrate, they may switch to other types of prey.

Predators may process a relatively small part of trophic energy. Recalling that primary production must first produce the organic energy available for grazers and detritivores, it appears that predators get only a small proportion of the energy fixed by primary production. Actually, predators may get more energy than appears to be available because a portion of the energy that has passed through the lower trophic levels to become detritus is incorporated into animals such as worms, which are eaten by predatory fishes.

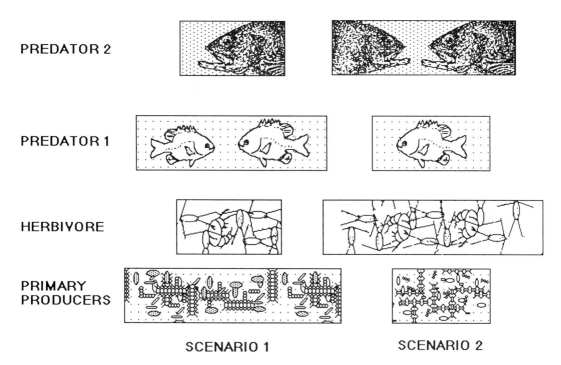

Figure 12.4. The abundance of top carnivores can have an influence on the standing crops of organisms in the lower trophic levels. As shown in this illustration, high bass density results in low prey (bluegill) density, with an increase in zooplankton which graze heavily on phytoplankton (right side of the diagram). When predator density is low, the impacts on the lower trophic levels are reversed (left side of diagram).

Top carnivores may be more important in structuring aquatic communities than is suggested by the amount of energy they process. In freshwater communities, top carnivores may control the abundance and species dominance of the lower trophic levels (Figure 12.4). Top carnivores are often piscivores, i.e., they eat smaller fishes, which themselves eat zooplankton. If the top carnivores are abundant, they reduce the abundance of the smaller fishes; thus, there are fewer fishes eating zooplankton, resulting in higher numbers of zooplankton, which in turn, eat most of the algae. Conversely, if the top carnivores are removed, for example, by overfishing or by a winter fish kill, more small fishes, fewer zooplankton and an algal bloom will result. Research has suggested that the abundance of algae, sometimes a problem in lakes that receive nutrient inputs from sewage, could be managed by controlling the fish populations.

Fish Production

In ponds and small lakes, the abundance of food of suitable size and type is useful in predicting fish production. For some lakes and reservoirs, the production and standing crop of fishes can be estimated by a few factors that relate to the total primary production. The original **Morphoedaphic Index** used only the total dissolved solids divided by the mean depth to predict fish production. Improved estimates have been made by including temperature, phosphate inputs and area-depth ratios.

In large water bodies, including the oceans, predicting fish production, especially for many economically important fisheries, cannot be done by knowing primary production alone. The food webs may be too complex, too many other species may be using the same foods, or too many other factors may control fish populations. Many adult populations of marine fishes do not appear to be limited by food supply. Fish growth is often near maximal, as judged by fish scales or otoliths, for the water temperature in which they live. In spite of the apparent abundance of food for adult fishes, the size of a year-class is highly variable. Some years, eggs may be plentiful, but few or no off-spring survive to reach maturity. Other years, survival of a year-class may be very good, and that year-class may sustain a commercial fishery for many years. Fishery managers would like to know the factors that control the year-class success of economically important fisheries.

Much research has been aimed at understanding the factors that determine whether eggs and larvae will survive to form a strong year-class or will die. Some of the evidence suggests that young fishes undergo a "critical stage" at the time they absorb the yolk sac and develop a mouth and gut. If food of the right size and type (small zooplankton and large algae) is available during this critical stage, the larvae will survive and grow enough to avoid predators. If food is not available during the critical stage, then most of the larvae will die, and that reproductive effort will be wasted. Small fishes are also vulnerable to predation—they represent a high energy source that is easy to see, catch and eat. The smaller the fish, the more predators exist that can eat them, such as large copepods and fishes a little larger than the fish in question. Thus, the availability of food and the number of predators may both determine year-class success.

Climate also influences food availability and predators. Calm weather promotes stratification of the water and allows algae and zooplankton to be concentrated in the surface waters, where their densities are high enough to feed larval fishes. Conversely, storms will mix these algae and zooplankton deep into the water column, where they may be too scattered to provide a dense enough food supply to larval fishes. Also, strong winds may blow the fishes away from coastal up-welling areas into areas where food is scarce. Studies suggest that the numbers of larval and young fry stages are so reduced that the older fish have a more ample food supply. Much more detail on this important life stage has been covered in the chapter on early life histories.

Disease and Parasitism

Diseases of fishes and other aquatic organisms can be caused by a variety of worms, protozoa, bacteria, fungi and viruses. The role of pathogens (disease-causing organisms) and parasites in natural populations is rarely documented unless a disease has been catastrophic. Given the importance of disease and parasites in fish culture, these interactions are likely important in controlling natural fish populations, at least at certain times and in some places. For example, between 1976 and 1977, perch (*Perca fluviatilis*) populations were markedly reduced (from 6,000 to 126 being caught in traps) by a fungal infection in Lake Windermere, Britain. As much as 34% of the perch in Lago Maggiore, Italy, were found to be infected with the worm parasite, *Diphyllobothrium latum*, although this fish/human/dog parasite was rare among humans living in the region (humans can contract the so-called dog tapeworm by eating raw fish). In this same lake, mass mortalities of bleak (*Alburnus alburnus alborella*) have been reported from fungal infections of the gills. Crayfish were common in Lake Hajalmaren, Sweden, until a plague in 1908. Diseases in hatcheries are most common during periods when fishes are stressed by handling, low oxygen or high ammonia, and it is possible that natural stresses, such as overcrowding and starvation, may also increase disease incidence in natural populations.

Habitats

Aquatic habitats share the same basic ecological processes, but the net result of those processes varies considerably from one habitat to another. Thus, some habitats such as small streams, which receive very little light energy, depend on organic inputs from outside, including leaves and

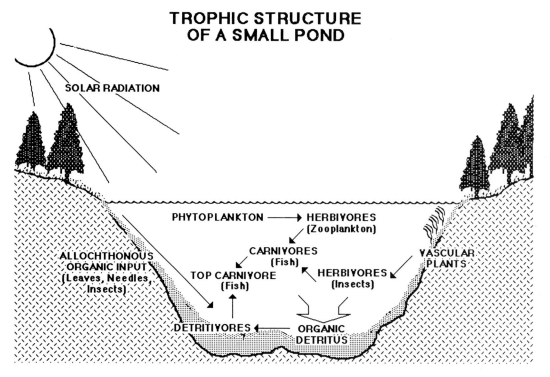

Figure 12.5. Small ponds may have significant levels of both autochthonous and allochthonous inputs.

wood from the terrestrial community (**allochthonous** inputs), while open ocean habitats must produce all of the organic materials from within by using sunlight and recycled algal nutrients (**autochthonous** inputs).

Ponds to Lakes

Small ponds are usually shallow and have high levels of primary production from plankton, rooted vegetation and, sometimes, benthic, attached algae if light penetrates to the bottom (i.e., high autochthonous inputs) as shown in Figure 12.5. Small ponds may also receive significant allochthonous inputs from the surrounding land, especially if the slope is steep, the drainage basin large and the rainfall intense. Shallow ponds subjected to strong winds often do not stratify, and the bottom sediments may be stirred up so that nutrients from the bottom waters and sediments continually resupply the lighted surface water. For this reason, ponds are often highly productive. Ponds usually also have rich detrital communities from the material that settles out of the water column and from the material that washes in from the terrestrial community.

Fish populations in ponds can be very dense. If no predators, such as bass, are present, sunfish and shiner densities can become so high that the fishes become stunted from lack of food. Predators, especially juvenile stages, require food smaller than their mouth size and may starve if only larger fishes are present. Thus, managing ponds requires coordination both between the hatching times and thus the sizes of predators and prey species.

Many studies on fish ecology have been pursued in ponds because they are amenable to manipulating nutrients and fish abundances to test hypotheses on ecological theories. For example, optimal foraging theory predicted the choice of foods and uses of habitats for bluegill (*Lepomis macrochirus*) foraging in open water, sediments and vegetation. Sunfishes will change their feeding location based on the "profitability" of the prey and the presence of closely related competitors. Pond studies also showed that young bluegill change from feeding in open water where food is more abundant, to feeding in vegetation if largemouth bass (*Micropterus salmoides*) are present; medium-sized and large bluegill, which are too large for bass to eat, do not change their feeding location.

Fish ponds are often used for aquaculture. Left to the natural productivity, fishes can be grown at little cost. Primary production can be increased by adding fertilizers or animal wastes, which simultaneously increases the detrital trophic level and algal nutrients. Truly intensive aquaculture can be accomplished by adding prepared food, which is consumed directly.

Lakes usually derive most of their energy from autochthonous primary production and less from the terrestrial community. Temperate lakes generally undergo the stratification described in a previous chapter. The fish communities of lakes are significantly affected by temperature and thermal stratification patterns. Cool lakes are likely to be dominated by salmon and trout, and warm lakes by bass, pike and walleye as top predators, and sunfishes and perch as carnivores on zooplankton as well as prey of the top predators. Stratified lakes tend to have warmwater fishes in the surface waters and coldwater fishes in the hypolimnion. If the water is turbid, catfish, carp and suckers will likely be common. Many species are often present if diverse habitats are available, e.g., many bays, some shallow areas with extensive rooted vegetation, some deep hollows separated by shallow sills. Such environments have complex competitive and predator-prey interactions. Much of the evidence for predator control of community structure has been derived from lakes, in which the abundance of top predators controls the standing crop and species dominance of the lower trophic levels (Figure 12.4).

Few temperate lakes still contain the fish species that were present before agricultural and industrial development. Habitat destruction, especially of wetlands and marshes, pesticides, introductions of exotic species and overfishing have all contributed to the shifts in species dominance. These changes have been well documented in the Great Lakes.

Streams to the Sea—The River Continuum

Small streams (Figure 12.6) often receive much of their organic energy input from the terrestrial community in the form of insects, leaves or pine needles (allochthonous input). Aquatic primary production (autochthonous input) is usually low because small streams tend to be shaded. The leaves and pine needles are processed by aquatic insects, which in turn are eaten by fishes and other predators.

As streams become larger and broader, more sunlight and plant nutrients reach the water, and the balance shifts to greater autochthonous production. Diatoms and other algae attached to rock surfaces are the major source of food for insect species that scrape such surfaces and are eaten by fishes. As streams become still larger, turbidity may reduce the light available for primary production, and the river may again depend largely on allochthonous inputs to support zooplankton.

Fish communities change from the small streams, where most fishes are feeding on terrestrial insects and invertebrate leaf processors, to more complex communities that include herbivores, predators, top predators and detritivores. Temperature is important, and cool streams are often dominated by trout and sculpins (family Cottidae) while warm streams often feature smallmouth bass (*Micropterus dolomieui*), green sunfish (*Lepomis cyanellus*), minnows (family Cyprinidae), darters (family Percidae) and catfish (family Ictaluridae). The number and species present will be determined by the location, flooding and drought frequency, and the introduction of exotic fishes that may replace local fishes. Predation and competition are important in controlling densities and species dominance. Resource partitioning is demonstrated by co-occurring fishes selecting a

TROPHIC STRUCTURE
OF A STREAM - RIVER

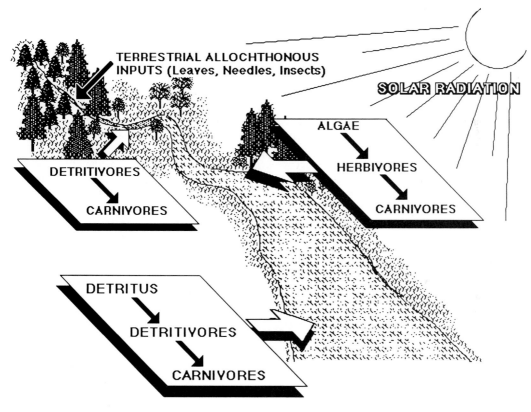

Figure 12.6. In small shaded streams, much of the input is allochthonous. In larger open streams, the autochthonous input by algae may be dominant. As streams become rivers, the allochthonous energy inputs may again dominate.

mutually exclusive range of prey. This presumably reduces competition and allows more species to coexist than if each species was a generalist.

Estuaries

Estuaries are very rich communities because they receive the allochthonous inputs from rivers and surrounding terrestrial environments and have extensive autochthonous production from rooted plants, attached macroalgae and the epiphytes that grow on these larger plants. Much of the base of estuarine systems is detritus because many of the rooted plants and macroalgae are not consumed directly, but die and decay before they are eaten. Light is more likely to limit primary production than nutrients, because of sediments being mixed by wind, tide and wave action, all of which tend to mobilize nutrients from the sediments into the water column.

The fish communities contain many species that represent freshwater, migratory, true estuarine and marine fishes. Some freshwater fishes that live in the upper reaches of estuaries are white catfish (*Ictalurus catus*), mosquitofish (*Gambusia affinis*) and common carp (*Cyprinus carpio*). Migratory fishes include salmonids (*Oncorhynchus* spp., *Salmo salar*), shad (*Alosa* spp.), striped bass (*Morone saxatilis*) and eels (*Anguilla* spp.). Estuarine fishes include smelt (family Osmeridae), seaperch (family Embiotocidae) and spotted seatrout (*Cynoscion nebulosus*). Some marine species use estuaries for breeding, but are otherwise found in open water. Included are herring (*Alosa* and *Clupea* spp.), croaker (*Micropogon undulatus*) and menhaden (*Brevoortia* spp.). Predation and competition may be less important than temperature and salinity in structuring estuarine communities because of the greater relative overlap of food habits of the species present as compared to other habitats.

Coastal Habitats

Most marine fishes are found near the continents in the widely diverse habitats that range from the intertidal zone to the edge of the continental shelf. Close to land, in shallow water, the food chains and fish communities differ depending on the substrate. That is, rocky intertidal, exposed sandy beaches, mudflats, salt marshes, mangrove swamps, seagrass beds, kelp beds, etc., all have different balances of food inputs and different communities of fishes. Offshore areas subject to upwelling of nutrient-rich waters into the photic zone are among the most productive ecosystems. Food chains tend to be short in upwelling areas, and much fish biomass can be harvested. This diversity of environments creates much of the fascination with studying nearshore marine communities.

Epipelagic Zone

The open ocean has an enormous surface area, but its great depth widely separates the surface from the sediments. Virtually all productivity must come from autochthonous production (Figure 12.7). The open ocean has fewer species of fishes as compared to the other more complex habitats. However, these few open-ocean fishes are extremely important in human commerce and include tunas, salmons, sardines, herring and anchovies. Bottomfishes are now being more heavily exploited because most of the more traditionally fished species are becoming overexploited. Many of the fishes that live in the ocean for most or all of their lives make enormous migrations between areas that have ample food supplies at different seasons of the year.

Coral Reefs

Although the area of the world occupied by coral reefs is small, they are particularly intriguing. The waters where coral reefs occur are usually very low in the nutrients required for algal growth, and plants have adapted to absorb nutrients as the water flows past. The ecology of reefs ensures that the nutrients available are recycled within the reef community. Many symbiotic relationships have evolved, such as the algae that grow within giant clams and corals. Those algae avoid the problem of low levels of nutrients in the water column by living within nutrient-rich animals.

The fishes of coral reefs are extremely diverse, and many are brightly colored and very beautiful. Many of the fishes are predators, and most have well-developed defenses against predators. Few fish species are herbivores, but the abundance and biomass of herbivores is impressive, and their activities change the surface of the reefs by preventing the filamentous and leafy algae from overgrowing and killing the coral.

Many other communities, such as the deep ocean, polar seas and hypersaline communities use the aforementioned processes, but to different degrees. Also, the differences in the species present make ecology fascinating. Ecology is important in managing fisheries and protecting fisheries from pollution and other adverse effects. These topics are considered in subsequent chapters.

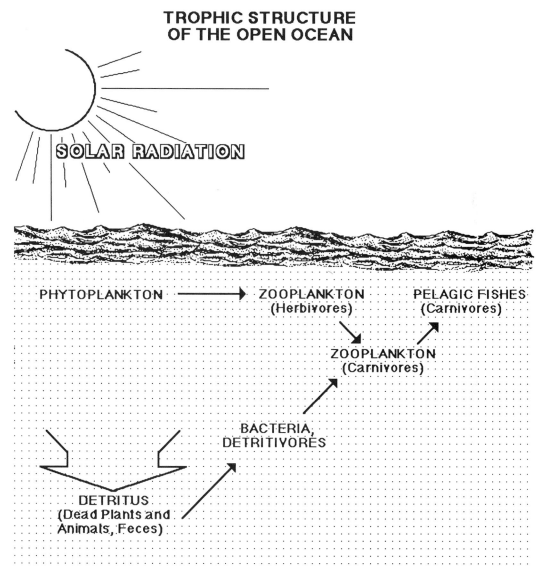

Figure 12.7. Open ocean productivity relies almost exclusively on autochthonous inputs since little or no nutrient runoff from land is available to the system.

13

Analysis of Exploited Populations

Frederick G. Johnson

Introduction

As we mentioned in the first chapter, a fishery involves both a harvested population (the fish stock) and a population of harvesters (the fishermen). In this chapter, we take a look at gathering information on how these two populations interact and how this information is used to manage fisheries. In practice, this information is analyzed quantitatively with mathematical and statistical techniques, and sophisticated computer technology. However, we will take a more qualitative look at what tends to be a very quantitative topic. In general, fishery scientists need to know how many fish or shellfish are available (stock size) and how fishing affects the stocks (population dynamics).

For many years, the ulterior motive behind these studies was to provide an estimate of the **maximum sustainable yield (MSY)** for each fishery. The MSY is an estimate of the greatest average harvest that can be removed from a fish stock each year and still allow the stock to rebound and produce maximum harvests in subsequent years. In this approach, it is assumed that fishing intensity and the fishery habitat remain stable. If more than the MSY is harvested in a given year, the remaining stock will be too small to produce a maximum yield the following year. Ironically, if less than the MSY was harvested in the first year, the remaining stock would still not produce as much as it could have for harvest the following year. In general, natural populations have the ability to increase their rates of production (generation of new biomass through birth and growth of individuals) when their populations suffer significant losses, and this ability is the foundation of the MSY concept. Until the 1970s, the purpose of most fishery management was to determine what population size gave the greatest average production for each fishery stock in question.

Since the 1970s, a new index has replaced MSY as the quantitative guideline for determining yearly harvests. The new index, **optimum yield (OY)**, takes into account other factors besides maximizing the catch.

The additional factors considered in OY include issues pertinent to the human side of fisheries, such as economics, politics and sociology. The OY estimate is used to establish the **total allowable catch (TAC)** for a given fishery in a given year. OY estimates are generally less than MSY estimates, although in unique circumstances OY could be set higher than MSY for a short time, perhaps due to economic considerations. Optimum yield replaced MSY as the official management policy for U.S. fisheries after the Fishery Conservation and Management Act of 1976.

Population Dynamics

Even if the size of a population is stable over time (meaning that the number of individuals in the population remains constant), the composition of the population nevertheless changes as some individuals enter the population and others leave. Categories of population gains and losses are indicated in Figure 13.1. The most important contribution to a fishery stock is the **recruitment**, which usually refers to the number of individuals that survive and grow to a size large enough to be harvested during the season. Many more individuals will enter a population by birth than will be

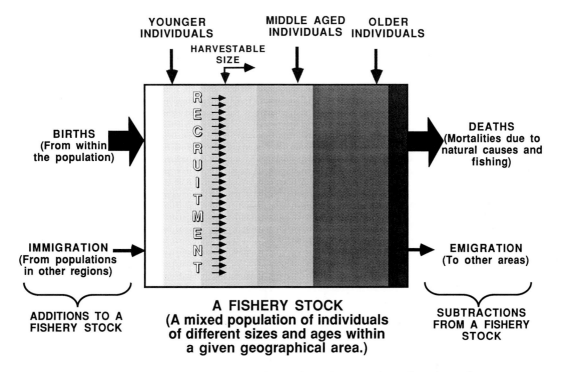

Figure 13.1. A hypothetical fishery stock, showing factors that add to and subtract from the stock.

recruited into a fishery, since most will die before they reach harvestable size. This is true for all stocks except those with extremely low fecundity, such as whales and live-bearing sharks.

The relative impacts of immigration and emigration on stock sizes are often minor, except perhaps for animals that spend a long time as planktonic larvae. These larvae (primarily invertebrates) may be transported long distances by ocean currents before they are ready to settle on the bottom and, if the prevailing currents carry them away from the coastline, they may even postpone their metamorphosis and remain planktonic until they reach a suitable habitat.

The implications of immigration, emigration and particularly migration may be of major importance to the management of fish stocks. For fishery managers to work effectively, they must know whether the stock in question is a **unit stock** or a **mixed stock**. Stock unity is maintained by reproductive isolation. Fishery scientists study many aspects of a fish stock to determine whether individuals in it belong to a single population or not: **morphological characteristics** (subtle differences in body form), **meristic characteristics** (differences in the number of countable structures such as scales, spines, gill rakers and vertebrae), differential **growth rates, migratory movements** established by tagging studies and biochemical characteristics established by **electrophoresis** .

For most fishery stocks, recruitment is mainly determined by birth rate (also called **natality**) and pre-recruitment mortality, rather than immigration or emigration. Among both fish and shellfish stocks, the older females tend to have a much higher fecundity than younger, recently matured females and, therefore, the older females in a population have a proportionately greater impact on recruitment than younger ones do. For this reason, fishery regulations for wild stocks that are depleted, such as sturgeon, may stipulate that the largest individuals have to be released following capture.

The major factor that reduces the abundance within fishery stocks is the death of individuals, or the **mortality** rate. We can divide the causes of mortality into natural mortality and fishing mortality. **Natural mortality** includes death due to so-called natural causes, such as disease, old age, parasitism, environmental extremes like freezing or catastrophic storms, starvation and predation by animals other than humans. **Fishing mortality** is associated with fishery harvests. Please keep in mind, however, that the predatory actions of humans are by no means unnatural. The reason we separate the sources of mortality to fishery stocks is to manage them effectively for human use. The basic difference between natural and fishing mortalities in this context is that fishing mortality can be controlled by adjusting **fishing effort**, while natural mortality is not so readily subject to control by fishery managers. However, in some instances management of a stock may include steps to reduce natural mortality, such as removal of a predator. Although mortality is usually partitioned into natural and fishing mortality, an increasingly important "gray" area of mortality is man-induced mortality, which is not fishing per se. Dams and pollution, for example, are caused by man and increase mortality, but the fish killed are not utilized.

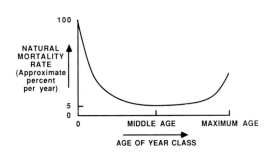

Figure 13.2. A common pattern of natural mortality rates by age.

The rate of natural mortality among individuals in a population is highest when they are very young, declines as the animals become older and then increases again as the animals approach the end of their average life span (Figure 13.2). Species of salmon represent a special case, in that they all die after they spawn. This situation is called **semelparity** and would be reflected by a more abrupt increase in mortality of older fishes than is shown in the figure. Even among semelparous fishes such as the salmons, though, far more individuals die at a young age.

The rate of natural mortality for an age group of fishes or shellfishes generally reflects the rate of predation upon that age group, because predation is usually the major cause of natural mortality for animals in nature (in aquaculture, however, we may be able to control losses due to predation). The very high rate of mortality of young, larval or juvenile aquatic animals therefore reflects the fact that predation upon these life stages is very high. As fishes grow older and larger, their chances of survival improve, primarily because they are no longer such easy prey. To give you an idea of the actual magnitude of these mortality rates and the degree to which they vary with age, consider the following: young fishes in the larval drift (zooplanktonic) state face mortality rates of 5–10% of the age group *per day*, whereas 5–20% of those fishes that survive to early or middle maturity die *per year*. The changes in mortality with age describe the year-to-year survival of a year-class, or cohort, as it ages. (Since an animal must either die or survive, the mortality rate and survival rate must add up to 1 at each age.)

The number of fish in each age group within the combined stock is usually quite variable from year to year, because the survival to recruitment age is better, by chance, for certain cohorts. These differences may be revealed by length-frequency analyses (discussed in Chapter 8). Length-frequency analyses of samples of a fish stock at successive points in time can tell us both how well each year-class grew (in length), and how well its members survived (Figure 13.3). This leads us to consider the major factors that affect recruitment.

Reproduction starts with spawning, so the number of mature individuals that spawn successfully is certainly of major importance to the abundance of the resulting year-class of juveniles. Since fisheries tend to concentrate on larger individuals, the number of spawners is usually more affected by fishing effort than the number of immature fish. If so many potential spawners are taken

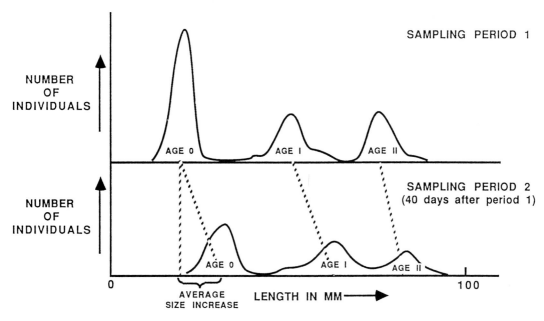

Figure 13.3. Lenth-frequency composition of a pond smelt population at two successive sampling periods, showing growth and survival of each year-class.

by a fishery that too few are left to foster ample offspring, it leads to **recruitment overfishing**. Another category of overfishing is called **growth overfishing**, which occurs when many fish are harvested too young and too small, before they enter their phase of rapid growth; this results in a very low yield of biomass for the number of fish taken.

Besides the number of offspring produced by a stock in its spawning season, the recruitment from a year-class may also be a function of population density (the number of individuals per unit area) and environmental conditions. Population density becomes a factor if there is competition among individuals for limited resources, such as space or food. If there is competition for food, fish often become **stunted** (for example, the maximum size of brook trout decreases when they are over-crowded), although deaths due to starvation are uncommon in large aquatic ecosystems. Competition for space may be keen on spawning grounds or on benthic substrates where sessile invertebrates find the habitat suitable for settlement. Higher population densities also make it easier for diseases and parasites to spread. When stock sizes are very large, cannibalism of younger individuals by older individuals may also limit recruitment. The impact of predation by other species tends to be a complicated, density-dependent and very important factor. If predatory animals maintain a constant number, they can be expected to take a certain number of individuals out of an age group. If a given year-class is small, the predators may take most of it, leaving little recruitment. If the new year-class is large because of highly successful spawning, the predators may become saturated with prey and leave many potential recruits. If several successive spawnings are large, the predators may also become more abundant, through either their own reproduction or immigration, and thereby exert a greater predatory influence thereafter. Fishery scientists are only beginning to appreciate the full significance of non-human predation on fishery stocks, but the more they learn, the more significant it seems to be.

A number of environmental factors affect recruitment (some of them are considered in Chapter 12). Among them are temperature, feeding conditions, ocean currents, wind patterns, floods,

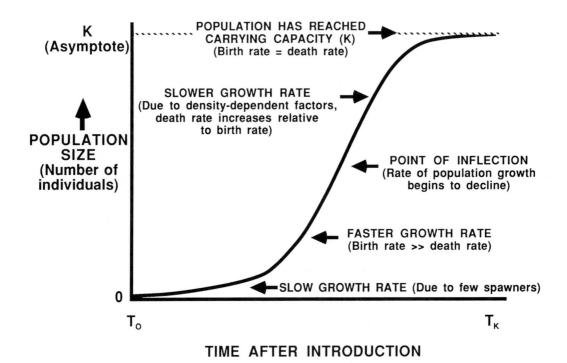

Figure 13.4. Growth of a population after it is introduced into a new and favorable habitat.

droughts and pollution. For us to understand the dynamics of a fish stock, leaving aside the impact of fishing for the moment, it is best to begin by restricting our attention to what happens when a population begins to grow in a new environment.

Let us assume that a lake or pond is **reconditioned** by the application of **rotenone**, a poison that kills all fishes in the lake but leaves the rest of the animals and plants alive (this is a common practice in inland fishery management (Chapter 19). Let us also assume that the lake is planted with a species of fish that can reproduce on its own (brook trout, *Salvelinus fontinalis*, is a good example). We can then follow the population growth with time as shown in Figure 13.4. At first, the growth of the population is slow because there are only a few planted fish that can spawn and bring new offspring into the stock. As the new offspring grow and become sexually mature, they too spawn and the rate of population growth increases accordingly. This continues until the older fish begin to die off (mortality rates increase) and the remaining fish begin to saturate the available resources of the habitat. As the population approaches the **carrying capacity** of the habitat (the asymptote of the graph at time T_k), the growth rate of the population declines as a result of **density-dependence**. At this point, birth rates equal death rates, and the population size is stable over time at the maximum level the habitat can support.

This kind of population growth also occurs after introducing an exotic species to a favorable habitat (exotic species are those that are not native, or indigenous, to a habitat). The introduction of striped bass (*Morone saxatilis*) into San Francisco Bay is a good example. In 1880, small numbers of bass were introduced. By 1900, the population had grown to support harvests of 500 tons per year. The recent introduction of coho salmon (*Oncorhynchus kisutch*) to Lake Michigan was also a spectacular success. (Introducing exotic species requires great care, since the resulting ecological manifestations are often unpredictable and may be harmful.)

During the early phase of population growth (that period which takes place before density-dependent factors reduce growth), fishery scientists can quantify the population's **instantaneous growth rate**. An "instantaneous" rate is one that occurs over small increments of time. Instantaneous rate functions include instantaneous growth (in number of individuals over time), **instantaneous individual growth** (in size or weight), **instantaneous mortality**, **instantaneous recruitment** and **instantaneous yield** (in biomass, which takes both numbers of individuals and growth of individuals into account). These rate functions are used in mathematical models that are designed to mirror changes in natural populations. It is important for these models to reflect actual population changes closely enough to make the impact of fishing on these populations predictable.

A relatively simple model that closely describes the early part of a population growth curve in a new habitat (a previously unoccupied ecological niche) is the **exponential growth** model:

$$N_{t+1} = N_t e^{rt}$$

where N_t = the number of individuals at the beginning of a time interval,
N_{t+1} = the number of individuals at time t+1 (one time interval later),
e = 2.71828,
t = the length of the time interval (e.g., 1 year), and
r = the instantaneous population growth rate (this represents a combination of birth and death rates).

Now let us put our model to work and see how it performs. First of all, assume we have studied the fish in our lake enough to know that their instantaneous population growth rate is 0.5 (we determined this by finding the number of fish at two times and then solving for r). Assume also that we planted 100 fish in our lake at time zero, the instantaneous growth rate will remain constant over the next 14 years, and the environment in the lake will not change. Using the equation described above, the model predicts that there will be the following number of fish at the end of each year:

Year	Number of Individuals	Year	Number of Individuals
0	100	7	3,312
1	165	8	5,460
2	272	9	9,002
3	448	10	14,841
4	740	11	24,469
5	1,218	12	40,343
6	2,009	13	66,514
		14	109,663

These data are plotted in Figure 13.5, and we can see the typical pattern of exponential population growth. Suppose we sample the new population of fish in the lake, and our population estimates are very close to the predictions based on the model for the first few years. What if we find, however, that by the fourteenth year there are only about 25,000 fish in the lake, only one quarter of the population we expected with this fish model. That means we need a new model that can incorporate the fact that the population growth rate actually decreases in a limited environment as the population increases.

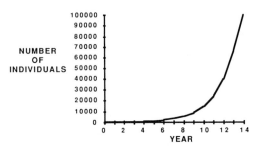

Figure 13.5. Exponential growth of a hypothetical population of brook trout in a lake after the fish are planted.

Our new model is for **density-dependent population growth** and includes a negative feedback factor that increases in importance as the population increases. We can assume that our lake can only support a finite number of fish and, when that number is reached, the lake's resources are used up. We call this number the **carrying capacity** (K) and use $\frac{K-N}{K}$ as our negative feedback factor. We can now hypothesize that the carrying capacity of our lake is 100,000 fish, for example, and then test that hypothesis by sampling the lake over successive years to determine the **goodness of fit** of our sampled data to the predictions from our new model.

Our new model is as follows:

$$N_{t+1} = N_t + rN_t \ \frac{K-N_t}{K}$$

where K = the carrying capacity of the lake, 100,000 (the other symbols were defined in our previous model).

Using the new model (plotted in Figure 13.6), we obtain the following predictions of population size:

Year	Number of Individuals	Year	Number of Individuals
0	100	14	24,304
1	150	15	33,503
2	223	16	44,642
3	337	17	56,998
4	505	18	69,254
5	756	19	79,900
6	1,131	20	87,929
7	1,690	21	93,237
8	2,521	22	96,390
9	3,750	23	98,130
10	5,555	24	99,047
11	8,178	25	99,519
12	11,933	26	99,758
13	17,187	27	99,879
		28	99,939

Since our prediction of 24,304 fish in the 14th year is very close to our population estimate of 25,000 from sampling, we can conclude that there is a good fit between our model and the observed values.

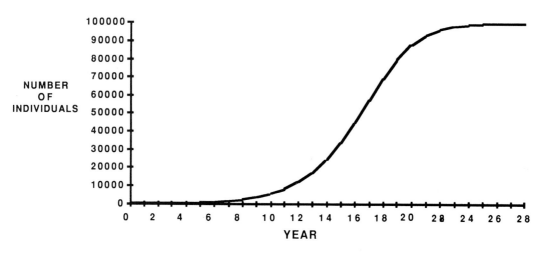

Figure 13.6. Density-dependent growth of a hypothetical population of brook trout in a lake after the fish are planted.

We can see that Figure 13.6 looks very much like Figure 13.4. The point of inflection in Figure 13.6 appears at approximately the 17th year, after which the rate of population growth slows down. That is also the point at which the population has reached half of its maximum size. When the population is small, the term $\frac{K-N_t}{K}$ is approximately equal to one, so population growth is largely unaffected by density-dependence. When the population is near the carrying capacity, $\frac{K-N_t}{K}$ is close to zero, so the population growth is very small. The change in the population is sometimes denoted as ΔN (pronounced "delta" N) and if Δt (the change in time) is 1, then $\frac{\Delta N}{\Delta t}$ is equal to

$$N_{t+1} - N_t$$

or

$$\frac{\Delta N}{\Delta t} = N_{t+1} - N_t = rN_t \frac{K-N_t}{K}$$

This is known as the **logistic growth model.** The solution for this "differential" equation (which is beyond the scope of this book) is called the **logistic equation** and expresses population number as a function of time, as shown by the "sigmoid" ("S" shape) curve in Figure 13.6. The sigmoid curve described by the logistic equation is special in that the point of inflection occurs at a stock size of K/2, or one half of the carying capacity.

Yield Models

It seems reasonable that if we want to have a trout fishery in our lake, we should try to keep the population of fish in the lake at a level where the production is greatest (the particular population size, N_t, where ΔN is largest). To find that point, we simply plot

$$rNt \ \frac{K-N_t}{K} \ \text{vs.} \ N_t$$

which shows production as a function of population size. The resulting graph is shown in Figure 13.7. We find that the maximum production occurs at a stock size of 50,000 fish, or half of the carrying capacity. The maximum yearly increase ($\Delta N = 12,500$ fish) is the maximum production and it equals $\frac{rK}{4}$. The quantity $\frac{rK}{4}$ also defines the maximum sustainable yield according to the **Schaefer surplus production model.**

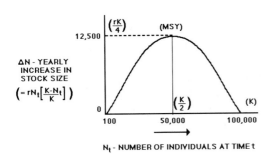

ΔN - YEARLY INCREASE IN STOCK SIZE

$\left(= rN_t \left[\frac{K-N_t}{K} \right] \right)$

N_t - NUMBER OF INDIVIDUALS AT TIME t

Figure 13.7. A surplus production curve, showing population production as a function of population size (based on the logistic growth model).

The Schaefer model brings the concept of harvest into the logistic growth model. If we decide to turn the yearly surplus production into fishery harvest, we need a new term for losses to the population caused by harvest, and we can designate this by C_t (the catch, or fishing mortality in time interval t), where $C_t = F_tN_t$ (F_t, the fishery harvest rate for time interval t, multiplied by the stock size in time interval t). The Schaefer model is usually specified in terms of biomass (weight), but is as follows for numbers of individuals (so we can compare it with our previous model):

$$N_{t+1} - N_t = rNt \ \frac{K-Nt}{K} - C_t$$

Note that when production

$$rNt \ \frac{K-N_t}{K}$$

equals catch, the change in stock size ($N_{t+1} - N_t$) is zero, so the stock is at **equilibrium** when all of the surplus production is going into harvest. We have already learned that maximum production occurs at a stock size of K/2 according to the logistic model, and the Schaefer model assumes that if harvests equal the surplus production each year, the equilibrium that results from this balance can continue indefinitely.

One of the problems with models that predict yields based on an equilibrium state is that a state of equilibrium may never actually exist in an exploited population. Another serious problem relates to fishing effort, because as a fishery develops over time, fishing effort tends to intensify and improve in effectiveness, and both of these factors lead to overly optimistic estimates of sustainable yield. Nevertheless, the Schaefer model was successfully applied to the management of the Pacific halibut fishery in the 1950s and has been refined to provide the basis of management for many other commercial stocks.

Another type of model used to predict fishery yields, particularly for salmonid fisheries, is the spawner/recruit model. This model is based upon the expected number of recruits from a year-class (the number of fish that return to the river of origin) that was spawned by some number of parent fish in those rivers earlier. This model is not complicated by multiple spawnings because salmon are semelparous. Some of the returning fish are caught, and the remaining number of recruits that are allowed upstream to spawn is called the **es-capement**. If one recruit spawns for each parent that spawned earlier, the relationship between the number of spawners and the escapement would be a simple linear one, represented by a 45° line (called the **line of equal replacement**, as shown by the dashed line in Figure 13.8). The line of equal replacement is a hypothetical relationship that is useful for management purposes. The actual form of the spawner/recruit curve varies, but its general shape is shown by the solid line in Figure 13.8.

In actual situations, the number of recruits that return from a small escapement of spawners generally exceeds replacement, but this excess of recruits over spawners declines as the number of spawners increases. This is another example of the density-dependence of population growth, but in this case, the density-dependence specifically applies to the car-

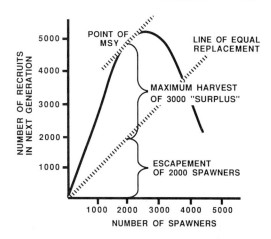

Figure 13.8. The spawner/recruit curve, which predicts yield on the basis of the escapement of the parent generation. This model is only applicable to semelparous species such as salmon.

rying capacity of the spawning habitat in the natal streams and the associated stream or lake rearing area. The reduction in number of returns versus the number of previous spawners at high spawning densities is easier to understand when one realizes that the suitable area available for the **redds** (spawning nests) is limited, as is the food available to the offspring after they hatch. At extremely high escapements, the spawning adults may even destroy existing redds in an attempt to build their own and, in addition, higher incidence of disease or attraction of additional predators may occur.

The spawner/recruit curve is used by fishery managers to estimate the level of escapement at which there is the greatest harvestable surplus of recruits. The point on the spawner/recruit curve at which the harvestable surplus is greatest can be found mathematically, but if you simply extend a line parallel to the line of equal replacement upward until it makes a tangent to the curve (i.e., meets the curve at only one point), the corresponding number of recruits above the line of equal replacement is the **harvestable surplus** (or MSY); at this MSY spawning level, the vertical distance between the spawner-recruit curve and the escapement curve is the greatest.

The spawner/recruit curve is not a general one, and the shape of the spawner/recruit curve for each species of salmon in each river system may vary. Besides that, considerable year-to-year variation in survival from the eggs within a river system further complicates the issue because the models assume a steady state. This results in disagreement among fishery managers about the exact form of the spawner/recruit relationship and the associated amount of harvestable surplus. Moreover, if factors other than fishing cause a reduction in the number of returning recruits, such as environmental degradation due to human actions or unexpectedly harsh weather, that reduction has to be subtracted from the ensuing harvest allotment rather than from the escapement portion if a maximum harvest is to be achieved in the following generation.

A somewhat different class of models includes the so-called **dynamic pool models**. These make use of data on individual growth, and natural and fishing mortality. In the most complex of these models, mortality rates may vary by age. With knowledge of the growth and natural mortality rates, in addition to stock size, the fishing rate and minimum vulnerable size can be adjusted to obtain a maximum yield from the stock. One weakness of classical dynamic pool models is that they assume fishing rates do not affect recruitment (i.e., reproductive success), at least over the range of fishing rates considered as management alternatives. Length-frequency studies are particularly beneficial for providing the data needed to formulate dynamic pool models. These models are particularly useful for oceanic stocks containing multiple age groups of harvestable fish. There are some newer models that attempt to combine spawner-recruit and dynamic pool yield concepts, but applications of such models await better data on the relationships between spawning stock size and recruitment. Such data are difficult and time consuming to accumulate for a fishery on several age groups of a stock.

One of the benefits of commercial fisheries is that they provide an economical source of necessary data for management models. Much of the information needed to develop a dynamic pool model, for example, can be taken from fish that are landed and enter the marketplace. To provide information on growth and age, scale samples and length and weight measurements can be taken at little cost and without harming the marketability of the product. Relative numbers of individuals among the different age groups can be compared over time to yield information on cohort survival and relative abundance. For some of the larger marine fisheries, as much as one third to one half of the stock might show up in the marketplace each year, thereby providing a huge sample from which to obtain data. Good historical data helps fishery scientists to better understand how the stocks respond to changes in the various parameters of concern, including the effects of fishing and environmental changes.

Ways to Estimate Stock Size

The fundamental parameter needed to assess how fishing affects the stock is an accurate estimate of stock size. Three general approaches to estimating the number of individuals present in a fishery stock are direct assessment, mark-and-recapture studies and catch-per-unit-effort studies.

Direct assessment implies that a stock is somehow being counted by direct means or en masse, either over an entire area or over a subsample. Examples of direct methods include fish counts taken at fish ladders, draining of ponds, flyover surveys at spawning grounds and hydroacoustic (sonar) surveys conducted from research vessels. In specific situations, such as when fish are heavily concentrated in schools, spawning in restricted areas or migrating through areas easily accessible to observers, direct assessment can be practical and efficient. In most cases, however, fish populations are too scattered to make direct assessment practical. A variation of direct assessment that is useful for stocks that are spread over larger areas is the **area-density** method (often called area-swept when trawls or seines are used), where numbers of individuals are only enumerated in certain subplots of the total geographic area that is occupied by the stock, and then the entire stock size is estimated based on an extrapolation of the subsample density over the entire range.

Mark-and-recapture studies are useful to enumerate small, relatively well-mixed stocks. One method, often called the **Peterson method**, involves capturing a number of individuals from the population, marking them with tags and then returning them in good health to mix with the rest of the stock. The assumptions that underlie this strategy are that the marked fish will mix randomly through the existing stock, no tags will fall off, tagged fish will survive and behave the same as untagged fish, and tagged fish are neither more nor less likely to be caught. If these conditions hold true, then the expected proportion of marked fish in a second sample of fish caught will be equal to the proportion of marked fish released originally into the entire stock. This relationship is expressed as follows:

$$\frac{R}{C} = \frac{M}{N}$$

where M = the number of fish marked,
  N = the number of individuals in the stock,
  C = the catch when the stock is fished the second time, and
  R = the number of marked fish present in the catch, C.

Multiplied out, this relationship yields:

$$\hat{N} = \frac{MC}{R}$$

where $\hat{N}$ = our estimate of N.

As an example, suppose we capture 160 fish in a pond by seining, tag all of them and then release them into the pond in good condition. After waiting a day or two for the fish to distribute themselves and recover from any effects of handling, we seine the pond again and catch 172 fish, 20 of which have tags. Our estimate of N is

$$\frac{(160)(172)}{20} = \hat{N}$$

or 1,376 fish in the pond.

The most commonly used index of stock abundance is **catch-per-unit effort** (CPUE). This method is useful for most fishery stocks, even those that are very large, diffuse and hard to assess by other means. We can understand the relationship between CPUE and stock size best by returning to our analogy of the lake that we planted with trout. Suppose that after the population reaches the carrying capacity, we go fishing. Regardless of how we choose to fish, it is important that we quantify our catches on the basis of a standard unit of fishing effort. Examples include:

- Catch per 8 hours of angling per person
- Catch per fishing pole per day
- Catch per hook per day
- Catch per 60 feet of gill net per hour
- Catch per seine set
- Catch per trap per hour

Whichever units we choose, we should stick with the same units over the course of our study for the sake of comparison.

What we find after opening the fishery is that catches are excellent at first, but if we maintain fishing effort at a high intensity without closing the fishery (and allowing the stock to recover), the CPUE will begin to decline. Those of us who fish popular sportfishing lakes on opening day and then again later in the same season know this only too well. The reason that catches are better per-unit-fishing-effort early in the season is that there are more fish. The reason that fishing success slacks off later in the season is that there are fewer fish remaining. If we measure fishing effort carefully by conducting a **creel census** (interviewing anglers) or by conducting our own **test fishing** or both, we can get a very good idea of the size of the harvestable stock.

CPUE indices alone can reveal trends in **relative abundance** (changes in relative population abundance over time, as increasing or decreasing) from year to year. In order to estimate **absolute abundance** (actual numbers of fish present in a stock) from CPUE, we need to first calibrate CPUE to abundance. This is done by estimating abundance using either a mark-recapture study, a direct assessment, or a series of catch and effort data for several years (where the decline in CPUE is compared with cumulative effort or cumulative catch). Consequently, catch-per-effort data are routinely compared with alternate means of stock assessment to provide greater certainty on the accuracy of the estimates and to show how CPUE reflects absolute abundance.

Effects of Fishing

Let us now return to our lake analogy and go fishing for trout after the population has reached the carrying capacity of the lake. We will fish intensively and remove all of the largest individuals

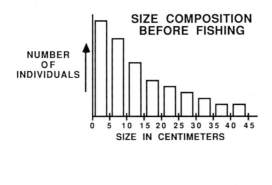

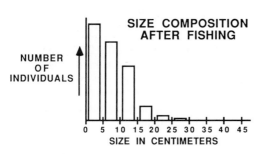

Figure 13.9. The effects of fishing on size composition of individuals that remain in the population.

Figure 13.10. Hypothetical consequences of contrasting harvest strategies based on the Schaeffer surplus production model.

by using nets with a mesh size that retains only the large fish and spares the smaller fish. If we consider some quantitative aspects of the fish stock before and after fishing, we find that by fishing we have both reduced the size of the population and changed the size composition of individuals within the population (Figure 13.9).

Our intention now is not only to demonstrate how the fishing affects the size structure and abundance of the stock, but to determine how the level of productivity changes as well. This is the key to managing a successful fishery—*after well-managed fishing, the stock that remains increases its production level,* and this was implicit in our discussion of the surplus production models. When we remove the largest fish from our lake, the remaining fish have more food available to them, more access to spawning areas, more refuges available and the like. Consequently, their growth rates and reproductive rates increase after fishing (provided we do not overfish the spawning stock to the point of impacting recruitment). If we wait an appropriate length of time before we fish again, we keep the stock at a high level of production. We can demonstrate the effectiveness of well-timed and properly applied effort by returning to the logistic curve and examining three different fishing scenarios (Figure 13.10). In scenario A, we adjust the fishing effort wisely and take bountiful repetitive harvests. In scenarios B and C, we fish the same number of times, but cumulative harvests are smaller. In scenario B, we overfish recruitment. The catch is good the first season but not in subsequent seasons. In scenario C, we underfish and leave the stock at a level too close to the carrying capacity for maximum productivity to ensue.

In this chapter, we have concentrated on the responses of fishery stocks to the environment and to fishing. However, there is another side to the story—how the harvest system responds to the presence of a fish stock. This issue is an important one and is examined in more detail in Chapter

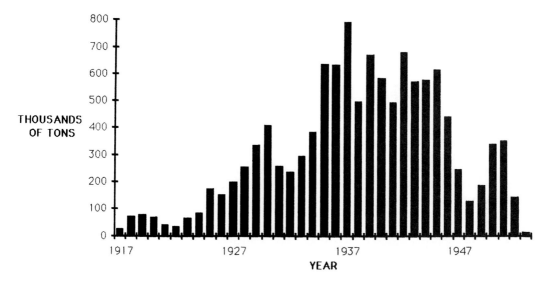

Figure 13.11. The classic pattern of a boom and bust fishery is exemplified by the Pacific sardine fishery that collapsed in 1950.

14. Far from remaining consistent, as is assumed in many yield models, fishing effort tends to evolve as a fishery develops, and this evolution of fishing intensity complicates the issue of quantitative fishery management. Another even less predictable complication arises with environmental extremes or anomalies, such as the El Niño wind pattern changes in the Southern Hemisphere. Complicating factors such as these tend to reduce the level of fishing effort that can safely be applied to a stock.

Most of the world's major fisheries have been overfished at one time or another, and many are currently overfished. The most spectacular and perplexing cases where overfishing is either wholly or partly responsible are the fishery collapses. Major collapses have occurred in fisheries for the Pacific sardine (*Sardinops sagax*), the North Sea herring (*Clupea harengus*), the Peruvian anchoveta (*Cetengraulis mysticetus*) and the Alaskan king crab (*Paralithodes camtschatica*). The record of yearly catches for these fisheries, beginning with their early developmental years, shows a typical pattern as indicated in Figure 13.11. There are four common stages in the collapse of a fishery. The first is the **discovery** stage, when the stock is found and means to harvest the stock is acquired. The second stage involves **expansion**, as more and more fishermen exploit the resource, markets and market demand intensify, and fishermen improve their harvest effectiveness. The third stage is **collapse**, when the stock is fished beyond its ability to bounce back. Collapses may be exacerbated by natural climatological phenomena, outbreaks of disease and other problems not related to harvest. The fourth stage, **resistance to conservation measures**, is an emphatically social one. By the time a fishery collapses, demand for the product is high, prices increase as supply falls off, and many fishermen are so heavily invested that each needs to be the one that catches those last precious fish. Fishery managers play a crucial and delicate role as they try to prevent a developing fishery from evolving in the manner described above.

14

Marine Fishery Management

Marc L. Miller and Robert C. Francis

Marine fishery management is a hedge against resource scarcity involving the exercise of authority and the regulation of human activities. The modern field of fishery management has evolved out of a late 19th century concern about overfishing in the common property fisheries of Northern Europe.* Because people—not fishes—abide by rules, and because people design and adjust rules, fishery managment is very much a social process.

As discussed in Chapter 2, fisheries around the world provide numerous sociological examples of what Miller and Gale have termed **natural resource management systems** consisting of four elements: **natural resources, profit-seeking industries, management bureaucracies** and **diverse publics.** The most important differences between fishery systems of the First and Second Worlds (i.e., the major industrial, non-communist or non-socialist nations; the major communist and socialist nations) as compared to those of the Third and Fourth Worlds (i.e., nations marked by widespread poverty; the most poverty-stricken nations) trace to the greater magnitude of governmental and scientific resources tagged for management in the more industrialized Worlds.

Regulation of fishing involves the design of licensing systems, the imposition of restrictions on gear and target species, and the implementation of season and area closures. These and other management tools are widely employed with the hope that a sustained harvest over a long period of time will result in a prosperous fishing industry and a permanent supply of food for society.

Miller, Gale and Brown have shown that the federal management of marine fisheries in the United States has its origins in the Progressive Era at the beginning of this century. At that time, and in response to short-term oriented and exploitative *laissez faire* attitudes towards the environment, a double-edged conservation ethic emerged. One kind of conservation—associated with Gifford Pinchot, first chief of the U.S. Forest Service—is termed **extractive conservation** and has been a basic theme in the management of the mineral, timber and commercial fishing practices. The other kind of conservation—associated with environmentalist John Muir—is termed **aesthetic conservation** and has long been central to park and wilderness management.

Importantly, neither aspect of conservation is "best." Today, managers accept the difficult task of reconciling the different priorities of commercial and recreational fishermen who advocate extractive conservation, and environmentalists who promote aesthetic conservation.

To do the job, modern fishery managers rely on a multidisciplinary community of fishery scientists for advice. In the United States, federal fishery management plans for marine fisheries are annually developed to mesh with the notion of **optimum yield.** This concept, embodied in the Magnuson Fishery Conservation and Management Act of 1976, invites fishery scientists—in particular, biologists, economists and anthropologists—to work together to inform managers, who ultimately must make decisions about conservation (how many fish can be caught) and allocation (which categories of fishermen get the catch).

---

*In the informal argot of fishery management, the terms "fish" and "fishery" are widely used, even when the resource in question is a marine mammal, crustacean, mollusc, coral, sponge, aquatic plant (e.g., seaweed and other algae), reptile (e.g., sea turtle) or some other life form.

Before moving to a discussion of the scientific framework for the study of fishes and humans, it must be stressed that fishery management policy decisions have more than a scientific heritage. While science is critically important in helping managers understand how fishes and people behave, it does not tell them what to do. For normative advice, managers take into account the preferences of industry, the opinions of the public and their own philosophies of resource management. Fishery management decisions in the United States, then, are the output of the social process of representative government where scientific input supplements that of other kinds.

Science and Fishery Management

Fishery science—whether the focus of inquiry is on animal or human behavior, or whether any particular analysis produces an abstract or an applied product—is ideally grounded in the ethic of objectivity. Accordingly, fishery scientists are trained to evaluate their research according to the standards of reliability and validity. A measurement is reliable to the extent another investigator using the same method would obtain the identical result. A measurement is valid to the extent the investigator has correctly labelled the object or feature under study.

Analytically, fishery scientists approach the study of fishes and people in exactly the same way. Assessment research looks at behavior, given an environmental constraint. For biologists, the behavior of fishes is of central interest and this is seen as constrained by the physical environment.* For social scientists, human behavior is of primary interest, and this is constrained by an institutional environment. In both cases, fishery scientists investigate the status of systems by studying structure and process.

Of course, environmental and institutional constraints are not, in fact, always constants. Changes in these do stimulate other changes in the behavior of fishes and people. Fishes, for example, are affected by variations in oceanic conditions, as during an El Niño southern oscillation event. People are affected by modifications of laws and bureaucratic and business practices, as when nations negotiate bilateral fishing agreements and firms cooperate in international joint venture operations.

Interdisciplinary work in fishery science nicely reflects the linkage of fishes to people in fisheries. What fishes do (which is manifest in their availability) influences society. In a symmetric way, what people do (which is manifest in fishing effort) influences the lives of fishes.

The applied scientific paradigms of fishery biology and what might be called the fishery social sciences (most visibly, natural resource economics and cultural anthropology) as these are aligned to respond to problems of fishery management can be introduced with comments on key *concepts* and issues of *measurement*. Concepts are the scientific abstractions which guide understandings of what is important to study. Measurement refers to how scientists proceed empirically to establish the quality and quantity of concepts, and the nature of relationships between concepts. In applying their expertise to policy problems, fishery scientists—whether they address themselves to the fishes or to the people side of the fishery management equation—rely on multivariate statistics, the computer manipulation of data and techniques of mathematical modelling.

Behavior of Fishes

The specialty in fishery science known as fishery biology has its genesis in the interplay of oceanography and marine biology in the last quarter of the 19th century. To contribute to fishery management, fishery biologists first assess the status of fish stocks and then try to comment on the consequences of alternative policy decisions for fishes. The job of fishery biology, then, is to ask and answer such questions as: How fast do fishes grow? How long do fishes live? Where do fishes

---

*In this chapter, the authors say they are interested in the behavior of fish and people using the same tone as, for example, physicists when they say they are concerned with the behavior of falling objects. For a discussion of specific behaviors of fish such as courtship, aggression, parental care, etc., which have only tangential relevance to fishery management, refer to Chapters 7 and 9.

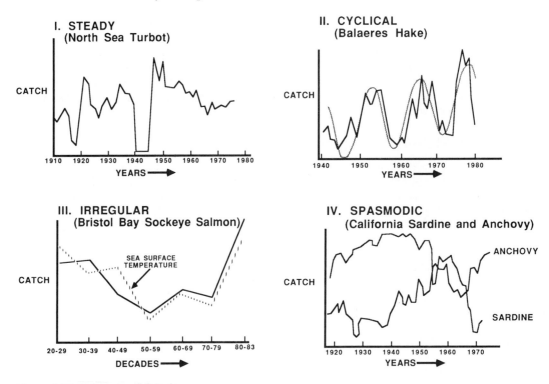

Figure 14.1. Categories of fisheries.

go? How many fishes are there, and of what ages? How many fishes of each age can be harvested to sustain (or increase, or decrease) the size of the stock?

Concepts

Three concepts dominate fishery biology: stock dynamics, production and effort. The study and estimation of these is generally called fishery **population dynamics**.

Stock dynamics refers to the manner in which fish populations vary over time and space. A classic paper by Caddy and Gulland describes the various ways in which fish stocks change and why a knowledge of these patterns is important to management. These authors divide fisheries into the four categories shown in Figure 14.1.

Steady fisheries are characterized by catches of about the same size year after year, with variations that generally remain within 20% to 30% of the long-term average yield. Major changes in catches that do occur are, in most cases, due to changes in patterns of fishing. Much of the standard stock assessment methodology has been developed for application to relatively steady fisheries. The North Sea turbot fishery (Figure 14.1) is one such example. With the exception of significant reductions in fishing effort due to World Wars I and II, the catch has remained virtually constant over almost a century.

Cyclical fisheries show cycles, oscillations of high and low catches, which are repeated at regular intervals. Many times, these intervals are of an interdecadal magnitude. The Balaeres hake fishery (Figure 14.1) is an example of a cyclic fishery which fluctuates on an interval of about 13 years. Obviously, attempts to manage or develop a cyclical fishery on steady state assumptions will be ineffectual.

Irregular fisheries are those in which catches vary greatly from year to year without any clear pattern. In many cases, major variations in harvest appear to be tied to fluctuations in the abiotic component of the ecosystem. Figure 14.1 shows the Bristol Bay, Alaska sockeye salmon fishery to be irregular. When landings are averaged over 10-year time periods and related to temperature, a possible causal relationship between catches in this irregular fishery and the marine environment is suggested.

Spasmodic fisheries, such as the California sardine and anchovy fisheries (Figure 14.1), are distinguished by periods of high abundance or availability alternating with collapse or depletion of the resource. Some of these fisheries (e.g., the Peruvian anchoveta fishery) are among the largest in the world during periods of high abundance. Figure 14.2 shows highly correlated trends in the three Pacific sardine fisheries over a 50-year time period. One might surmise that environmental phenomena occurring on a global basis have a major impact on the rise and fall of these stocks. It is interesting to note (Figure 14.1) that in many cases, when one spasmodic stock (e.g., sardines) declines in a particular region, another (e.g., anchovy) increases.

One other form of irregular fluctuation occurs when the entire structure of a fish community changes, in many cases due to the effects of human activities (e.g., fishing, pollution, introduction of exotic species). Longhurst and Pauly discuss community turnovers and replacements in tropical marine ecosystems. For example, these authors note the often repeated explosive growth of squid populations after the reduction by fishing of associated fish components of a particular ecosystem. When this type of species succession occurs, high-value steady-state fisheries are often displaced by low-value, fluctuating, generally irregular fisheries.

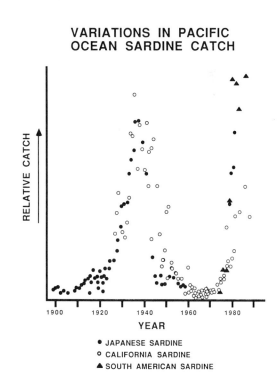

Figure 14.2. Relative sardine catches.

Production refers to the rate at which populations grow and decay. Fishery biologists describe fish population production with reference to the interrelationships between four basic processes: growth, mortality, recruitment and migration. Growth refers to the process by which individual members of a stock alter their biomass. Mortality refers to processes—predation, aging, fishing—that cause fishes to die. Recruitment refers to the process by which new individuals are added to the exploitable population. Migration refers to the processes by which individuals are added to or removed from a population through movement, into or out of, the physical area encompassed by the fishery. Production is routinely described by mathematical models which relate these four basic processes. Note that production is a rate and not a quantity. A small, productive stock can generate the same sustained yield as an unproductive large stock.

Francis has employed mathematical models to characterize production and its management implications for several "typical" species types encountered by fishery managers. He characterized populations as slow-growing, long-lived (e.g., 30- to 80-year life span) at one extreme and fast-growing, short-lived at the other. The former have a very low ratio of production to biomass (sometimes referred to as "turnover"), whereas the latter have a very high ratio of production to biomass. For example, Pacific ocean perch (*Sebastes alutus*), a common slope rockfish of the Northeast Pacific, is capable of turning over no more than 5% of its biomass per year, whereas yellowfin tuna, *Thunnus albacares,* is capable of turning over 20% of its biomass in the same period. Kleiber et al. estimate that skipjack tuna (*Katsuwonus pelamis*), the most abundant and productive of the world's tunas, are capable of annually turning over more than 100% of their biomass. From the fishery management point of view, Francis warns that fisheries that depend on long-lived, low production stocks (e.g., most rockfish) can quickly generate a harvest pressure that vastly exceeds the long-term productive capacity of the resource.

Effort is a concept used to describe the intensity of fishing on a particular stock. Typically measured by an amount of capturing gear (e.g., number of vessels), effort is a complex human variable which has proven difficult to study in a fully satisfactory way. Theoretically, the amount of effort expended for a fixed catch is inversely related to the underlying stock abundance. (Fishery scientists also investigate how catch varies with a fixed level of effort; hence, the acronym CPUE—catch-per-unit-effort.) The more fish there are, the less effort is needed to realize a fixed harvest; it is this idea that has driven many of the most widely used stock assessment methods. If one can accurately monitor catch and effort, one can assess both the current status and the long-term productive capacity of a fishery. The work of Schaefer on the yellowfin tuna fishery of the eastern Pacific Ocean is a famous example of the application of this concept. Unfortunately, as is pointed out by Hilborn and others, it seems to be quite rare that this relationship between the effectiveness of fishing effort and the abundance of an exploited population holds true.

Measurement

As was stated earlier, measurement refers to the quantification of concepts. Insofar as the behavior of fishes is concerned, measurements are obtained in two ways. When fishes are tagged, caught, or otherwise monitored in controlled studies by biologists, measurement is of the direct type. By contrast, indirect measurement refers to those instances in which scientists get their data from government or industry records of commercial and recreational harvests.

While stock dynamics and production are commonly measured both directly and indirectly, a weakness of fishery science is that effort is usually measured only indirectly. To correct this situation, social scientists are beginning to work with biologists in the conceptualization and direct measurement of how effort is apparent in the behavior of fishermen.

A reliable time series of stock assessment is a prerequisite for an understanding of species dynamics and interspecies relationships. Much of what fisheries agencies around the world do is to carry out the collection and processing of information essential for both direct and indirect stock assessments. Direct surveys of fishery resources run the gamut from low-cost directed fishing by commercial vessels to high-cost fishing and hydroacoustic surveys by expensive government research vessels. A major objective of surveys is to estimate the absolute biomass level of exploited stocks of fishes.

A second method of direct assessment of fish stocks involves the marking and subsequent recapture of individual fishes. This method is applied to migratory fishes (e.g., salmon, tuna), whose movements during most stages of their life histories preclude direct assessment by survey. Both direct survey and tagging information, if properly collected, can be used to estimate population size as well as important population variables such as growth, mortality, rates of migration and recruitment.

Indirect methods of stock measurement involve the collection of fishery statistics on the landed catch—concerning both absolute biomass and biological (e.g., age or size) structure—and fishing intensity or effort (e.g., vessel days at sea, number of hooks fished per period of time).

In many fisheries, particularly those in the Third and Fourth Worlds, indirect or fishery-based information is all that is available for stock assessment. In other cases, millions of dollars are spent for sophisticated surveys of fishery resources where meager, if any, information is collected on catch and effort. One often wonders why fisheries are so poorly managed. From the scientific point of view, one explanation is that fisheries tend to be poorly measured.

It stands to reason that production can only be known if the right variables and parameters of a stock are properly measured. In practice, production per se is not measured. Rather, it is estimated using mathematical and statistical models employing data generated by direct and indirect measurements. In their simplest forms, these models fall into three categories: **stock production models**, **age and size structured models**, and **tagging models**. Stock production models estimate the rates of production of population biomass directly from either direct or indirect measures of stock biomass. As noted above, a crucial assumption underlying these models is that the ratio of catch to effort is somehow related to levels of stock biomass. Age and size structured models use both direct and indirect measures of the biological structure of a stock to infer stock production relationships. Many of these models are also used to estimate stock abundance over a protracted time period. Tagging models are used to translate mark-recapture data into estimates of all aspects of production.

Measurements of effort range from direct and detailed observations made by observers aboard commercial fishing vessels, to rough counts of marine recreational anglers, to incomplete records of the number of vessels operating in a fishery. On the surface, effort would seem to be the easiest of all concepts to measure or estimate. In practice, it presents many difficulties. At the root of this is the fact that human activity in most fisheries is very competitive and therefore carried out as secretively as possible. Fishermen simply do not relish their activities being measured or monitored. As a result, meaningful indices of effort are rare. So, although measures of effort are the bread and butter of indirect stock assessment, they are frequently unavailable for the assessment of biological production.

Behavior of People

For nearly all of the first half of this century, fishery science was exclusively populated by fishery biologists. After World War II, however, the field began to attract economists. With the passage of the Magnuson Fishery Conservation and Management Act of 1976, which mandated the formal study of the social dimension of U.S. fisheries, cultural anthropologists, sociologists, and some political scientists have joined the fray.*

To contribute to fishery management, fishery social scientists first assess the status of elements of society and then try to comment on the consequences of alternative policy options for categories of people. This job is somewhat more complex than that which faces the fishery biologist for three reasons. First, unlike that of fishes, the behavior of people is really a three-dimensional puzzle. The ABC's of social science are **affect** (what we feel), **behavior** (what we choose) and **cognition** (what we think), and these are causally interrelated in ways which are far from well understood.

Second, while the adaptation of fishes to the oceanic environment is typically achieved over a long period (e.g., decades), the adaptation of people to an institutional environment (e.g., a new law, or form of government) can occur much more quickly. People, unlike fishes, can change both the physical environment and the institutional environment.

---

*For an introduction to the Regional Fishery Management Council system established by the Magnuson Act, see the work of Kelly, Gale and Miller, Miller, and Marasco and Miller.

Third, human populations, again unlike fish stocks, are stratified by overlapping social variables—education, class, kinship and religion, to name a few. Few quarrel with the assumption that an individual fish is analytically the same as the others with which it shares species, age, class, size and gender. Analogous treatment of people, however reasonable to a scientist, invites controversy. Many people prefer to reach their own subjective conclusions about the human condition of a fishery, and then have their lawyers assert it, than to have social phenomena objectively assessed.

Indeed, the reality that people have differences of opinion can be a serious impediment to continuity in the fishery social science research agenda. Which fishermen—those who are captains or crew, those who own vessels, those who fish full-time—should be studied? Which non-harvesting populations—processors, retailers, environmentalists, recreationalists, tourists, consumers, our children's children—have interests that should be studied and considered in policy decisions? Should analysis focus on fishing activities found in a small strip of coast, a state, a region or a nation?

Questions such as these and their biological equivalents—Which species should be managed? Which species should be given "incidental" status? What defines a fish stock?—underscore that applied fishery science of all kinds is more dependent than many of its practitioners would like to believe on managers and industry for direction as to what problems merit study. Because marine fishery policies are social policies, we would do well to keep in mind Fuller and Myers' classic definition of a social problem:

> Social problems are what people think they are and if conditions are not defined as social problems by the people involved in them, they are not problems to those people, although they may be problems to outsiders or to scientists . . .

Concepts

As noted above, fishery social science has a shorter history than fishery biology and, as a partial result, has made a smaller contribution to fishery management. Today, federal budgets and manpower for fishery social science in the United States lag embarrassingly far behind those for fishery biology. The National Marine Fisheries Service in the Department of Commerce, for example, employs many biologists, but only a small number of economists—and virtually no other social scientists. The situation is much the same in state fishery agencies.

Scientific concepts useful in fishery management must meet two criteria. First, they must have theoretical value, they must be essential to the solution of an intellectual riddle. Second, they must have practical value, they must be essential to the resolution of a real world policy issue.

Fishery social science meets the first criterion. Economics, anthropology, sociology and political science, like all sciences, have developed and debated countless theories, and the supply is inexhaustible. The second criterion, that of addressing a policy concern, has been dealt with more successfully by economists than by their social science colleagues.

Key concepts in fishery economics include **value** and **efficiency**, in addition to the cornerstone notions of supply and demand. Anderson introduces fishery economics as "the study of the optimal allocation of (scarce) resources to a fishery in such a way that the value of production is maximized." Economists generally seek to adjust capital and labor inputs to fisheries in a manner that permits attainment of maximum economic yield, the point on a production curve where the difference between total cost and total revenue is the greatest. Efficiency is strictly defined by economists as the allocation of resources that maximizes the value of production. With this thinking, efficiency is not said to have been reached if anyone can still be made better off without making someone else worse off. (The majority of economists who are concerned with marine fishery management are natural resource economists, a subset normative, or welfare economists. All of these, together with positive [i.e., non-normative] economists, are versed in neoclassical microeconomics.)

The other fishery social sciences differ from economics in that paradigms are not organized to concentrate on the maximization of any one variable, such as efficiency. Anthropologists and sociologists, for example, do not assume that efficiency is desirable. Instead of focusing on prescription, they put their effort into describing the social processes that hold society together (and, sometimes, break it down). Economists may know how to measure efficiency, but all of us, including economists, know that equity (fairness) is in the eye of the beholder. This leads anthropologists to show great diversity in the questions they ask about fisheries. Some, for example, might study the distributions of attitudes of American fishermen toward the idea that individual transferable catch quotas should be implemented in the management of commercial fisheries. Others might be inclined to study how kinship functions to regulate the behavior of small-scale fishermen in the Third World, or how lifestyle motivates people to become fishermen, or what other factors cause fishermen to participate in different fisheries (often in different states, with different gear and in different statuses) in the course of a year. Finally, anthropologists might study the religious function of fishes in subsistence and ceremonial fisheries. The findings from all of these kinds of sociological studies, and many others as well, have the potential to help managers to anticipate the likely social, cultural and political consequences of policy alternatives.

Measurement

Indirect measurement of the behavior of fishermen by fishery social scientists is made possible by the same government and industry statistics (landing and ex-vessel price data) utilized by fishery biologists, and also by market, census, demographic and other data compiled by numerous government agencies with no interest in fisheries. Direct measurement of human behavior—as, for example, is standard in social surveys, ethnographic field work and numerous methods of interviewing—is intermittently conducted by university-based scientists as contract research for an agency, or as independent abstract research. Social science methodologies that have potential to help managers decide between policies that sustain the sociological status quo and those that initiate social change include the benefit-cost and sectorial input-output accounting techniques of economists, and social impact assessment methods of sociologists and anthropologists.

Inference

Inference refers to how the findings of science are transformed into advice for fishery managers. And it is in this process where values, ambition and naïveté influence the conduct of scientists. This is the phase of the management process where science meets politics, where things tend to break down. Biology, anthropology and economics suddenly become adversarial, part of a debate not so much about the correctness of science as about the role of science in societal decisions. Apples are compared with oranges and often chaos results.

In fishery science, as in science elsewhere, analysis boils down to finding a pattern in data. This done, scientists, when speaking to their peers, customarily comment on the confidence they have in their work, usually invoking the standards of reliability and validity. But there is no consensus about how scientists should convey their results—not to mention their hunches and visions—to a complex, non-academic audience.

Fishery scientists face a real dilemma when they try to communicate what they know (and the degree to which they are sure about it) to non-scientists. Part of the problem is of their own making; scientists have too long neglected how to effectively display quantitative patterns and probabilistic qualifications. But managers are also at fault. Managers—many of whom are not scientists, and all of whom are doing something more complicated than science—ask for the "best available" scientific information without realizing how loaded this seemingly benign request is. How can we tell whether the "best available" science is more than good enough, just good enough, not quite good enough, or worth nothing? The problem is exacerbated when managers demand a simple answer (a "yes" or a "no," or a single number), or press for a quick answer, or want scientists to make management decisions for them.

Too often, fishery scientists have been unable to resist the temptation to move from the arena of scientific thinking to that of managerial policy making. The warning sign that this is happening is evident when scientists stop saying "If the policy is x, the consequence is y" and begin saying "Do this!" If there cannot be a clear-cut separation between the work of science and the work of management, one should, at the very least, be explicit about which activity one is engaged in at the moment.

Management Philosophy and Process

Marine fishery management is now a social process strongly influenced by the interaction of a scientific ethic with a conservation ethic. However, it has taken 100 years for this to happen. In the first half of this century, significant progress was made in the understanding of biological processes and in the development of an ideology of resource conservation, but fishery management policies were not driven by either.

In the aftermath of World War II, fishery science and a philosophy of conservation began to play more important roles in fishery management systems. In the theoretical realm, major advances in fishery science appeared in the work of W.E. Ricker, R.J.H. Beverton, S.J. Holt and M. Graham. In the political realm, coastal states, beginning with Peru in 1947, moved to declare sovereignty or jurisdiction over fishery resources in adjacent seas. A number of conservation agreements were signed in the post-World War II years, notably the International Convention for the Northwest Atlantic Fisheries (1949), the Inter-American Tropical Tuna Convention (1950), and the International North Pacific Convention (1953).

In 1948, the United Nations Food and Agriculture Organization (FAO) began to compile world-wide fisheries statistics. Seven years later, FAO convened the International Technical Conference on the Conservation of the Living Resources of the Sea. This conference provided the stimulus for a variety of conventions emanating from three United Nations Conferences on the Law of the Sea between 1958 and 1982.

The global implementation of zones of national jurisdiction extending 200 nautical miles from shore during the last 40 years has profoundly altered the responsibilities of coastal states in the utilization of fisheries resources. Most strikingly, the new ocean regime presents governments with a greater opportunity and motivation to influence the destinies of fishes and people.

That coastal states have responded differently to this challenge is illustrated in the contrasting styles of fishery bureaucracies around the world. Fishery management bureaucracies can be compared by reference to philosophy and process.

Philosophy

Philosophies of fishery management arise from legal mandates for regulation of fishing and from the professional training of managers. Insofar as mandates for management are concerned, Brewer and Burke have shown that fisheries objectives, where they exist at all, are often inconsistent, unordered, unfocused and weakly connected to policy decisions.

When legislation is vague, fishery managers can find justification for almost any kind of policy. At one extreme, a fishery might be closed to protect an endangered fish. At the other extreme, a stock might be harvested to extinction to preserve a human community.

Consequently, the background of managers becomes a critical variable in the fishery policy process. With the exception of managers of subsistence, artisanal and other small-scale systems, fisheries authorities have not traditionally been recruited from the ranks of harvesters. Exceptions to this rule are found in the United States, where fishermen can be members of Regional Fishery Management Councils established by the Magnuson Act, and in Japan, where fishermen's cooperatives have management authority. More often, managers have backgrounds as marine biological and fishery scientists, as career government bureaucrats or as political appointees and diplomats.

Unfortunately, there is no consensus about the best way to prepare to be a fishery manager. Bureaucrats and appointees may be unfamiliar with theoretical or methodological changes in fishery science (for example, the expansion of fishery science to involve the social sciences). Scientists may be unfamiliar with political or programmatic constraints on management. Royce has suggested that this educational problem can be avoided for future generations of fishery scientists with changes in curricula.

Process

The social consequences of fishery management (as, for example, measured by the timeliness of decisions, the minimization of conflict generated among constituencies, the attainment of specific welfare objectives) are sometimes directly related to the nature of the policy making process. When bureaucratic uncertainty is high, the legitimacy of the management authority is called into question by those managed. Collaboration among managers, scientists and those managed is not a universal feature of marine fishery management systems. Unfortunately, and because fishery management bureaucracies have rarely documented the way decisions have been made, the literature concerning the fishery policy-making process is small.

It is conceivable, although not necessarily desirable, that the policy process in a fishery management system could exactly mirror the political process sustaining government. To the extent this is the case, fishery management reduces to a political exercise.

Alternatively, the policy process in a fishery management system could be influenced as well by the scientific ethic. Fishery management in this case solves problems by recourse to the objective measurement of the biological and social world. When management takes place in the context of representative government, opportunities exist for fishermen and other interested parties to be usefully brought into the management arena as participants in the policy making process.

Several Examples

Marine fishery policies are rarely pure products of a single line of reasoning. More often, they reflect a blend of scientific, conservation and business ethics. Larkin suggests that a Martian visiting earth would discover **ichthyocentric** and **anthropocentric** approaches to fishery management. He observes that ichthyocentric managers, as found in the Canadian bureaucracy, are careful to ensure the long-term viability of fisheries and accordingly are attentive to "all the biological bogeymen that scientists may generate from their speculations." By contrast anthropocentric managers, as found in the Japanese bureaucracy, look first to social and economic issues, "serenely confident that the fish will somehow look after themselves."

Managers in the United States, as exemplified by the members of eight Regional Fishery Management Councils, fall somewhere in between, but probably tend to favor people over fishes. Overall, the most compelling criticism of U.S. marine fishery management holds that the beneficiaries of policies have been fishermen, rather than the public who pays for management. Indeed, it is difficult to argue that provision of the "highest quality fish product, on a regular basis, at the lowest price" has been a higher management priority than support for the domestic fishing industry.

Such a brief exposure to these systems cannot reveal any "best" philosophy of fishery management any more than it can isolate a best set of cultural values. The great majority of debates about the relative risks of policies to fishes and people are destined to be metaphysical. The fishery scientist can do no more than insist that managers' guesses about the statuses of fish systems and human systems be empirically tested.

Discussion

"I know of no more important practical lessons in this earthly life of ours—which, to the wise man, is a school from the cradle to the grave—than those relating to the employment of a sense of vision in the study of nature."

—George Perkins Marsh

This chapter has introduced the multidisciplinary field of applied fishery science and has remarked on some of the philosophies and processes of marine fishery management. Given the power of the written word, some readers will doubtlessly take the observations and opinions in the text to constitute a definitive statement. Perhaps more than a few readers will extract a thesis along the lines that scientists, if they do not already have the situation under control, are well endowed to march to the rescue of the people and fishes, who have so unwittingly become entangled.

To counter any such conclusion, this last section flags a few marine fishery management problems which will be hard to handle.* These problems are categorized according to whether they concern **conservation, allocation,** or **institutional design.** Conservation problems have both philosophical and scientific dimensions. Philosophically, what is meant when we say, for example, we respect species diversity? Does it make sense to stop fishing on mixed stocks in order to guarantee the survival of one stock, say, wild salmon? (The costs of treating diversity as an absolute value come into perspective if one extrapolates smallpox virus as an endangered species.) The strictly scientific conservation problems are just as irksome. When will the sophisticated concepts, theories and measurements of science be enough to satisfactorily account for the tremendous biological variability in marine fisheries? Presently, fishery scientists are more than uncertain about how to model uncertainty.

A major trend in marine fishery management is that allocation problems, rather than conservation problems, increasingly signal where the action is. The reason for this, of course, is that people talk and fishes do not. In the First World, allocation problems arise from the common property feature of fishery resources. As fishery economists have long pointed out, the creation of property rights in fisheries would greatly simplify management. Then, too, a tax on commercial fishing would help pay for management, which directly benefits the fishing industry. However, these ideas, and the related notion that entry to fisheries be limited, are abhorrent to many who want to continue to treat the ocean as a frontier, a place where anyone has a chance to succeed or fail.

It seems unlikely that a move to establish property rights in U.S. fisheries will find broad-based industry and public support in the near future. (An even more radical and unpopular—but also economically justified—alternative calls for the nationalization of the fishing industry.) This being the case, managers contend with two types of allocation problems. The first, allocation of access, has to do with the possibility that increases in the number of fishermen will result in economic overfishing. The imposition of limited entry regimes (in which the number of fishing licenses is fixed) does not permanently solve this problem because overcapitalization of fleets is not discouraged.

The second problem, allocation of catch, is equally troublesome. Here managers must decide how to distribute the available catch equitably across competing categories of commercial, recreational and subsistence fishermen. Needless to say, this Solomonesque assignment requires more than scientific insight.

Finally, fishery management must face problems of institutional design. Three, in particular, come to mind. First, there is the problem of international cooperation in the management of highly migratory and trans-boundary stocks. How can institutions be linked to allow the timely exchange of scientific information and the extension of authority? Second, there is the problem of ecosystem management. How can fishery managers collaborate with forest, habitat, wildlife and other natural

---

*For collections of essays on the contemporary problems of fishery science and management, see the work of Rothschild, Gulland and Wooster.

resource managers to address questions of how natural systems are interrelated? Third, there is the problem of a shortage of scientific knowledge. How can managers improve their performance in fisheries where either the behavior of fishes is too complex to model, or where for a number of political or fiscal reasons no fishery science is conducted at all?

These, then, are a few of the truly maddening problems which make the management of marine fisheries challenging. Solutions will require fishery managers, scientists, fishermen and the public to be imaginative and confront not only each other, but themselves.

Additional Reading

Anderson, L.G. 1977. The Economics of Fisheries Management. The Johns Hopkins University Press, Baltimore.

Brewer, G.D. 1983. The management of world fisheries. Pages 195–210 *in* B.J. Rothschild (ed.), Global Fisheries.

Burke, W.T. 1983. Extended Jurisdiction and the New Law of the Sea. Pages 7–50 *in* B.J. Rothschild (ed.), Global Fisheries.

Caddy, J.F. and J.A. Gulland. 1983. Historical patterns of fish stock. Marine Policy 7:267–278.

Francis, R.C. 1986. Two fisheries biology problems in West Coast groundfish management. North American Journal of Fisheries Management 6:453–462.

Fuller, R.C., and R.B. Meyers. 1941. The natural history of a social problem. American Sociological Review 6:320–328.

Gale, R.P. and M.L. Miller. 1985. Professional and public natural resource management arenas: forests and marine fisheries. Environment and Behavior 17:6:651–678.

Gulland, J.A. 1988. Fish Population Dynamics. The Implications for Management. John Wiley and Sons, New York. 2nd ed.

Hilborn, R. 1985. Fleet dynamics and individual variation: why some people catch more fish than others. Canadian Journal of Fisheries and Aquatic Science 42:2–13.

Kelly, J.R. 1978. The Fishery Conservation and Management Act of 1976: Organizational framework and conceptual structure. Marine Policy (January):30–36.

Kleiber, P., A.W. Argue and R.E. Kearney. 1983. Assessment of skipjack (*Katsuwonus pelamis*) resources in the central and western Pacific by estimating standing stock and components of population turnover from tagging data. South Pacific Commission, Tuna and Billfish Assessment Programs, Tech. Rep. 8. Noumea, New Caledonia.

Larkin, P.A. 1988. Comments on the Workshop Presentations. Pages 287–289 *in* W.S. Wooster (ed.), Fishery Science and Management: Objectives and Limitations. Springer-Verlag, New York.

Longhurst, A.L. and D. Pauly. 1987. Ecology of Tropical Oceans. Academic Press, San Diego.

Marasco, R.J. and M.L. Miller. 1988. The role of objectives in fisheries management. Pages 171–183 *in* W.S. Wooster (ed.), Fishery Science and Management: Objectives and Limitations. Springer-Verlag, New York.

Marsh, G.P. 1965. [1864]. Man and Nature: Or, Physical Geography as Modified by Human Action. Harvard University Press, Cambridge, MA.

Miller, M.L. 1987. Regional fishery management councils and the display of scientific authority. Coastal Management 15:309–318.

Miller, M.L. and R.P. Gale 1986. Professional styles of federal forest and marine fisheries resource managers. North American Journal of Fisheries Management 6(2):141–148.

Miller, M.L., R.P. Gale and P.J. Brown. 1987. Natural resource management systems. *In* M.L. Miller, R.P. Gale, and P.J. Brown (eds.), Social Science in Natural Resource Management Systems. Westview Press, Boulder, Colorado.

Miller, M.L., R.P. Gale, and P.J. Brown (eds.). 1987. Social Science in Natural Resource Management Systems. Westview Press, Boulder, Colorado.

Rothschild, B.J. (ed.). 1983. Global Fisheries. Springer-Verlag, New York.

Royce, W.F. 1984. A professional education for fishery scientists. Fisheries 9(3):12–17.

Schaefer, M.B. 1957. A study of the dynamics of fishing for yellowfin tuna in the eastern tropical Pacific Ocean. Bulletin Inter-American Tropical Tuna Commission 2:247–285.

Wooster, W.S. (ed.). 1988. Fishery Science and Management: Objectives and Limitations. Springer-Verlag, New York.

15

Commercial Fishing

Bill High

Introduction

For hundreds of years fishermen have hooked, trapped, surrounded or otherwise captured their prey. Even now, the same basic capture methods account for virtually all aquatic life taken by man. Nonetheless, efforts to improve gear efficiency, safety and selectivity have resulted in significant changes.

The Changing Face of Commercial Fisheries

Change has been most rapid since the technological revolution following World War II. Electronics, nylon and hydraulic power are among the key developments.

Electronics aid in "seeing" underwater and navigating. Sonar developed to track submarines also locates fish schools, and electronic echo returns from the sea bed provide both depth and details of the sea bed. Today, all modern vessels carry one or more fish-finding and depth recorders that display, sometimes in vivid colors, signals from fishes and the ocean floor. Some electronic devices even reveal a trawl's position while it is fishing (Figure 1.4, Chapter 1).

Radar and radio direction finders allow vessels to return with precision to longlines or pots set hours or days before. LORAN, an electronic reference grid, improves determination of vessel location and is an essential piece of gear on fishing vessels. With the more accurate but less common satellite navigation systems, vessels can return to within a few yards of their target.

Nylon, a strong, durable and rot-resistant material, allowed the development of large fishing nets. Other synthetic materials have also been used because of their special characteristics of buoyancy, abrasion resistance, color and elasticity. Development of giant high seas purse seines and enormous pelagic trawls were only made possible through the use of these synthetic materials.

The importance of new technology depends on the particular fishery. However, few fishermen question the value of hydraulic systems that allow energy to be routed through rigid or flexible pipes to powerful motors, reducing the need for human labor. While hydraulic motors lift loads, turn the vessel and pump liquids, their most valuable contribution to the fishing process is their ability to pull huge fish-laden nets onboard (Figure 15.1).

Mario Puretic is perhaps the best known advocate of hydraulic-assisted fishing systems. Introduced in 1955, his forerunner of the Marine Construction and Design (Marco) power block helped to dramatically change methods of retrieving purse seines and was followed by a wide range of hauling devices, including pot line haulers and net reels.

Gear and Methods

If you tried to describe a standard gear for any fishery, you would be subjected to the criticism of most fishermen, who would point out how their own special difference sets the standard. There-

Figure 15.1. Hydraulic motors provide power to the trawler's stern-mounted net reel, deck-mounted dual-trawl winches and two lifting winches at the base of the boom.

Figure 15.2. Sablefish strike at a baited hook attached by a short gangion to the groundline.

fore, the gear descriptions that follow are general and only intended to provide the novice with an understanding of the basic techniques. Unless dictated by regulations, identical construction is rare for gear of the same name, even when the gear is used in a particular region for a particular species. Fishermen incorporate a combination of science, experience, tradition, imagination and intuition in attempting to make the gear (1) catch more target species or fewer unwanted species, (2) blend better with their vessel design, size and power, and (3) more simple for handling and repairs.

Terms used to describe fishing gear or methods likewise vary widely across the United States and internationally. For example, the terms longline and setline, used interchangeably in the U.S. northwest, describe essentially the same gear northeast U.S. fishermen call a trawlline and southern watermen refer to as a trotline. A great number of terms describe or identify a relatively few basic fishery methods.

Hook and Line

Hooks represent one of the earliest capture forms. Typically, the hook is baited or made to simulate natural food. Infrequently, a bare multiple hook may be jerked through the water to impale densely schooled species. Hooks may be fished singly, at the end of a weighted line, attached at intervals along a heavy mainline fished either on the bottom or near the surface, or pulled through the water as part of a lure.

Handline

A handline is usually directly controlled by the fisherman. It may be held to sense a striking fish and retrieved hand over hand, or wound on a hand-turned or power reel. The fisherman may operate several lines alternately. Because handlining is labor intensive, this fishing method is usually restricted to relatively high value species destined for specialty markets.

The electrically powered auto jig is a form of handline machine developed in Norway that automatically sets and retrieves a line having multiple baited hooks. When hooked fishes apply a preselected resistance, the line is reeled in.

Gulf of Mexico fishermen seeking red snapper and grouper incorporate up to 10 electrically or hydraulically powered line spools called Bandits along their vessel rail. Multiple baited hooks are affixed near the weighted end of the line. At water depths of 100 to 200 feet (25.4 to 50.8 m), the striking fish can be felt with a hand on the line and are retrieved when several are likely to be hooked.

Longline (Setline, Trotline)

Longlines, whether fished on the sea bed for tilefish, cod, sablefish and halibut or near the surface for shark, billfish and tuna, are typically composed of a mainline (groundline), short leaders (**gangions**) with hooks at the loose (terminal) end and a buoyline. The mainline, which may consist of steel cable, nylon, or synthetic rope depending upon the specific fishery, may run for several miles.

Demersal (bottom) gear is anchored at each end with a buoyline between the anchor and surface floats. In some fisheries, small weights are attached at intervals along the groundline to reduce line drift in currents or to prevent hooked fishes such as the Pacific halibut from pulling the usually slack groundline many feet across the bottom or up into the water column. Hook-bearing gangions are affixed at intervals along the groundline, either securely tied in place (fixed gear) or snapped on with a metal clip (snap gear). The leaders may be short and close together, as are the 10-inch (25-cm) long leaders at 3- or 4-foot (0.8- or 1.2-m) intervals for sablefish (Figure 15.2), or longer and farther apart, like the 3-foot (1-m) long leaders at 13- to 32-foot (4.0- to 9.8-m) intervals for Pacific halibut.

Steel, rope and monofilament groundlines may be stored on large spools and gangions clipped on at selected intervals while the groundline is set. Rope groundlines with gangions tied in place are

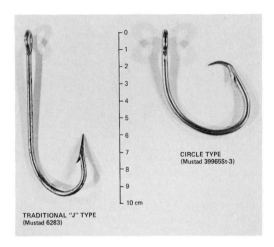

CIRCLE TYPE
(Mustad 39965St-3)

TRADITIONAL "J" TYPE
(Mustad 6283)

Figure 15.3. Traditional ''J'' style halibut hook and the more efficient circle hook.

often coiled by hand with the hooks carefully placed in the coil center. Small vessels, or those participating in longlining only part time, may coil the groundline into a wood, plastic or metal tub. Baited hooks are draped around the tub lip to prevent snags as the mainline is deployed. Traditional demersal longline hooks have been of the J design. In 1983, the International Pacific Halibut Commission and the National Marine Fisheries Service conducted experiments that clearly demonstrated the vast superiority of a more circular hook (Figure 15.3). By 1984, most halibut fishermen had converted to the circle hook and by 1986 most longline-caught sablefish were taken on circle hooks.

In international waters, longlines are fished near the surface for salmon, tuna and a variety of billfish species. Within the U.S.-managed waters of the Atlantic Ocean and Gulf of Mexico, hundreds of vessels use surface gear to fish for shark, tuna or swordfish. As with bottom gear, specifications vary depending on species, fishing strategy and region. The mainline for shark fishing is usually steel cable, which can withstand abrasion from sharp teeth and rough skin. Seven-hundred pound (318 kg) breaking strength monofilament nylon line with weaker gangions is fished in lengths up to 30 miles (50 km) or more for tuna and swordfish. Buoylines between the mainline and surface floats are attached at intervals to achieve the desired depth of the drifting gear. Gangions are much longer than those used in demersal longline fisheries, extending 30 to 100 feet (10 to 33 m) below the mainline at about 200-foot (61 m) intervals. Both conventional J hooks and circle hooks are baited with herring or squid. Some fishermen dye the squid flesh various colors to improve performance.

Chemical light sticks, usually attached a short distance above tuna and swordfish hooks, greatly increase the catch at night. Fishermen believe the light attracts nearby squid, which in turn attract target species to the vicinity of the baited hooks.

Several automated gear operating systems have been introduced into the demersal longline fisheries in an effort to reduce the labor-intensive activities of baiting, setting and hauling. One system retrieves gear, separates fishes from hooks, clears old bait from hooks, untwists gangions snarled by active fishes and rebaits the hooks at a rate of up to four hooks per second. However, not all fishermen are convinced such expensive mechanical devices, with their inherent problems, result in increased income.

The trotline, a variation of the longline system, is used by Chesapeake Bay watermen to catch blue crab. Baits consisting of beef and poultry parts or eel are tied by short gangions to the groundline or may be tied directly onto the groundline. Crab hold tenaciously to the bait as the line is lifted and are netted by hand at the surface.

Trolling

Baited hooks or artificial lures may be pulled (trolled) through the water by a vessel underway. Primary species captured by trolling include king mackerel off Florida, albacore throughout the Pacific Ocean and two salmon species along the Pacific coast of northwestern North America. Multiple lines are fished simultaneously and various techniques to keep the lures separate are employed.

Figure 15.4. A salmon troller with one fishing line deployed from each forward pole and two lines from each midship pole.

Figure 15.5. Caribbean-style fish trap.

Salmon troller vessels range in length from about 20 to 50 feet (6.1 to 15.2 m) and deploy 4 to 6 mainlines, each with numerous lures clipped at various distances above the terminal end weight (cannon ball). Stainless steel mainlines are set and retrieved from power- or hand-wound reels called **gurdies**. Lures are widely spread by temporarily securing the mainlines to the ends of poles set out from the vessel (Figure 15.4).

Hundreds of lure designs are marketed to attract fishes or, sometimes, just the fisherman. Lure effectiveness depends on target fish species, time of year, time of day, water color, weather conditions (sea state and cloud cover), geographic location, prey species present, etc. Spoons are fabricated from several bright metals in numerous shapes. Usually a large single hook is affixed at the terminal end. Wood or plastic plugs are shaped to resemble fishes and may have two single or treble hooks. Plastic lures called **hoochies** resemble squid, a popular, natural fish food. The lure is secured by a snap and swivel to a monofilament leader that can be clipped directly to the mainline.

After the cannon ball weight is lowered over the moving vessel's side, a lure attached to a leader is clipped onto the mainline. Additional lures are clipped on the line every few fathoms* as the weight is lowered. The multi-lure array is then lowered farther to a depth where the fisherman expects to find the target species. Finally, a heavy tag line from the pole is attached securely by a clamp to the mainline. As the mainline is released farther, the entire array swings outward and is supported by the pole. Each mainline is similarly set to tow from a pole.

Hooked fishes cause the pole to bounce. Fishermen wishing to relax within the cabin place small bells on the poles to alert them to a hooked fish. When one or more fish are on a lure array, the fisherman retrieves the mainline. As the reel winds in the stainless wire, it is drawn back to the vessel side from the pole tag line, which is removed and carefully set aside. As lure assemblies reach the surface, those without fishes are quickly unclipped from the mainline and set on deck or allowed to trail behind the vessel. Leaders having fishes attached are pulled in by hand. Skilled fishermen using a club with a gaff hook stun the vigorously fighting salmon without tearing the hook out of the fish's mouth and, almost in the same motion, hook the fish and lift it aboard. Small fishes may be lifted from the water by the hook and leader, though occasionally a poorly hooked fish will pull free.

Albacore tuna trollers, sometimes called jig boats because of the weighted feather lures they fish, operate far from shore. For that reason, fishermen need larger boats, usually vessels 45 to 70

*1 fathom = 6 feet or 1.8 m.

feet (13.7 to 21.3 m) in length. Since trolling methods for salmon and tuna are similar and some gear is interchangable, the larger salmon trollers commonly switch to albacore late in the summer in anticipation of higher earnings. The albacore fishery begins off California in early June and progresses northward toward British Columbia as offshore water temperatures rise above 58°F (14°C). A growing fleet of large high seas trollers follow the migrating albacore throughout the western Pacific.

Main trolling lines for albacore tend to be short and are not deployed from gurdies. Several mainlines are spread by tag lines along two midship poles and are then pulled by hand or by a power line-hauler. The demands on a single fisherman in retrieving several hundred 10- to 25-lb (4.5- to 11.4-kg) hyperactive fishes in often rough seas are considerable, so some vessels carry 2 or 3 crew members.

Traps (Pots)

Like hooks, fish and invertebrate traps have been used since recorded history. Although there is no universal rule as to which term to use, **pot** is commonly used for devices that target invertebrates such as crab and lobster, and **trap** is used when finfish are the target species.

Trap designs vary widely depending primarily on fishing conditions and species behavior. For example, octopus seek the protection and seclusion of a dark interior, so octopus pots may be deployed unbaited. On the other hand, crab, lobster and shrimp are attracted primarily by chemoreception to bait placed within the pot. Wood and chicken wire fish traps (Figure 15.5) used in the Caribbean have a tunnel of molded wire mesh that provides an entrance for the fishes. Extensive underwater observations show that escape efforts by small fishes attract larger predator species such as grouper, producing a live bait effect. Similar fish traps were used successfully in Florida until banned by state regulation.

Pots are set individually with a hauling line and surface marker buoy or may be set at intervals along a groundline. Occasionally, heavy boat traffic or thieves prompt watermen to set strings of pots without surface buoys that reveal gear location. Of course, recovering the pot string depends on the fisherman's explicit local knowledge, accurate bearings and a grappling hook. Although uncommon, timed release devices may be used to temporarily hold buoys beneath the surface until the fisherman returns.

Sablefish Trapping

During the 1970s, a significant trap fishery developed along the North Pacific coast for sablefish (also known as blackcod), a high-value, deep-water species. The traps, of several designs, are attached by short bridles at 25- to 50-fathom (45- to 90-m) intervals along a 3/4-inch (1.9-cm) polypropylene groundline. Bait, consisting of chopped herring or squid pieces, is placed in perforated plastic jars and hung within the traps, which are then set at depths of 200 to 500 fathoms (360 to 900 m), usually on steep slopes or along the heads of submarine canyons. Fishing, or soak time, is limited to 1 or 2 days as trapped sablefish held longer begin to die. Soon after the catch is dumped from the trap onto the deck, each fish is headed, eviscerated and packed in ice or frozen. Much of the sablefish catch is exported to Japan where the flesh, with its high oil content, is highly regarded.

Sablefish trap fishing is popular in British Columbia, Canada. However, competition and conflicts between longlines and the trap strings resulted in closing most of Alaska's coastal waters to sablefish trapping by the mid 1980s.

American Lobster

One of North America's largest nearshore fishing fleets deploys a variety of lobster pot styles from the U.S. mid-Atlantic states to the Canadian Maritime Provinces. The traditional wood lath

Figure 15.6. Traditional-style American lobster pots.

Figure 15.7. Popular Dungeness crab pot style.

pot (Figure 15.6) remains
popular, but both molded plastic and plastic-covered wire mesh pots have gained advocates because of their added durability and simple construction.

Although offshore lobstermen deploy larger lobster pots in long strings, most are set individually by a single fisherman in a 20- to 35-foot (6.1- to 10.7-m) boat and are hauled by leading the buoylines through an electrically or hydraulically powered line hauler. At the height of fishing activity, it may be difficult for vessels to navigate through the vast concentration of lobster pot buoys set in harbors, channels and along the relatively shallow coastal waters. Lobsters are most active and come within the influence of pots at night. Therefore, pots are hauled, rebaited and reset during daylight.

Crab Pot

Several crab species are captured in pots, including stone, blue, Dungeness, king and snow (Tanner) crabs. The stone crab fishery of Florida is unique in that the large pincer claws are removed and the animals, capable of reproducing the claws, are returned to the water. This intense fishery begins in the fall with more than two-hundred, 30- to 50-foot (9.1- to 15.2-m) boats setting a total of 400,000 pots. The annual catch of claws is near 4 million pounds (1.8 million kg).

Chesapeake Bay produces nearly one-half of the U.S. blue crab catch in a season that begins in mid-March. The 4-foot (1.2-m) square, mesh-covered pots have several tunnel openings and are set in shallow water. Both the hard- and soft-shell blue crab are fished with these pots, as well as by other methods.

Dungeness crab are fished in nearshore Pacific waters from California to the Alaska Peninsula. The circular pot style (Figure 15.7) is composed of a welded steel frame covered by hand-woven, monel wire mesh. All possible contact points between the frame and mesh metals are insulated with rubber wrappings to reduce galvanic corrosion. Two opposing tunnels are usually fitted with hinged vertical wires (triggers) across the tunnel gate that only allow passage into the pot. Chopped razor clams and herring pieces placed in perforated containers serve as bait.

Captured male crab, meeting the regional size requirement, are stored alive in circulating seawater tanks aboard the vessel. Legal sizes for crab are determined by the width of the carapace.

Figure 15.8. Adult king crab walk at speeds of 1 knot or more when pursued across the sea floor.

Some shell crushing is inevitable within the crowd of crabs because, unlike American lobster catches, crab pincers are not pegged or otherwise held closed to protect shell appearance and integrity. Only viable crab may be sold, since rapid deterioration after death makes them unfit for human consumption.

King crab (Figure 15.8) are fished primarily in the western Gulf of Alaska, in the Aleutian Islands, in the Bering Sea, and west to the Soviet Union and Japan. Alaska's resource was identified in the 1950s and resulted in an explosive fishery that peaked in 1980, when 190 million pounds (86.4 million kg) were taken. By 1987, the Alaska catch was only 27 million pounds (12.3 million kg) and even less in 1988. Factors contributing to the declining catch are overfishing, natural fluctuations, disease and, to some degree, damage to the crab stocks resulting from incidental catches by large bottom trawl fisheries.

In the early king crab fishery, vessels came from the collapsed California sardine fishery and were modified for holding live crab and large, bulky pots. The wooden boats ranged in size from 50 to 80 feet (15.2 to 24.4 m) and were generally inadequate for the demands of the poor weather encountered during Alaska winters. Later, a new generation of large steel vessels emerged, ranging from 85 feet (25.9 m) to over 130 feet (39.6 m) with large open decks to carry the pot inventory (Figure 15.9). In turn, as the crab fishery declined in the 1980s, some of the large modern, rough-sea crab vessels were modified for the joint-venture pollock trawl fishery.

King crab pots are large (from 6 by 6 feet [1.8 by 1.8 m] to 8 by 8 feet [2.4 by 2.4 m] by 30 inches [76 cm] high) and weigh up to 800 lbs (364 kg). The design allows numerous crab, which may reach weights over 20 lbs (9.1 kg) and a length of 5 feet (1.5 m) or more, to be held in one pot. Vessels can usually carry only a portion of their several hundred pot inventory; thus, more than one trip to the fishing grounds is necessary. Several stackable pot shapes are used, but many fishermen claim catch rates are less than those for the rectangular style. Each pot is set individually (inset, Figure 15.9) and marked at the surface with several large plastic floats attached to 5/8-inch (1.6 cm) or 3/4-inch (1.9 cm) synthetic rope lifting lines. Floats on each assembly must be marked with the vessel's identification symbols.

Chopped pieces of herring placed in 1 or 2 perforated plastic jars bait the pots which, at times, incidentally capture cod, halibut or other fish species. Large pieces of these fishes are likewise hung in the pot for bait. Pots are soaked 1 to 3 days depending on anticipated catch, weather and other factors.

King crab vessels use the same gear and methods to harvest Alaska snow crab; only the season and, generally, the depth vary. (The distributions among the several king and snow crab species overlap considerably.) A wooden slat placed across the pot tunnel gate (the rigid metal opening through which the crab pass into the pot) reduces the large king crab entrance dimensions

Figure 15.9. Modern king crab fishing vessel. Inset: Crab fishermen on a small nearshore vessel deploying a load of king crab pots.

to better retain the smaller snow crab. Most king and snow crab fisheries take place during the fall and winter in remote regions, making them perhaps the most difficult and dangerous of American fisheries.

Whelk

Channel and nobbed whelk, snail-like animals, are captured in small, wood lath pots set individually or in strings. Vessels in New England's Narragansett Bay and Vineyard Sound harvest over 1 million lbs (450 thousand kg) of whelk during the May to November fishery. Baited with horseshoe crab, each of the vessel's 100 to 600 pots may produce 5 to 20 lbs (2.3 to 9.1 kg) during a 1- to 3-day soak.

Pound Net (Fish Trap, Weir)

Far removed from the small fish traps set from a vessel in large numbers are the large, semi-permanent net configurations designed to entrap entire fish schools. Although large floating or pile-driven Alaska salmon traps were legislated into virtual extinction in 1959, juvenile herring, mackerel, squid and other species are taken by pound net, primarily along the Atlantic coast.

The structure is composed of a web or sometimes other types of fence (lead) material beginning at the shore and leading directly seaward 300 to 600 feet (91 to 183 m) or more to a large web enclosure. Fishes migrating nearshore swim offshore in an attempt to bypass the obstruction. Poles driven into the sea bed support a web corral of several chambers. Where the water is deep, both the lead and the trap are supported by anchored logs. Fishes are guided by web walls through the wings and heart to a **spiller** (pot) to await the drying up and brailing process in the spiller section. (Brailing involves using a large dip net attached to a crane; the net is lowered into the spiller and, when filled, is lifted onto the vessel where the net can be quickly opened and the catch spilled into the hold.) Large fish traps can be extremely efficient and are precisely regulated to manage the resources that they impact.

Along the Atlantic coast from the Maritime Provinces south to Pamlico Sound, North Carolina, as well as in the Great Lakes, more than 400 pound nets intercept migrating fish schools. The Maine traps use leads of driven stakes, with brush and twine out to a depth of about 30 feet (9.1 m). Trapped sardines are held overnight until their digestive tracts have emptied, thereby enhancing quality. The few Rhode Island floating traps take a variety of spring migrating species, as do pile-driven traps in Chesapeake Bay and Pamlico Sound.

Seines

A seine is designed to be set around a fish school and form a webbing wall between the opposing forces of a weighted bottom line (**leadline**) and floats at the top line (**corkline**). The net may be set either blind (that is, on an assumption fishes are within the encircling net) or when fishes are sighted near the water's surface. In some seine fisheries, pilots of spotter aircraft will locate fish schools for cooperating fishing vessels.

Beach Seines

Beach seines fished in relatively shallow water use the sea bed to bar escape beneath the leadline while one or both net ends are pulled back to shore. For large nets, an open skiff may be used to carry the stacked seine directly away from shore as an attached haul line is paid out to the shore crew. The net is set parallel to the shore, and then the second haul line is returned to the beach. Each net end is simultaneously pulled ashore and stacked by the fishing crew to concentrate the catch for sorting. Species taken by beach seines and the similar **long haul seine** of Long Island Sound include striped bass, bluefish, mackerel, weakfish, butterfish, porgies, blackfish and freshwater carp. Small beach seines operated entirely by human power are used in some artisanal fisheries.

Purse Seines

Purse seines are usually set around fish schools in waters too deep to allow the net wall to reach the sea bed. In order to prevent the fish school from descending beneath the web, the weighted net bottom is pulled together (pursed), creating a net floor, thus barring escape.

Purse seines produce the greatest volume of fishes in the United States, primarily because of the extraordinary Atlantic and Gulf of Mexico menhaden fishery, which accounts for 40% of the entire U.S. catch. A menhaden seine is distributed between two large open boats and set as the boats draw away to encircle the school.

Fishing techniques and net configuration for single boat purse seines vary in detail, depending on the target species, regulations and ocean conditions. Distant water tuna seiners must be large enough, sometimes in excess of 200 feet (61 m) long, to safely manage the heavy 600 fathom (1,080 m) long net, carry catches sometimes in excess of 1,500 tons and endure the many storms that may find a vessel on the high seas.

In all seine fisheries, the captain must predict the school's probable course and speed in order to place the net across its path and close the net before the fishes divert around the ends or descend beneath the unpursed web wall. A skiff or sea anchor holds the net lead end (Figure 15.10) as the

Figure 15.10. Seine skiff slides down stern of a tuna seiner to begin pulling the net from the deck.

vessel deploys the net in a circle around the target school and returns to the lead end. When the circle circumference is greater than the net length, rope or cable is paid out until both ends come together. **Warp** threaded through large rings along the leadline is pulled back to draw the entire leadline aboard ship, thereby closing all avenues of escape. Because the vessel tends to be pulled into the net circle during haulback, the skiff, having returned its net end to the vessel, may tow the vessel away from the seine while a hydraulic power block high on the boom or an on-deck net reel retrieves the web to concentrate the catch (Figure 15.11). Salmon seiners often hold the net open with both vessel and skiff towing slowly toward oncoming migrating fishes (Figure 15.12) in order to take a larger catch. Timing must be precise so that fishes will not bypass the net.

Figure 15.11. Seine skiff prepares to pull vessel from its side before the net is retrieved. Note second skiff near the shore holds a seine end for another vessel.

Tuna net sets begin only after crew members, searching the horizon, discover evidence of fish. The school may **breeze**—that is, cause a visible water disturbance as fish breach—or move just below the surface. Birds may hover above the fish school, attracted to small food species disturbed by the passing school. Also, some tuna species aggregate near floating objects or animals such as porpoise, whales and turtles.

Figure 15.12. Holding open a salmon seine to intercept fish schools migrating near the shore.

When encompassed by a net, whales strike the web with enough force to tear holes and pass through. Usually, the adjacent tuna fail to find the same escape route. On the other hand, encircled porpoise (dolphin) cannot escape without crew assistance. U.S. tuna vessels are required to employ a process called **backdown** and a modified net to promote porpoise escape. When the tuna net is nearly retrieved with both fish and porpoise concentrated into a narrow web channel, the vessel pulls the corkline underwater in timed sequence (Figure 15.13). The objective is to safely spill out the mammals while retaining the rapidly moving tuna school (Figure 15.14). If the fish reach the submerged corks, they will also swim out of the net. Retained high-value species like tuna and salmon are brailed into the hold, while menhaden and herring may be pumped directly onto high capacity transport carriers.

Fishing regulations influence vessels, gear and fishing methods. Alaska limits seine vessel length to no more than 58 feet (17.7 m) and prohibits using the efficient drum seine technique. Washington State and the Province of British Columbia allow the drum seine, but Canadian seine skiffs may not have engines. Most fishery regulatory agencies prescribe allowable fishing areas, times, duration, net length, web size, species, etc.

Figure 15.13. Porpoise mill about within a tuna seine as vessel pulls corkline underwater to facilitate their escape.

Trawl

Trawls are large, cone-shaped nets similar to an airport wind sock that are pulled (towed) through the water by 1 or 2 vessels. Slow-moving fishes and invertebrates are engulfed and directed to the net's terminal end. At intervals up to several hours, the entire net is recovered to dump the confined species for sorting and processing. Catches may be small, such as a few pounds of high value prawn-size shrimp, or enormous, as in a 50-ton or more haul of Pacific whiting or Alaska pollock.

Otterboards (doors), rigged forward of the net, tend to spread the net wings horizontally while opposing forces of large, pressure-resistant metal or plastic floats on the headrope and chain secured to the footrope spread the trawl mouth vertically. Nets with large openings are needed for vertically distributed fish schools and species that tend to escape up into the water column, as do herring and numerous other roundfish species. On the other hand, flounder nets strive for maximum horizontal spread, with the headrope overhanging the footrope. The net may be fished in close contact with the smooth sea floor, held off rocks and other damaging substrate with rolling devices along the footrope, or pulled at given distances above the sea bed using special rigging configurations and electronic sensing devices.

Figure 15.14. Yellowfin tuna and a porpoise mill around within the seine as the backdown process begins.

Net size primarily depends on the power of the vessel to pull and control the net as well as handle the necessary tow cable and otterboards. On large trawlers, the entire net and catch is commonly retrieved up a sloping stern ramp (Figure 15.15) onto the main deck for sorting or transfer to a processing area below decks. Vessels less than about 80 feet (24.4 m) usually retrieve the net and its bridles onto a reel mounted on deck. The codend, a heavy web terminal bag, is routed to the vessel side, where the catch is brought aboard in a series of lifts weighing up to about 5,000 lbs (2,270 kg) each.

Figure 15.15. Soviet factory trawler retrieves a net up its stern ramp.

Most shrimp trawlers operating throughout the Gulf of Mexico and other coastal waters fish two smaller nets simultaneously. Each net is deployed from a boom extended outward from the port and starboard sides. Both the doors and net body are retrieved to the boom so that the codend can be lifted onto the deck.

Trawl designs may incorporate special features to exclude unwanted species. Shrimp trawls fished in the United States, where turtles may be encountered, may include a **turtle exclusion device** (TED) that diverts turtles safely out through the webbing. Several nations have developed

Figure 15.16. Near total absence of small fishes mixed with this catch of pink shrimp attests to the effectiveness of an experimental sorting trawl.

specialty nets to divert finfish from shrimp catches, thereby greatly reducing the hand sorting process before the shrimp are placed in automatic peeling machines (Figure 15.16).

During the 1960s, foreign high seas catcher-processor vessels developed large volume trawl fisheries along the U.S. and Canadian east and west coasts. At the time, there was little U.S. demand for certain so-called industrial grade flounder, whiting and pollock. However, passage of the Magnuson Fishery Conservation and Management Act of 1976 established a means for U.S. and Canadian trawlers to catch those species for direct at-sea sale to foreign processor vessels. Many large vessels caught in the decline of king crab fishing were able to modify their fishing systems and enter the rapidly expanding Alaska joint-venture trawl fishery. Other U.S. vessels such as the 334-foot (102-m) F/V *Arctic Storm* were converted to factory trawlers that fish and process the catch into **surimi**, a minced fish product that is the base for numerous consumer-oriented fish products. In 1986, the F/V *Artic Storm* converted 100 million lbs (45.5 million kg) of whole fishes into 20 million lbs (9.1 million kg) of surimi.

Gill Net

Although widely modified for various freshwater and marine fisheries, gill nets are generally designed to intercept and entangle passing fishes. Gill nets are often set in water of low visibility or during the night to be less visible. The net forms a wall of loose, fine, low-visibility meshes. Selected mesh dimensions allow only the head and perhaps forward portion of the body to pass through. Web having a depth appropriate to the fishery is hung from corkline to leadline at a ratio of about two lengths of web to one length of line.

Gill nets are set at the surface, suspended just below the surface to discourage accumulation of floating debris, or up from the bottom, depending on the behavior of the targeted species. Surface nets for Florida king mackerel and most Pacific salmon drift in river or ocean currents with one end tied to the fishing vessel; for shad and, in some cases, salmon river fisheries, nets are secured onshore or to anchors. Sunken nets have anchored leadlines with pressure-resistant floats that lift the web up into the water to catch California halibut, cod, haddock and dogfish.

Perhaps the most intense and concentrated U.S. gillnet fishery takes place each summer for several weeks near the mouths of rivers entering Bristol Bay, Alaska. As sockeye salmon concentrate prior to their migrations into nearby rivers where they will spawn, hundreds of nets literally block river passage during highly regulated fishing periods. Nets are set end-to-end and side-by-side, sometimes only a few feet apart.

In recent years, restrictive legislation, especially in the Great Lakes and California, has reduced gillnet use. Although the catch, species and size can, to some extent, be selected by net length, depth, mesh size, area and time of fishing, many non-target species, including marine mammals, may be captured. Additionally, when lost on the fishing grounds, a net may continue to entangle and kill a variety of animals for several years. This phenomenon is known as **ghost fishing** and is also a problem with lost pots. A single salmon gill net that was lost in Boundary Bay, Washington, was retrieved a week later with nearly 1,000 trapped Dungeness crab.

Some New England fishermen deploy gill nets in a manner somewhat similar to a purse seine by encircling bluefish schools. As the fish push into the web during escape efforts, they become entangled.

Trammel nets, set near the sea bed for California halibut and, in the 1950s, for Alaska king crab, are a form of entangling gill net. The barrier wall is formed by three layers of web, each of a different mesh size. Fishes or crab became trapped between the layers or in web pockets formed as the animals push the small mesh through larger meshes. Crab are difficult to remove and the gear causes injury to protected females. Also, the gear poorly discriminates between species and sizes captured.

Diver-operated Fisheries

The earliest underwater harvesters of marine animals breathed from air hoses and wore bulky, hard-hat diver dress to gather sponges in Florida and a few abalone along the rocky California coastline. Today, **SCUBA** (Self-Contained Underwater Breathing Apparatus) as well as surface-supplied divers participate in a variety of North American fisheries. The best established, with a peak production in the 1960s of 5 million lbs (2.3 million kg), was the red and pink abalone harvest. Extensive overfishing and a drop in catch by the 1980s to only about 1 million lbs (0.5 million kg), primarily of the less desirable black abalone, prompted the California Department of Fish and Game to limit diver-held licenses. A limited abalone fishery developed in southern Alaska and British Columbia.

During the 1970s, divers began harvesting large sea urchin species, primarily along the Pacific Coast, but also to some extent in upper New England. Multiple roe sacs were removed from the urchin shell, processed to meet rigid consumer demands and shipped fresh by air to specialty markets in Japan. These historically unwanted species were soon overfished in readily accessible areas, thus requiring stringent conservation measures. The resource has now become more stabilized, making the fishery viable once again.

The geoduck (pronounced ''gooeyduck''), a large clam, is harvested by divers in Washington State, British Columbia and Alaska, 1 to 2 feet (0.3 to 0.6 m) deep in the substrate and at water depths greater than those of other commercial clam species. As conventional harvest systems could not reach the well-buried, 1- to 4-lb (0.5- to 1.8-kg) bivalves, divers created their own fishery by developing a hand-held, high-pressure water jet that was used to flush out individual clams. Washington State divers annually produce about 3 million lbs for fresh sale, mostly to restaurants.

Numerous small, regional fisheries are pursued by divers. Up to 400 divers in the mid-Atlantic states harvest oysters, while 100 or more divers in Maine gather scallops. Herring roe deposited on kelp fronds during spawning are valued highly in Japan. Consequently, the egg-laden kelp is the object of a small but intensive Alaska fishery. The harvest is rigidly controlled, and competition is fierce among the divers, who usually have only a few hours to gather the allowable catch. Other species taken by divers include clams, American and spiny lobsters, mussels, seaweeds and sea cucumbers. Most states prohibit commercial spearing of fishes, though at one time it was relatively popular among Florida skin divers.

Shellfish Harvesters

All coastal waters and most estuaries at one time supported commercial harvests of wild or reared shellfish. Pollution has led to shutting down some traditionally harvested regions. Gathering methods vary widely, from hand picking at low tide to hydraulic dredging with escalator belts that mechanically lift the dislodged shells to the surface.

Oysters and some clam species lie on or just below the surface of the sea bed. Within a protected intertidal zone, the molluscs may be gathered by hand, rake, or shovel during low tide. Filled bags or baskets can be left until the rising tide permits watermen on a shallow-draft barge to retrieve the catch. A novel harvest method employed by over 200 North Carolina fishermen is called clam kicking. In water 10 feet (3 m) or less deep, small boats are equipped with deep propeller shafts that cause the propeller wash to dislodge clams from the sediment so they may be picked up in a small dredge pulled closely behind the boat.

Chesapeake Bay supports an extensive oyster fishery composed of 1,000 or more small boats operated by watermen working small dredges or patent and shaft tongs. The long-handled scissor tongs are repeatedly lowered to the bottom and shellfish are picked up as the two rake-like halves are drawn together. Traditional, sail-powered craft called skipjacks continue to operate in the fishery, especially during harvest periods and in areas set aside specifically for sail power. The Gulf of Mexico states of Louisiana and Texas are major oyster producers, with about 1,500 small dredge boats. Much of the catch is taken from privately owned shallow sand bars and mud banks.

Dredge designs are selected to serve the species sought, boat size, water depth and bottom type. A dredge incorporates a heavy, steel, sled-like frame and is towed slowly across the sea bed. Affixed to the leading edge are short steel rods or water jets, which dig into the substrate to dislodge shellfish. The water current propels shellfish back into a heavy web or steel mesh bag; however, for some fisheries the bag is replaced by a continuous escalator belt that carries the catch directly to the surface.

Throughout the mid-Atlantic and New England states, hydraulic-assisted clam dredges are operated. Water is pumped at high pressure from the vessel through multiple nozzles into the substrate. The 500 to 1,000 gallon (1,900 to 3,800 l) per minute water stream breaks open the sediment and lifts clams into the path of the rigid dredge frame. Up to 200 bushels (7,040 l) of surf clams per day may be gathered in a series of 5- to 10-minute tows by a dredge. Large volumes of dead shell at times may also be collected, requiring considerable sorting for live clams.

Scallop drags are generally lightweight dredges. Scallops usually lie atop the sea floor in depressions made in a mud substrate or on rocks and other coarse material. Small, sturdy trawls are employed to gather the small but abundant calico scallop along the U.S. coast from the Carolinas to Florida. Several tickler chains strung across the net mouth just ahead of the trawl footrope dislodge the scallops or cause the bivalves to swim up into the water column, where they will be enveloped by the oncoming trawl.

Alaska halibut longline fishermen were for many years aware of regions where large scallops inhabited the sea bed. While the fisherman's groundline was moving across the bottom during gear retrieval, it often passed into the open shell of a feeding weathervane scallop, causing the opposing shells to close on the line. Many scallops reached the vessel deck still tightly closed around the fishing line. In 1968, several New Bedford scallop dredge vessels began commercial harvest of the Alaska species. The limited resource was quickly overfished, and within 3 years most vessels returned to the Atlantic or entered other fisheries.

The Pacific coast oyster fishery is primarily one of privately owned beds carefully managed and harvested on a farm-like basis. Production is balanced with demand and has increasingly been the result of hatchery-reared larvae or spat.

Summary

Few living aquatic resources are not utilized. For many, overfishing continues in spite of extensive management plans. Some fisheries have recovered from disastrous impacts by man and, with careful management, continue to provide a bountiful catch. Pollution not only affects the viability of commercial species, but also can make them unfit for human consumption. Innovative fishermen continuously improve their harvest efficiencies. Therefore, further regulation of gear, seasons, species, geographic regions, participants and allowable catch is inevitable.

16

Aquaculture

Robert R. Stickney

Introduction

Aquaculture is the rearing of aquatic organisms under controlled conditions. More simply, aquaculture is underwater agriculture. The concept is not a new one—fish culture began in China perhaps as much as 4,000 years ago. The Egyptian tombs have pictographs indicating that certain types of fishes were being reared during the days of the Pharoahs, and oyster culture was being practiced under the Roman Empire.

Most people think of aquaculture as the production of aquatic animals for human consumption, and many of the aquaculture efforts around the world are being conducted for that purpose. There are, however, other purposes for which aquatic organisms are grown. Examples include producing minnows for bait, rearing tropical fishes and goldfish for the aquarium trade and producing ornamental aquatic plants (e.g., water lilies). In addition, the often large-scale hatchery programs that exist in the various states and provinces of North America produce fishes for release into streams, lakes, reservoirs and the marine environment to enhance commercial and recreational fisheries as well as to repopulate water bodies with endangered and threatened species.

Aquatic plants are also produced for human consumption. In the Orient, for example, seaweed production (e.g., red and brown algae, division Rhodophyta and Phaeophyta, respectively) involves the labor of several hundreds of thousands of people. The seaweeds may be consumed directly by man (e.g., *Porphyra* spp., division Rhodophyta), or extracts may be obtained that become components of a variety of substances each of us uses every day. Ice cream, toothpaste, cosmetics and a wide range of other household items contain extracts from seaweeds. Some of those businesses are very large, and they are legitimate aquaculture enterprises. In this chapter, however, our discussion centers on the production of aquatic animals for human food.

Under natural conditions, as much as 100 pounds per acre (kg/ha) of fish might be produced in a lake within a year. Aquaculture systems, by contrast, can produce several thousand to even a million pounds per acre (kg/ha) in a year. The difference relates to the definition of aquaculture presented above. The aquaculturist exerts control over the species being reared. That control may include, but is not limited to the following:

1. *Design, construction and maintenance of the culture system being employed.* Aquaculturists utilize ponds, cages, net-pens, raceways, tanks and other units. They do not attempt to convert unaltered natural water bodies into culture systems.
2. *Maintenance of suitable water quality.* In order for an aquaculture venture to be successful, the water must be of a quality suitable for the species being reared. Of most importance are the levels of dissolved oxygen and ammonia, water temperature and, in the case of marine species, salinity. Each of these and other water quality variables may or may not be controlled, depending on the nature of the culture system.
3. *Control over reproduction.* Unless the culture species can be reproduced in captivity, there is no way to undertake genetic selection and the improvement of the species with

respect to its suitability for culture. Successful aquaculture of any species ultimately depends on captive breeding and producing broodstock from animals that are hatched in captivity.

4. *Provision of nutritionally complete feeds.* Species being reared by aquaculturists are usually fed prepared feeds, similar to the feeds used by livestock producers. Such feeds contain the nutrients necessary to meet the daily requirements of the species under culture. In some instances, natural foods are relied upon (e.g., oyster and mussel culture), but for most species, manufactured diets are employed.

5. *Control of diseases.* Aquatic organisms, like other plants and animals, are susceptible to diseases of various kinds. The aquaculturist must be able to recognize the diseases that affect the particular species being cultured and know how to treat them properly.

The underlying concept for commercial aquaculture is that the species being produced must have sufficient economic value that the aquaculturist can realize a profit once the plant or animal is ready for market.

Feeding the World through Aquaculture

Commercial aquaculturists throughout the world are in business to make a profit. When that profit is associated with providing food for other people, the commercial aquaculturist is not generally interested in feeding the hungry, but in providing the best quality product possible and selling it for the highest possible price. In some places, **subsistence culture**, which involves the rearing of a modest amount of aquacultured product for consumption by a family, group of families, or a village involved in the enterprise, is being practiced. One or a few small ponds can often produce enough fish, for example, to meet the local animal protein demand, perhaps even with enough left over to sell profitably. Subsistence aquaculture, as is true of all aquaculture, depends on the availability of water in suitable quantity and quality. Water of that nature is often not available, so the aquaculture option may not exist.

Commercial aquaculture is not, in general, a means of providing inexpensive animal protein to the masses. Aquaculture can create jobs, enhance the overall economy of a region and thereby help to improve the plight of underprivileged peoples. It is not a panacea, however, and should not be looked upon as the means to prosperity for large numbers of those who live in the developing nations of the world.

The Big Two in Fish Culture

The most widely cultured groups of fishes in the world are the carps (family Cyprinidae). While we in the United States are most familiar with the common or European species (*Cyprinus carpio*), carp culture is dominated by production of a variety of Chinese carps, including the common, bighead (*Aristichthys nobilis*), silver (*Hypophthalmichthys molitrix*) and grass carp (*Ctenopharyngodon idella*). Carp culture is widely practiced in the People's Republic of China, and with over one billion people in residence, it is little wonder that carp dominate world fish culture. Add to that some additional species of Indian carps (cultured on the subcontinent of India and represented by fishes in the genera *Catla*, *Labeo* and *Cirrhina*), and the dominance of the carp family in aquaculture becomes even more understandable. The common carp that we have in the United States was introduced from Europe during the 19th century. There is virtually no aquaculture of that species in North America, but it continues to be cultured in parts of Europe, in China and elsewhere (Figure 16.1).

The Chinese have developed carp culture to a fine art. They use a system known as **polyculture** (two or more noncompeting species are reared in the same water system) in which at least four

Figure 16.1. One form of the common carp is a strain in which most of the scales have been removed through selective breeding.

Figure 16.2. Various species of tilapia are reared in tropical regions around the world. The blue tilapia, *Oreochromis aureus* (shown here), is among the most popular for aquaculture.

species of carp are grown in the same pond. Ponds in China are often fertilized with organic fertilizers (night soil [human sewage] and livestock manure), which produce plant and animal food for the fish. Agricultural wastes may also be used. Common carp feed on benthos, silver carp on phytoplankton, bighead carp on zooplankton and grass carp on rooted aquatic vegetation. Thus, various food supplies are used by the various culture species, allowing several niches to be filled and greater production to be realized than if only one species were used. Stocking rates are related to the food supplies. In recent years, prepared feeds have become more common in China, though pelleted diets may still be used in combination with fertilization.

Indian carps (various species) and common carp are more commonly reared in **monoculture** (only one species present in the culture system). Depending on expected production, the ponds may be fertilized, or prepared feeds may be offered. In Europe and Israel, common carp are maintained at high densities and are fed pelleted rations that meet their nutritional requirements.

Carp will spawn naturally in ponds, though hatcheries are often maintained. Hormones may be injected into the adults to induce spawning. Eggs and milt may be obtained by manual stripping. The eggs are maintained in a hatchery and the young fish are stocked into nursery ponds. The system is relatively simple. Carp are able to tolerate fairly wide ranges in environmental conditions, so the technology required for their culture is not highly sophisticated.

The second most widely cultured group of fishes in the world today are the tilapias, which are members of the family Cichlidae (Figure 16.2). Over the past several years, the taxonomists have recommended that many of the foodfish species of tilapia be changed from the genus *Tilapia* to *Sarotherodon* and, subsequently, *Oreochromis*. All three generic terms appear in the scientific literature, which can lead to confusion.

Tilapias are native to the Middle East and north Africa, but have been introduced throughout the tropical world and into many subtropical and temperate areas. Most species die when water temperatures fall below about 50°F (10°C). Therefore, culture in temperate climates depends on production of a crop during the warm months and maintenance of broodstock in warm water (often in indoor heated holding facilities) during the winter.

Various species of tilapia are under culture around the world today, primarily in the tropics. All of the popular ones are known for their rapid growth, ease of production and heartiness. Tilapias are extremely tolerant of poor water quality, reproduce readily in almost any environment and reach market size within several months. Most species feed on a combination of plants and animals, and they do not require high-cost prepared feeds unless they are being reared at high densities where natural food supplies become exhausted.

Tilapias are popular in subsistence culture in much of Africa, the Far East and Latin America. Commercial production is highly developed in Israel, Indonesia, the Philippines, Thailand, Taiwan, Jamaica and various other nations.

While not miracle fish, tilapias are important to aquaculturists. Their production does not come without problems, however. Some species begin to reproduce as early as 3 months of age. Also, a female may breed once a month upon reaching maturity. Mature females are typically of submarketable size, and once they begin to produce eggs, their growth is drastically reduced. Thus, part of the population in a pond does not reach market size within a reasonable time. This is compounded by the fact that the offspring compete for available food. When the food supply is limited, growth of the population is negatively impacted and the result is a pond with large numbers of very small fish.

Various schemes have been developed to overcome the overpopulation problem. Many of those schemes are not particularly useful to the subsistence aquaculturist because they involve a level of technology not generally available to that group of culturists.

Overpopulation with small fish can be reduced or eliminated by controlling reproduction. One way of doing that is to stock predators that can eat the young fish but cannot consume the parents. This has been successfully accomplished in various countries. More sophisticated methodology involves stocking only males, which can be accomplished by hand-sorting adult fish, or by feeding male hormones to young fish and converting the females to males. Another method of producing all-male populations involves hybridizing certain species.

Leading Aquaculture Species in the United States

Finfish aquaculture in the United States is dominated by channel catfish (*Ictalurus punctatus*) culture in the south and trout production (primarily rainbow trout, *Oncorhynchus mykiss* [formerly *Salmo gairdneri*]) in the north. Aquaculturists are also raising salmon (Atlantic salmon, *Salmo salar* and coho salmon, *Oncorhynchus kisutch* are the primary species involved) in marine net-pens, and a few other fishes like red drum (*Sciaenops ocellatus*), buffalo fish (*Ictiobus* spp.), grass carp (*Ctenopharyngodon idella*) and sturgeon (*Acipenser* spp.) are also being cultured. Marine invertebrate culture involves the production of oysters (American oyster, *Crassostrea virginica* and Japanese oyster, *C. gigas*), mussels (blue mussel, *Mytilus edulis*), clams (southern quahog, *Mercenaria campechiensis*, northern quahog, *M. mercenaria* and others), shrimp (marine shrimp in the genus *Penaeus* and freshwater shrimp, *Macrobrachium rosenbergii*) and crayfish (e.g., *Procambarus clarkii* and *P. acutus acutus*).

Channel catfish culture is centered in Mississippi, where ideal conditions exist for the construction of ponds and where large volumes of ground water are available. Since the early 1960s, when commercial catfish culture began to develop in the United States, annual pond production has gone from an average of about 1,000 pounds per acre (kg/ha) to in excess of 4,000 pounds per acre (kg/ha) on well-managed farms. Good quality control has produced a widely accepted product that has been finding increasing favor among the consuming public outside of the south, where catfish has been a traditional favorite.

Catfish farming is also practiced in various other states, though the farther north one goes, the shorter the growing season and consequently, there are few catfish producers in the north. Besides Mississippi, some of the key catfish producing states are Arkansas, Louisiana and California. There is also production in the remaining southern states and a few others. One producer in Idaho has been growing catfish for several years by using geothermal wells that provide water of ideal temperature year round (Figure 16.3). The catfish industry is presently valued at over $400 million annually.

Figure 16.3. Channel catfish are produced in large numbers in concrete raceways in Idaho, where warm groundwater supports fish growth throughout the year.

Figure 16.4. Concrete raceways such as those shown here in the Snake River Canyon of Idaho are used for the commercial production of rainbow trout.

Over 90% of the trout produced for human consumption in the United States are grown in the immediate vicinity of Twin Falls, Idaho. In the Thousand Springs region of the Snake River Canyon near Twin Falls, major underground rivers emerge from the canyon walls and enter the Snake River. The water from those underground rivers is of excellent quality and nearly ideal temperature for trout production. Trout farmers divert the water through raceways (Figure 16.4) before allowing it to enter the Snake River. The trout industry is somewhat smaller than the catfish industry at the present time and does not appear to be growing appreciably, though there is significant potential for growth if new markets for the product can be developed.

Atlantic and coho salmon have been produced in net-pens in Puget Sound, Washington for a number of years (Figure 16.5), and a complementary Atlantic salmon industry is rapidly developing in the northeastern United States. Salmon net-pen culture has developed into a major industry in Norway, and Norwegian fishes are being air freighted to United States markets where they command a high price. Spurred by that success, other nations have been actively developing salmon culture industries, most notably Scotland, Chile and Canada. New Zealand and other countries involved in salmon culture also see the United States market as one target for their product.

Oyster and clam farming is largely practiced on leased intertidal beds (Figure 16.6), though some raft culture is practiced. Larval oysters, called **spat**, are either produced in hatcheries or collected in the wild on substrates to which the spat attach when they leave the zooplankton and become benthic. In hatcheries, spat are allowed to settle on dead oyster shell, called **cultch**, which is then placed on the sea bed (Figure 16.7). United States oyster farming is perhaps best developed in the state of Washington, which currently ships a large percentage of its production to the east coast and Gulf of Mexico regions, where disease and pollution problems have severely limited the availability of local oysters. Most of the oyster production in Washington is based upon the Japanese oyster, which is not native to the Pacific Northwest but was brought to Washington, Oregon and California from Japan many years ago.

Freshwater shrimp culture received a great deal of attention in the United States during the 1970s, but it has never developed to the extent once thought possible. On the basis of culture of the so-called Giant Malaysian prawn, *Macrobrachium rosenbergii* (Figure 16.8), freshwater shrimp culture was perhaps most successful in Hawaii, though a good deal of research and development took place in South Carolina, Texas and a few other states. Freshwater shrimp are relatively easy to produce in hatcheries, after which the postlarvae are stocked into culture ponds. In tropical climates, aquaculturists can potentially produce two crops per year.

Figure 16.5. Net-pens in the marine environment are used in several countries for the production of salmon. The net-pen complex shown here was located in Puget Sound, Washington and covered about 5 acres (2 ha). (Courtesy of Ronald Hardy.)

However, there are some major problems associated with freshwater shrimp culture. For example, freshwater shrimp are cannibalistic. When a shrimp **molts** (sheds its exoskeleton to provide an opportunity for growth), it is vulnerable to attack by other shrimp for a period of several hours until the new exoskeleton hardens. Cannibalism can cause significant losses. Other problems involve short shelf-life of the product after harvest, and difficulties in marketing have caused many shrimp culturists to move away from freshwater shrimp and toward marine species.

United States species of marine shrimp such as the white (*Penaeus setiferus*), pink (*P. duorarum*) and brown (*P. aztecus*) shrimp were subjected to intensive study during the 1960s and 1970s, but problems with their culture were never completely overcome. In the meantime, culturists in the Far East and Latin America began having better success with various other species. The Japanese developed the means to commercially culture Kuruma shrimp, *P. japonicus*, and culture of the tiger shrimp, *P. monodon*, was developed in Taiwan and elsewhere. *Penaeus stylirostris* and *P. vannamei* culture developed at an almost subsistence level in Latin America, where those species are native. In Ecuador, where shrimp culture has grown into a huge enterprise, postlarval shrimp have been historically stocked when ponds were filled by the rising tide. Once full, gates were closed to keep water in the ponds. The shrimp were not fed or managed in any way but were harvested after a few months. Typical yields were a few hundred kg/ha, but there was little investment except for the ponds. As the industry developed, feeding was introduced and the level of management increased. Hatcheries also went into production, and ponds were constructed at elevations

Figure 16.6. Oysters are commonly cultured on intertidal or subtidal beds such as the one shown here in Puget Sound, Washington. (Courtesy of Ken Chew.)

Figure 16.7. Strings of shell in a larval settling tank prior to the introduction of oyster spat-laden water. After the spat settle, the shells will be planted on oyster beds. (Courtesy of Ken Chew.)

where pumping was necessary to fill them. Today, Ecuador leads the world in shrimp production. Other Latin American countries such as Panama, Costa Rica and, increasingly, Brazil, are also producing large quantities of shrimp.

In the United States, Hawaii, Texas and a few other states have developed commercially successful marine shrimp culture. While high costs for land, labor and energy have undoubtedly impeded the development of shrimp farming in the United States, a few successes have been achieved, and we anticipate continued interest and consequent expansion of the domestic industry.

Culture Systems

Ponds

The primary culture system used in the world today is the earthen pond. A typical pond is about 3 feet (1 m) deep at the upper end and 6 feet (2 m) deep at the drain. The bottom and sides are made from compacted earth. Sides slope at a ratio of 2:1 or 3:1 (Figure 16.9). Steeper slopes make entry and exit from the pond difficult and promote erosion of the banks, while shallower slopes promote the establishment of aquatic weeds. Pond size varies greatly—some are only a few square yards (square meters) in area, but most are an acre (0.4 ha) or larger. Management and harvesting become difficult if a pond is too large, so most culture ponds do not exceed about 20 acres (8 ha).

Figure 16.8. The freshwater shrimp, *Macrobrachium rosenbergii*.

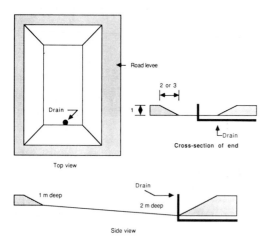

Figure 16.9. Diagrammatic views of a typical earthen pond with 2:1 side slopes, the drain at the deep end and a road levee. (Diagrams in Chapter 16 courtesy of the author.)

A well-designed pond should have a drain that allows the water to be completely removed within a day or two and should be provided with an inflow pipe of sufficient size that the pond can be filled within a reasonable period of time (a few days at most). At least one of the levees adjacent to each pond should be sufficiently wide that it will allow vehicular traffic for feeding and harvest (Figure 16.9).

Ponds can be stocked at various densities. If water is exchanged continuously or frequently, higher densities of culture animals can be maintained than when static conditions are employed. Catfish farmers in Mississippi are annually producing over 4,000 pounds per acre (kg/ha) in ponds, whereas a few years ago maximum yields were less than 3,000 pounds per acre (kg/ha). Improved feeds and feeding techniques, careful water quality management and other factors have led to the steady increase in levels of maximum standing crops, but further increases in most earthen pond situations are unlikely unless oxygen is added or water is continuously exchanged.

Raceways

Raceways are linear channels or circular tanks (Figures 16.4 and 16.10, respectively) through which water continuously flows at a rate that will provide a minimum of several exchanges each day. Raceways are commonly used in hatcheries for the rearing of young animals and are employed by the trout industry for production from fry to harvest size. Densities of up to a million pounds per acre (kg/ha) can be maintained in a raceway, so long as sufficient flow can be provided to maintain water quality. In a pond, fish are exposed continuously to the same water during the growing period (with some additions to replace evaporative and seepage losses), while in raceways, the water may be exchanged completely every several minutes to every few hours.

Cages and Net-Pens

Culture cages and net-pens (Figure 16.5) are structures placed in a natural environment and stocked with fish for growout. The only real difference between the two is size. Cages tend to be relatively small (a typical cage has a volume of 1 or 2 m^2), while net-pens are often several meters on a side and 10 to 20 meters deep. Cage culture has been practiced to a limited extent by catfish farmers. For example, in Arkansas (where leasing of state lakes for cage culture operations is possible), catfish have been commercially produced in cages (Figure 16.11). Cages have also been used by researchers. Most cage culture is conducted in freshwater environments, while net-pens are most commonly used in the marine environment. Net-pen culture has been largely restricted to protected waters, but recently, net-pen engineering has advanced to the stage that open sea pens are now available that can withstand most storms without damage.

Net-pen salmon farming is well developed or under development in Norway, Scotland, Canada, the United States, Chile, Japan and elsewhere. Objections of upland landowners, commercial fishermen and others to net-pen salmon operations have limited development of the industry in the United States, though several farms are in operation. Some people are concerned about visual impacts, the use of antibiotics in feeds, environmental impacts from waste feed and feces and transmission of diseases from net-penned to wild salmon, among others. While extremely high densities of net-pens within a given area can cause certain types of problems, scientific evidence and the experience of various countries have demonstrated that by placing net-pens where flushing by currents and tides is good, and establishing proper pen and animal densities, we can conduct environmentally sound net-pen culture. The visual impact question can be resolved by comparing the sight of a net-pen with that of a pulp mill, factory, or even a marina. Net-pens protrude above the water line by little more than a meter. Most farms have small numbers of net-pens (10 or less), so the surface area impacted is generally small. Net-pens are generally strict in terms of maintenance of navigation, water quality deterioration, numbers of pens that can be placed in a given area and so forth.

Recirculating Water Systems

A recirculating water system is basically a raceway culture system in which the water is reused after passage through a treatment unit. Recirculating water systems are expensive to operate because they require the input of significant amounts of energy (usually supplied by electricity). They have been used in regions where water is scarce or expensive, in indoor facilities where water must be heated and that heat conserved, and in research laboratories.

As water passes through a culture tank containing fish or invertebrates, it picks up fecal material and waste feed particles, loses oxygen because of respiration and collects ammonia released from the aquatic animals as a waste product. After the water leaves the culture chamber of a recirculating system, it typically passes through a settling tank where solids are removed (this may also be done by filtration of the water through sand or other types of fine material). The water then

Figure 16.10. Circular raceways such as the one shown here can be used for commercial aquaculture production and are widely used by researchers.

Figure 16.11. Cage culture can be used in large water bodies. In some Arkansas lakes, such as the one shown, areas in public lakes can be leased for fish production.

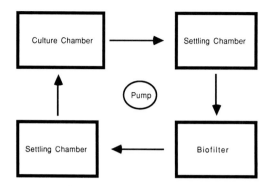

Figure 16.12. Diagram showing components of a typical recirculating water system.

Figure 16.13. A type of seaweed called nori is cultured extensively in Japan. Nylon nets are placed in large vats where the embryonic seaweeds become attached to the mesh. The nets are then taken out into large embayments and tied in place. Within several weeks, the seaweed will grow to harvestable size (several centimeters).

enters what is known as a **biofilter,** where ammonia is removed by bacteria; oxygen may also be added to the water in the biofilter. The treated water may then be passed through another settling chamber or filter, after which it is reintroduced to the culture tank. A typical recirculating water system is shown diagrammatically in Figure 16.12. In order to keep the water flowing through such a system, at least one pump must be included between two of the units. Gravity can be used to keep the water flowing through the other components.

The key to efficient operation of a recirculating water system is the biofilter. The biofilter is composed of material such as gravel or plastic that provides a large amount of surface area per unit volume and will support the growth of bacteria. Styrofoam, polyvinyl chloride, polypropylene and various other materials have been used effectively. Bacteria colonizing the biofilter substrate material remove ammonia and other nutrients from the water. As the bacterial mats grow, pieces will slough off the plastic and become suspended in the water flowing through the biofilter. The second settling tank (Figure 16.12) removes those bacterial mats from the system before they enter the culture tank.

When a recirculating water system is first filled with water and stocked, the biofilter will not function because nutrient levels at that time will be insufficient to support high levels of bacteria and, thus, the bacteria will not have colonized the system. It takes several days in freshwater systems and up to several weeks in seawater systems to establish proper bacteria levels. Therefore, it is important to introduce only small numbers of fish into the system until the biofilter becomes functional; otherwise, the ammonia level could become sufficiently high to kill the fish.

There are really no hard and fast rules as to how many kilograms of fish one can raise per cubic meter of biofilter medium. Functioning of a recirculating system depends on the species being raised, the density of the organisms, the rate at which water flows through the system, the type of biofilter medium and various other factors. Most aquaculturists have designed recirculating systems with little knowledge of how all the factors interrelate. In recent years, engineers and biologists have begun to work together to evaluate the critical parameters influencing the design of recirculating water systems and have developed guidelines for sizing the various components to meet the needs of the culturist.

Miscellaneous Culture Systems

The preceding subsections have outlined the major types of culture units currently being employed. Various modifications of them are possible. For example, a system may operate under the recirculating mode for a period of time, after which the biomass within the system exceeds the capacity of the biofilter to treat the water. At that time, a semi-recirculating mode of operation may be employed wherein new water is added continuously (with concomitant release of water down the drain), but the bulk of the water continues to be recirculated. The amount of new water added each day may be increased as biomass continues to increase. However, such systems have rarely been used.

Shellfish species such as oysters, mussels, scallops (e.g., *Pecten yessoensis* and *Argopecten irradians*) and clams may be grown in a variety of methods, including free on the bottom or in bags suspended from floating lines. Species that attach to substrates, like oysters and mussels, can be grown on lines suspended from rafts.

Seaweeds are sometimes grown on nets that float within the water column (Figure 16.13). Japan has developed nori (*Porphyra* spp.) culture on nets into an industry that employs thousands of people and produces enormous amounts of seaweed. Other seaweeds are also grown in Japan and various other countries.

Backyard aquaculture systems in which a family maintains a recirculating water system where they produce fish for home consumption have been advocated for a number of years. Designs are available for systems which are effective in producing fish, such as tilapia, where a high level of expertise is not required. Such systems require a substantial initial investment and may require supplementation with energy for aeration, water heating and so forth. If the operation of such a system is looked at as a hobby it may pay dividends. If the goal is to produce inexpensive fish, backyard systems are often a disappointment.

Management of Culture Systems

Once the species for culture has been selected and the culture system has been constructed and stocked, aquaculturists must address various management concerns. An overview of some of the most important areas is provided below.

Water Quality

In terms of things that should be monitored by the aquaculturist, perhaps the most important water quality variables are temperature, dissolved oxygen and ammonia. Other variables can be important under certain circumstances, but the three mentioned generally provide a good indication of the performance of the animals in the culture system.

Temperature

Aquaculture species are all "cold-blooded" or **poikilothermic**. That means that their body temperatures are virtually the same as the temperature of the water that surrounds them. Basically, there are two primary types of culture species with respect to temperature: **warmwater** species and **coldwater** species. Carp, tilapia, channel catfish and freshwater shrimp are examples of warmwater species. Trout, salmon and American lobsters (*Homarus americanus*) are examples of coldwater species. The optimum temperature for warmwater species tends to be about 86°F (30°C), while that for coldwater species is often 59°F (15°C). Some species of aquaculture interest, such as the yellow perch (*Perca flavescens*), have temperature optima between the warm and coldwater species and are known as **mid-range** species. Few mid-range species are currently being cultured.

Some aquatic species cannot survive when ambient temperature is not within or close to an optimum range. Tilapia, for example, typically die below about 50°F (10°C), and their growth is severely retarded below 68°F (20°C). Channel catfish, on the other hand, have a temperature optimum of 79°F to 86°F (26°C to 30°C), but can survive under the ice in nature when the water temperature is no warmer than 39°F (4°C). Trout cannot survive at temperatures much in excess of 68°F (20°C). Similar to channel catfish, trout can survive in extremely cold water.

When temperature changes dramatically and, in particular, when it moves out of the optimum range, aquatic animals are placed under stress. It is at such times that disease resistance is lowered and problems often arise. Knowledge of the temperature requirements of the species under culture and of the temperature at any given time will not only provide the culturist with valuable information about how well the animals are growing and how much to feed them, it will help to establish the disease resistance status of the animals.

Dissolved Oxygen

Oxygen enters water by dissolution from the atmosphere and through the release of that element by plants during photosynthesis. Animals with gills respire by absorbing oxygen that has been dissolved in water directly into the bloodstream through diffusion. As a general rule, if the water contains 5 parts per million (ppm or mg/l) of oxygen, it will support aquatic organisms. Some fish, such as tilapia, can survive at very low concentrations of oxygen, while others, such as trout, are stressed if the concentration falls below 5 ppm.

Daily changes in temperature are very small relative to the changes that can occur seasonally, particularly in temperate climates. Daily changes in dissolved oxygen, on the other hand, can be substantial. If one takes dissolved oxygen readings throughout a typical summer day when there is a significant amount of phytoplankton in a culture pond, one might observe a pattern similar to that shown in Figure 16.14. The range of oxygen values might be as little as 1 or 2 ppm during the 24-hour period, or it could be 10 ppm or more. Thus, the vertical axis in the figure shows only relative changes.

As shown in Figure 16.14, dissolved oxygen begins to increase at about dawn, when photosynthetic production of oxygen by the plant community begins. As the sun rises, photosynthetic oxygen production increases with the increasing amount of light energy available. While both plants and animals respire continuously, the rate of oxygen production exceeds respiration and there is a net increase in the dissolved oxygen level. At dusk, when there is insufficient light for photosynthesis, the oxygen level begins to drop because of respiration demands, and the drop continues through the night. As long as the lowest morning dissolved oxygen level is not below about 5 ppm, there should be no problem. However, the lowest level of dissolved oxygen can change dramatically from one day to the next. Daily production of oxygen can be influenced by the weather (cloudy days don't support as much photosynthetic activity as clear days) and by the biomass of culture organisms present. As the fish or shellfish being raised grow, they extract more oxygen from the pond each day.

Aquaculturists typically check dissolved oxygen in their ponds each morning before sunrise, because it is then that the lowest levels will occur. Emergency aeration can be employed to raise the oxygen level in ponds where problems exist. In most cases only one or a few ponds will be critically low in oxygen on a given morning, so each pond does not need its own emergency aeration equipment. Today, paddlewheel aerators are commonly used (Figure 16.15), but various other types of emergency aeration systems have been developed. One way to overcome the problem is to add large quantities of oxygen-rich water. That is a good solution if such water is available.

Ammonia

Ammonia occurs in two forms, **unionized** (NH_3) and **ionized** (NH_4^+). The ratio between the two in water depends on temperature, pH and a few other factors. Unionized ammonia is the more toxic form and tends to be present at increasing levels relative to ionized ammonia as temperature increases.

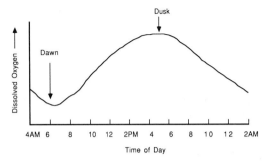

Figure 16.14. Typical daily pattern of oxygen fluctuation in a culture pond.

Figure 16.15. Paddlewheel aerators may be driven by electricity, hydraulic pumps, or they may be attached to the power takeoff of a tractor. The slashing water and creation of water movement in the pond rapidly increase the level of dissolved oxygen.

Ammonia is rapidly converted to **nitrate** (NO_3^-) by plants and bacteria in aquatic systems. Thus, in ponds where there are plenty of plants and bacteria present, ammonia toxicity is not usually a problem. In raceways, recirculating systems and other water systems where animals are reared at high densities, ammonia removal is often not as efficient as in a pond, and toxicity can occur. As is true of dissolved oxygen, different species of aquaculture interest have different tolerances for ammonia. Tilapia can tolerate high concentrations of total ammonia (several ppm), whereas trout are highly susceptible to levels well below 1 ppm.

Toxicity from ammonia can occur if a biofilter fails for one reason or another, if the biomass of a highly intensive system exceeds the carrying capacity of that system and in a few other instances. In the chemical reaction that carries ammonia to nitrate, there is an intermediate toxic chemical, called **nitrite** (NO_2^-) that forms, usually only for an instant before being transformed into nitrate. Ponds and other types of culture systems have experienced nitrite toxicity when ammonia production is high and the bacteria needed to convert the ammonia to nitrite are present in sufficient quantities, while those required to convert nitrite to nitrate are not. This has been a problem in some catfish ponds in recent years, for example. It usually occurs late in the growing season (early fall) when the water is warm, fish biomass is at its maximum and the feeding rate is very high.

Nutrition and Feeding

Under natural conditions in ponds, lakes, rivers and the ocean, fishes rely on natural productivity for their nourishment. Some aquaculturists also use natural food organisms to provide nourishment for the culture species. In the People's Republic of China, for example, ponds are stocked with various species of carp that feed on different parts of the food chain. The fish ponds may be fertilized to help promote growth of phytoplankton, zooplankton, rooted aquatic macrophytes and benthic organisms, each of which is fed upon by a different type of carp. In Japan and a few other countries, ground raw fish is often used to feed aquacultured animals (the fish may be mixed with small amounts of dry ingredients).

Oysters, mussels and clams are among the various shellfishes that feed by filtering algae and other organic nutrients from the water. The culture of those animals requires the presence of large algae concentrations. As it is not usually economical to culture the levels of algae needed for growout of shellfish, sites for the culture of shellfish are often selected on the basis of high natural levels of primary productivity

Most fishes and invertebrates of aquaculture interest are fed **prepared feeds**. Such feeds are composed of various ingredients in proper combinations so that the final product will meet the nutritional requirements of the species being fed. Diets vary considerably from one type of aquaculture animal to the next because of differences in nutritional requirements. For example, many crustaceans are unable to synthesize cholesterol, so that chemical must be provided in the feed. Fishes, on the other hand, do not require dietary cholesterol.

Determining the nutritional requirements of an aquaculture species can require many years of research. Diets are prepared in which various ingredients of interest are varied with respect to quality and quantity. Then the feeds are presented to the aquaculture species over a period of several weeks to months and the growth response is evaluated. Experimental diets may be prepared to examine the responses of the animals to variations in dietary protein, fat, carbohydrate, minerals, vitamins or energy.

Typical aquaculture diets are relatively simple. They usually contain some type of animal protein (fish meal, poultry byproduct meal, meat and bone meal) and other proteins supplied by plants (soybean meal, wheat, corn meal, peanut meal and cottonseed meal). The plant products also supply high levels of carbohydrates (sugars and starches). Some species, such as channel catfish, can tolerate levels of 40% carbohydrate in the diet, whereas others, such as trout, tolerate only low carbohydrate levels. Fat is supplied by the various ingredients mentioned, but supplemental fat is often added in the form of corn oil, fish oil or a variety of others. A mixture of required vitamins and minerals is also usually added. In some instances, wet, ground fish is used in the United States as a dietary ingredient. This is particularly true in the Pacific Northwest, where waste products from fish processing plants are readily available. The resulting formulation contains a high level of water and when pelleted (as described below) results in what is known as a "moist pellet."

An old expression is "You are what you eat." With current interest among Americans to consume more healthful diets, there is renewed interest in providing aquaculture animals with feeds that produce fishes and shellfishes that have the same qualities that are seen as healthful in wild fishes. A primary example is highly unsaturated fat containing the so-called omega-3 fatty acids. Wild fishes often have high levels of those fatty acids relative to the less desirable saturated fatty acids. On the other hand, cultured fishes reflect the fats in the prepared feeds they consume. By altering the fatty acids in the feed ingredients, the producer can alter the flesh of the fish in response to consumer demand.

Once a diet has been formulated and the ingredients have been mixed together in the proper proportions, the material is usually made into a pellet. Pellets are made by exposing the material to high pressure in a **pellet mill** or **extruder**. Pellet mills may use steam to help bind the ingredients together. Extruders use supplemental heat and extended exposure to high pressure to make pellets. Pressure pellet mills and extruders pass the feed mixture through a small aperture which leads to a product which, if not cut to short lengths, would be much like spaghetti strands. The diameter of the pellets varies, but is typically 0.2 to 0.24 inches (5 to 6 mm). A knife cuts off the strands as they exit the pelleting equipment. Lengths vary, but pellets are not typically longer than about 0.8 inch (2 cm).

Feeds produced by pressure pelleting are more dense than water; thus, they sink. During the extrusion process, on the other hand, the high heat used causes changes within the ingredients so that starches expand when the pellets leave the machine and come into contact with air. This rapid expansion of the material traps air within the pellets, which float when placed in water. Because of the higher temperatures and other factors, extruded pellets are more expensive than pressure pellets. Also, some nutrients are damaged by the intense heat of the extruder, so those ingredients (certain vitamins in particular) must be overfortified in the initial mixture of feedstuffs.

Advantages of floating pellets are that the aquaculturist can see that the fish eat the feed. By watching the fish eat, the producer can control the amount of feed offered and thereby avoid overfeeding. This can save money and will also prevent the buildup of organic matter from unconsumed feed. If the fish develop a disease, the aquaculturist may be able to identify the problem by observing changes in the behavior or appearance of the fish and can treat the problem before it progresses very far.

Floating feeds should not be used on all aquaculture species. Shrimp, for example, feed on the bottom and will not swim to the surface for pellets. However, nutrients will be quickly lost from sinking pellets which may dissolve in a few minutes (floating pellets may take 24 hours or more to disintegrate), so valuable sources of nutrition can be lost if the animals do not consume the feed quickly. Also, bacterial and fungal growth on feed particles that are not quickly consumed can lead to disease or toxicity problems.

In regions where aquaculture is routinely practiced, obtaining good quality feed specifically designed for a particular aquaculture organism may be as simple as buying dog food at the local supermarket. The culturist merely calls the local feed mill and has the feed delivered to the fish farm. In areas where aquaculture is new, feed may have to be shipped long distances; thus, feed costs will be high. Various good feed formulations are available for an array of culture species, though the actual nutritional requirements are known in detail for only a few species.

Genetics and Reproduction

If an aquaculturist wants to undertake selective breeding in order to improve the performance of the species under culture, it is necessary to have all components of the life cycle under direct control. For some species it has been necessary to obtain young animals for stocking by collecting them in the wild since reliable means of producing young in a hatchery have not been developed. In other cases, wild broodstock are obtained because we do not have the knowledge required to grow fish to adult size in captivity or, more often, the cost of growing and maintaining brood fish may be too high. However, for most successful aquaculture species, the life cycle from egg through adult is controlled by the aquaculturist.

Each aquaculture species requires certain conditions for reproduction. Those conditions vary widely. For example, some species spawn in fall or winter, some in spring, and some spawn almost continuously. The key to inducing spawning may be changing temperature (such as falling temperatures late in the year or rising temperatures in the spring), increasing or decreasing the amount of daylight present (also known as **photoperiod**) or a combination of the two. These types of environmental stimuli cause changes in the hormone activities within the aquaculture species and lead to the development of eggs and sperm and, finally, induce the behavioral activity that accompanies the actual spawning act.

Many aquatic species broadcast their eggs and sperm into the water. The fertilized eggs will become members of the plankton community. Upon hatching, the larval animals may continue to swim about in the plankton until they grow sufficiently large to enter the benthos or nekton community. This type of reproductive scenario is typical of shrimp, crabs, lobsters, oysters and certain types of fishes (for example, red drum [*Sciaenops ocellatus*] and striped bass [*Morone saxatilis*]).

More specialized reproductive modes are used by many of the species of aquaculture interest. Channel catfish produce adhesive masses of up to about 30,000 eggs (Figure 16.16). Tilapia males in the genus *Oreochromis* (the genus in which most culture species are found) construct shallow nests in pond bottoms into which eggs are deposited and fertilized. After fertilization, the female picks up the eggs in her mouth and retains them until after the fry hatch and are able to survive on their own. Trout and salmon lay their eggs in shallow nests (known as **redds**) constructed in the gravel bottoms of lakes and streams. The newly hatched fish remain in the gravel for a considerable amount of time after hatching.

In order to control reproduction in the hatchery environment, the aquaculturist must understand how the species of interest behaves under natural conditions. Those conditions can then be replicated. In some instances, however, culture environments have been set up to duplicate natural conditions, but the brood animals refuse to spawn. Under those circumstances, hormone injections can sometimes be used to induce spawning. Various natural hormones have been used for that purpose, with one of the most common being human chorionic gonadotropin (HCG), which is obtained from the urine of pregnant women. Small doses of HCG are injected into fish (either into the abdominal cavity or muscles), which can result in spawning activity within 24 hours of the injection. Striped bass and carp are routinely spawned with the aid of hormones, and the technique has been used on various other species, though it is usually not required for trout, salmon and channel catfish.

Once fertilized eggs are obtained, they need to be incubated under the proper conditions (temperature and light may be important, as are dissolved oxygen level and, in the case of marine fishes, salinity, among others). Eggs of trout, salmon and catfish are relatively large

Figure 16.16. Channel catfish lay eggs in adhesive masses. The male fish fans the eggs with his fins to keep oxygenated water flowing past each developing embryo.

and resilient, so high water flow rates are tolerated. The eggs of many species, however, tend to be susceptible to mechanical damage, so care must be taken to provide a calm environment to prevent the eggs from bumping into each other or into the walls of culture tanks. Various marine fishes fall into the latter category, as do many invertebrates. At the same time, some exchange of water is required to dilute waste products like ammonia which are produced by developing eggs and larvae.

Some eggs hatch into larvae within 24–48 hours after fertilization, while others may require several weeks or even months of incubation. Red drum fit the former category and salmon the latter. Generally, warmwater species develop and hatch quickly, while coldwater species require considerable time periods for development. A newly hatched fish is called a **larva** or **sac fry**. These young fishes are not capable of eating since their mouth parts remain undeveloped. They obtain their nourishment from a yolk sac located on the abdomen. After periods of a few days to several weeks, again depending upon species, the yolk sac will be depleted of nutrients and a critical stage in the life cycle is reached; that is, the time of first feeding. The young fishes, or fry, need to have suitable feed available to them when they have exhausted the nutrients in the yolk sac. If such food is not available, the young fishes will quickly starve.

For trout, salmon, catfish and a few other species, fry are sufficiently large that finely ground practical feeds can be used. For many fishes, the size at first feeding is only a few millimeters and prepared feeds that will provide all the nutrient requirements of the fishes cannot be readily manufactured. Therefore, live food may be required until the fry become large enough to accept prepared diets. In some cases, it may be necessary to train the young fishes to accept prepared feeds.

Most attempts to improve performance in aquaculture species through genetic alteration have involved selective breeding. In recent years, the production of animals with extra sets of chromosomes (**polyploidy**) has been used in a few cases. One of the most significant examples involves the production of polyploid oysters (*Crassostrea* spp. and *Ostrea* spp.). When eggs from oysters are subjected to certain types of chemical treatment at the proper point in their development, they will develop with two sets of female chromosomes and an additional set provided by the sperm of a male. Polyploid individuals not only grow faster in many cases, but they are sterile. In oysters, this means that glycogen (sugar) that is normally used to provide energy for the production of gametes (eggs and sperm) is not lost. Thus, the oysters retain their sweet taste throughout the year. In the past, oysters have often not been available during months without an ''r'' in them (May, June, July, August) because the animals were in poor condition. Polyploid oysters can be harvested and sold during those months without affecting consumer acceptance.

Polyploidy has also been used experimentally to improve growth of fishes, but commercial employment of the technique seems to be some years in the future. Growth hormones have been fed to various aquaculture species, sometimes with good results. Limitations on the use of hormones as feed additives in many countries has restricted the use of that technique.

Genetic engineering may allow us to produce fishes and shellfishes that have better growth rates, more desirable flesh characteristics and other attributes in the future. At this time, we know very little about the potential for applying that biotechnological discipline to aquacultured organisms.

Diseases and Parasites

Like other organisms, aquaculture species are susceptible to a broad array of diseases and parasites such as viruses, bacteria, parasitic protozoans, helminths (worms), copepods and others. When a disease or parasite outbreak occurs in a population of aquatic organisms, it is known not as an epidemic (a term used for the same problem in human populations), but as an **epizootic**. In most instances, epizootics only result after a population has been stressed.

Stress can occur in a number of ways. If water quality deteriorates, even for a short period, the animals exposed to that water will undergo stress. Handling is another cause of stress in aquaculture species, as are overfeeding or sustained underfeeding. Following exposure to a stressful situation, an epizootic may occur within as little as 24 hours or as long as 2 weeks. The period between the stress episode and the onset of disease or parasite infestations depends on the time involved in building the numbers of disease and parasitic organisms to a high enough level that signs of the disease are observable. Disease and parasite organisms are almost always present in the culture environment, but at very low levels. They only produce disease signs when they are promoted by a lack of resistance in the aquaculture species. An analogy can be drawn in humans. Students in a classroom may be exposed to people with the flu, but not everyone is infected. The immune status of the various individuals plays an important role in who will ultimately show flu symptoms. Students who are stressed (for example, by getting too little sleep) are often much more susceptible than those who are not.

The number of chemicals that can be used to treat disease and parasite problems in aquatic animals is small. Only about 10 compounds have been approved for species that are being reared for direct human consumption. Among them, few are effective at controlling bacteria, some work on a few parasites but not others, and many are themselves toxic if given in improper doses. Some treatment chemicals are effective when added to the water, but some must be ingested by the aquaculture organisms. Animals that are experiencing disease or parasite problems often refuse to eat, making treatment difficult.

Good overall management of the culture system is perhaps the best way to avoid disease and parasite problems. However, even the best managers experience epizootics on occasion. Treatment chemicals of various kinds should be available, and the culturist should know how to use those chemicals. Chemicals such as antibiotics should not be used routinely, but should only be employed when there is a problem requiring treatment, or when there is a very strong probability that a disease or parasite epizootic is imminent.

Harvesting and Processing

The fish or invertebrates being reared by an aquaculturist may all be harvested over a short period of time, or there may be intermittent periods of harvesting throughout the year to ensure that a constant supply of product reaches the market. The channel catfish industry, for example, used to be based on a system wherein the bulk of the harvesting occurred in the late fall when water temperatures became cold and fish growth was retarded. This meant a glut of fish on the market and consequent low prices to the producer. Now, the technique employed by catfish farmers is to stock each pond with several sizes of fish and harvest marketable individuals from a given pond by collecting them at intervals of several weeks year-round. If the producer has several ponds, harvesting might be a weekly or even daily occurrence, with no single pond being partially harvested more often than once every 3 or 4 weeks. Peaks and valleys in the availability of fish in the marketplace are thereby avoided, and the price is much more stable.

Harvesting of ponds is typically accomplished with seine nets. The pond may be partially drained and then a seine is pulled through that pond and fish are collected. If the mesh size of the seine is proper, only marketable individuals will be taken. Raceways can be harvested by crowding fish into a confined area and dipping them out with nets, a technique also used for harvesting cages and net-pens.

Following harvest, fish are typically loaded into hauling tanks (enclosed boxes filled with water and supplied with air or oxygen to help keep the animals alive) on trucks. Invertebrates may be transported in sacks (oysters and crayfish), in boxes on ice (shrimp), or in hauling tanks (lobsters, crabs). The animals may be carried considerable distances to a central processing plant, or they may be processed on the farm.

For fishes, processing may result in a number of forms. For example, channel catfish may be processed into steaks or fillets, skinned and eviscerated, prebreaded, or rendered into several other forms. Trout may be stuffed, gutted only (head and fins are left on), or prepared in other ways. The product may be shipped to the marketplace alive, on ice or frozen.

Economics and Marketing

For many aquaculture products, the profit margin (the difference between cost of production and amount received for each fish) is very small. The channel catfish industry has typically operated on a margin of a fraction of a cent to a few cents per pound. The same is probably true for other types of aquaculture products, though much higher margins have been obtained from salmon, shrimp and a few other species. When profit margins are low, aquaculturists need to grow large quantities of fish or shellfish so that the overall profit is sufficient to keep the business in operation. That is one reason that many channel catfish farms occupy several hundreds or even thousands of hectares.

Fixed costs (those which do not recur on an annual basis) for aquaculturists include the cost of the culture system (pond or raceway construction, costs associated with purchase and installation of net-pens or cages); land; hatchery, office and feed storage buildings; vehicles such as tractors; and various types of equipment (pumps, feeding machines, water quality analysis gear). Those costs are typically in the hundreds of thousands to millions of dollars and are usually obtained, at least in part, through loans. Therefore, profits from the operation must be sufficient to pay the principal and interest on such loans.

Variable costs (those which recur annually) include costs of fingerlings, larvae, or juveniles of the culture species; feed costs; costs of chemicals used in disease treatment; labor; insurance; taxes; fuel; and electricity. While such costs may occur on an annual basis (for example, the cost of purchasing fingerling fish), many of them occur monthly or at shorter intervals (electricity, feed). Sufficient cash flow must be available to pay recurring bills. Such bills can be substantial. A channel catfish farmer, for example, may provide up to 100 pounds per acre (kg/ha) daily of feed near the end of the growing season. The cost of feed will be a few hundred dollars a ton. Thus, if the farmer has a few hundred hectares of ponds, the feed costs could amount to many thousands of dollars daily!

The final test of the success of an aquaculture venture comes when the ultimate consumer eats the product. A primary goal of the aquaculturist should be to ensure that the fish, shellfish, or algae produced meets the most demanding requirements of the consumer. Wholesomeness, good flavor, attractiveness, color and various other characteristics of the product are important.

Additional Reading

Bardach, J.E., J.H. Ryther, and W.O. McLarney. 1972. Aquaculture. Wiley-Interscience, New York. 868 p.

Brown, E.E. 1983. World Fish Farming: Cultivation and Economics. Avi Publishing Co., Inc. Westport, Connecticut. 516 p.

Brown, E.E., and J.B. Gratzek. 1980. Fish Farming Handbook. Avi Publishing Co., Inc. Westport, Connecticut. 391 p.

Piper, R.G., I.B. McElwain, L.E. Orme, J.P. McCraren, L.G. Fowler, and J.R. Leonard. 1982. Fish Hatchery Management. U.S. Department of Interior, Washington, D.C. 517 p.

Stickney, R.R. 1979. Principles of Warmwater Aquaculture. Wiley-Interscience, New York. 375 p.

Stickney, R.R. (ed.). 1986. Culture of Nonsalmonid Freshwater Fishes. CRC Press, Boca Raton, Florida. 201 p.

Tucker, C.S. (ed.). Channel Catfish Culture. Elsevier, New York. 657 p.

Wheaton, F.W. 1977. Aquaculture Engineering. Wiley-Interscience, New York. 708 p.

17

Marine Mammals

Thomas R. Loughlin and R. V. Miller

Introduction

Marine mammals constitute three unrelated taxonomic categories within the Class **Mammalia** and include a broad assortment of animals incorporating a variety of morphological, physiological and behavioral traits that allow them to live in aquatic environments (Figures 17.1 to 17.3). The Order **Cetacea** contains about 75 species of whales, dolphins and porpoises. It is formally divided into those living cetaceans with teeth (Suborder **Odontoceti**), such as the killer whale, and those that lack teeth but utilize a system of "horny" plates called **baleen** (Suborder **Mysticeti**), such as the California gray whale and blue whale, the largest animal known to have lived. The Order **Sirenia** includes the manatees, dugongs and the extinct Steller's sea cow, which was slaughtered in the late 1700s. (Animals in this order are said to have led to the legends of mermaids.) The order **Carnivora** contains three families of **pinnipeds** ("feather-foot," or with flippers), which comprise 34 species of seals, sea lions and walruses, and the family Mustelidae, which contains the sea otter. Polar bears are also included in this order, but the designation of polar bears as marine mammals is equivocal, and they are not discussed further except for a brief mention under International Treaties regarding protective legislation. Most of the discussion in this chapter pertains to the cetaceans and pinnipeds.

The Cetacea and Sirenia are totally aquatic, whereas the Pinnipedia and sea otters are semi-aquatic—they feed at sea but haul out onto land or ice to breed, rest and molt (although in parts of their range sea otters rarely come ashore). The rear appendages of cetaceans and sirenians have been reduced and reside inside the abdomen as small vestigial pelvic bones. The body has attained a **fusiform** (torpedo-like) shape that tapers gradually from the chest to the tail. Within cetaceans, the caudal vertebrae extend down to the end of the tail, which bears two flattened and laterally expanded cartilaginous extensions called **flukes**. Swimming is accomplished by vertical undulations of the flukes (unlike fishes, which, in general, use horizontal movements of the vertical tail). Sirenians and sea otters use a method of propulsion similar to cetaceans. Fur seals and sea lions propel themselves in the water using powerful, paddle-like front flippers, which are modified arms and hands; they are able to raise themselves while on land using their front flippers. True seals and walruses swim by powerful sideways movement of the rear flippers and, except for sea otters (which have typical otter-like front legs), do not raise themselves on land with the front flippers. The 63 marine mammal species that occur in the waters of North America are listed in Table 17.1.

Evolution

Land-dwelling **vertebrates** (animals with backbones) have invaded the sea many times in the past. These groups include several marine reptiles in the Mesozoic Era (230 to 65 million years ago [mya]) and marine mammals in the Cenozoic Era (65 mya to present). Evolution of present-day marine mammals is poorly understood because there is little fossil evidence for their immediate ancestry, and their appearance in the fossil record is sudden. Although the direct ancestry for most

Figure 17.1. Photographs showing a baleen whale and an odontocete whale: (a) minke whale with baleen shown in the inset; (b) Pacific white-sided dolphin with its teeth shown in the inset. (Photos by T. Loughlin, NOAA and J. Harvey.)

Figure 17.2. The three types of pinnipeds: (a) an otariid (northern sea lion); (b) an odobenid (Pacific walrus); and (c) a phocid (bearded seal). (Photos by T. Loughlin, K. J. Frost, ADF&G, and NOAA.)

major marine mammal groups is equivocal, it appears that terrestrial mammals have successfully invaded the marine environment as many as five times.

Most of the fossil remains of early cetaceans were found in the Mediterranean Sea and Gulf of Arabia region in what was the Tethys Sea during the Eocene Period (Figure 17.4). Early cetaceans, the extinct Archaeoceti, first appeared in the Middle Eocene some 50 mya. None survived beyond the Miocene (7 mya), but some forms may have given rise to our present-day cetaceans. The Archaeoceti are believed to share a common mammalian ancestry that gave rise to the ungulates (e.g., elephants, horses, deer). Researchers have identified mysticete and odontocete fossil remains from

Figure 17.3. (a) A female West Indian manatee with calf, and (b) a female sea otter with pup. (Photos by T. Loughlin, K. J. Frost, ADF&G, and NOAA.)

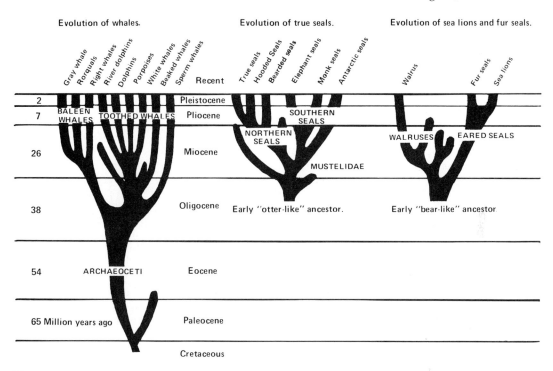

Figure 17.4. A generalized depiction of the evolution of cetaceans and pinnipeds. Note that many extinct forms are not shown. Some authorities suggest that otariids and phocids arose from a common ancestor and that their early lineage should be joined in the Oligocene. (Modified from Macdonald 1984.)

the late Oligocene and early Miocene. Such remains are characterized by progressive adaptation for aquatic modes of life, such as migration of the nostrils toward the top of the head.

Sirenians probably evolved about 50 mya (about the same time as the Archaeoceti) in the shallow tropical waters of the western Atlantic Ocean and Caribbean Sea. The first recognizable sirenian, *Protosiren*, was closely related to the ungulates and descended from an ancestor shared by elephants. Sirenians feed almost exclusively on plants, and their evolution and present distribution may be linked to this unique aquatic feeding niche. The order includes four extant (living) species in two genera.

The pinnipeds evolved from early carnivore ancestors during the Miocene Period (15 to 25 mya). The group is divided into three extant families: the **Otariidae** or fur seals and sea lions (14 species), **Odobenidae** or walrus; and 19 species of **Phocidae,** the true or hair seals and elephant seals. The otariids and walruses probably had a similar ancestry from a "bear-like" carnivore stock about 25 mya and evolved in the cool marine waters of the North Pacific Basin. Forms similar to those present today were scattered throughout the North Pacific by 8 to 11 mya. No otariids are known from the North Atlantic Ocean, but walruses are known to have crossed into the Atlantic, probably through the Arctic Ocean. The Phocidae arose in European and Mediterranean waters, probably from an "otterlike" ancestor, in the Miocene Period. By 12 to 15 mya, modern forms had crossed to the eastern United States and had dispersed into the Pacific Ocean through the Arctic Ocean and Central American seaway before the emergence of Central America.

Table 17.1. Marine mammals known to occur in waters of the United States.[1]

Scientific Name	Common Name
Class Mammalia	Mammals
Suborder Odontoceti	Toothed whales
Family Delphinidae	Delphinids
Orcinus orca	Killer whale
Pseudorca crassidens	False killer whale
Feresa attenuata	Pygmy killer whale
Tursiops truncatus	Bottle-nosed dolphin
Steno bredanensis	Rough-toothed dolphin
Stenella attenuata	Pantropical spotted dolphin
Stenella coeruleoalba	Striped dolphin
Stenella clymene	Clymene dolphin
Stenella longirostris	Spinner dolphin
Stenella frontalis	Atlantic spotted dolphin
Delphinus delphis	Saddle-backed dolphin
Lagenodelphis hosei	Fraser's dolphin
Lagenorhynchus acutus	Atlantic white-sided dolphin
Lagenorhynchus obliquidens	Pacific white-sided dolphin
Lagenorhynchus albirostris	White-beaked dolphin
Grampus griseus	Risso's dolphin or grampus
Globicephala macrorhynchus	Short-finned pilot whale
Globicephala melas	Long-finned pilot whale
Lissodelphis borealis	Northern right whale dolphin
Family Monodontidae	Monodontids
Delphinapterus leucas	Belukha or white whale
Monodon monoceros	Narwhal
Family Phocoenidae	Porpoises
Phocoena phocoena	Harbor porpoise
Phocoenoides dalli	Dall's porpoise
Family Ziphiidae	Beaked whales
Berardius bairdii	North Pacific bottle-nosed whale
Ziphius cavirostris	Goose-beaked whale
Hyperoodon ampullatus	North Atlantic bottle-nosed whale
Mesoplodon bidens	North Atlantic beaked whale
Mesoplodon carlhubbsi	Arch-beaked whale
Mesoplodon densirostris	Dense-beaked whale
Mesoplodon europaeus	Gervais' beaked whale
Mesoplodon ginkgodens	Ginkgo-toothed whale
Mesoplodon hectori	Hector's beaked whale
Mesoplodon mirus	True's beaked whale
Mesoplodon stejnegeri	Bering Sea beaked whale
Family Kogiidae	Pygmy sperm whales
Kogia breviceps	Pygmy sperm whale
Kogia simus	Dwarf sperm whale
Family Physeteridae	Sperm whales
Physeter macrocephalus	Sperm whale
Suborder Mysticeti	Baleen whales

1. Modified from Jones, J.K., Jr., D.C. Carter, H.G. Genoways, R. S. Hoffman, D. W. Rice and C. Jones. 1986. Revised check-list of North American mammals north of Mexico, 1986. Occas. Pap., The Museum, Texas Tech. University, No. 107. 22 p.

Table 17.1.—*Continued*.

Scientific Name	Common Name
Family Eschrichtiidae	Gray whale
Eschrichtius robustus	Gray whale
Family Balaenopteridae	Rorquals
Balaenoptera acutorostrata	Minke whale
Balaenoptera borealis	Sei whale
Baleanoptera edeni	Bryde's Whale
Balaenoptera musculus	Blue whale
Balaenoptera physalus	Fin whale
Megaptera novaeangliae	Humpback whale
Family Balaenidae	Right whales
Balaena glacialis	Right whale
Balaena mysticetus	Bowhead whale
Order Carnivora	Carnivores
Family Mustelidae	Mustelids
Enhydra lutris	Sea otter
Family Otariidae	Eared seals
Callorhinus ursinus	Northern fur seal
Eumetopias jubatus	Northern or Steller's sea lion
Zalophus californianus	California sea lion
Arctocephalus townsendi	Guadalupe fur seal
Family Odobenidae	Walrus
Odobenus rosmarus	Walrus
Family Phocidae	True seals
Phoca largha	Spotted seal
Phoca vitulina	Harbor seal
Phoca hispida	Ringed seal
Phoca groenlandica	Harp seal
Phoca fasciata	Ribbon seal
Halichoerus grypus	Gray Seal
Erignathus barbatus	Bearded seal
Cystophora crista	Hooded seal
Mirounga angustirostris	Northern elephant seal
Monachus schauinslandi	Hawaiian monk seal
Order Sirenia	Sea cows
Family Trichechidae	Manatees
Trichechus manatus	Manatee

Distribution

Cetaceans are distributed throughout the world's oceans and seas. In addition, a few species reside exclusively in rivers and estuaries of China, India and South America. Land barriers, water temperature and food supplies seem to be the prevalent factors limiting their distribution. Except for the narwhal, bowhead whale and belukha, which are polar breeders and occur strictly in Arctic waters (32°–50°F), all other cetaceans give birth in **temperate** (50°–68°F) or warm, tropical (68°–86°F) water. This suggests that most cetaceans originated in warm or temperate environments. Interestingly, the population levels and species diversity of the polar breeding cetaceans are low compared to the temperate and tropical breeders, a situation opposite to that for pinnipeds. Population levels for the narwhal and belukha range between 30,000 and 70,000 animals, whereas some tropical and temperate dolphin populations are in the hundreds of thousands. Bowhead whales, which were decimated by commercial whalers, are an endangered species and currently number between 5,000 to 7,000 individuals.

The distribution and feeding areas for many cetacean species vary by water depth and distance from shore. In the Gulf of Alaska, for instance, three feeding zones can be defined: coastal, coastal to continental slope and seaward of the continental slope. Some species are restricted to just one zone, whereas others may feed in all three. Cetaceans in the coastal zone typically have small populations made up of individuals that feed in small groups (e.g., killer whale, harbor porpoise and belukha). Their principal prey include a variety of small schooling fishes such as herring, as well as rockfishes, squids and octopuses. Cetaceans in the coastal to continental slope zone are more numerous in overall abundance and can be found in groups of a few individuals to many thousands of individuals (e.g., pilot whale, Dall's porpoise). These animals typically prey on fishes that live near the bottom (called **demersal**) such as walleye pollock, Pacific cod, Pacific sand lance, and some fish and squid that are higher in the water column (**mesopelagic**). For those species feeding seaward of the continental slope (e.g., sperm whale and beaked whales), populations and group sizes vary and they principally prey on fishes and oceanic squid near the top and middle of the water column (termed **epi-** and mesopelagic). For example, the principal food of sperm whales feeding in deep offshore waters worldwide is generally squids; however, in the Gulf of Alaska, fishes are more prevalent in their diets (68% fishes and 32% squid).

The sirenians are distributed in warm waters. The West Indian manatee is the only representative in the United States. It is found in decreasing abundance along the southeastern United States in coastal rivers and protected bays, and its distribution extends southward into Central America and northern South America. Two other manatee species exist, one in the Amazon River and the other in western Africa. One species of dugong exists and is found in the southwest Pacific Ocean and Indian Ocean, including Australia, New Guinea, Indonesia and Sri Lanka. Sirenians are not abundant anywhere, and perhaps because they are slow-moving plant eaters, they are vulnerable to excessive harvesting and poaching. In the United States, they are frequent victims of boat collisions and propeller injuries.

Pinnipeds occur in polar and temperate waters of both hemispheres. Their occurrence in cool environments is related to their dependence on land for breeding and productive marine waters for feeding. Their adaptations to retain important body heat while feeding in cool marine waters work against them when on land, where their thick fur or blubber may cause them to overheat. Consequently, the land areas where they breed (**rookeries**) are generally exposed to cool marine winds and are away from excessively high temperatures. Pinnipeds are rarely found in tropical areas and, where they do breed in warm land areas, they have access to cool marine waters and have developed special behavioral and anatomical adaptations that allow them to cope with the heat.

Fur seals and sea lions can be found in the North Pacific Ocean from Japan to Mexico; in the Galapagos Islands; along the coast of South America from Peru to southern Brazil; along the southern coasts of Africa, southern Australia and New Zealand; and in the oceanic islands circling Antarctica. There are no fur seals or sea lions in the North Atlantic Ocean. Two species of sea lions (California sea lion and northern or Steller sea lion) and one species of fur seal (northern fur seal) occur along the west coast of the United States. In the United States, California sea lions breed principally in California, northern sea lions breed from California to Alaska (but principally in Alaska) and northern fur seals breed principally in the Bering Sea on the Pribilof Islands (a small colony breeds on San Miguel Island, one of the California Channel Islands). All three species are abundant. U.S. population levels for California sea lions are about 70,000 animals and increasing; for northern sea lions the level is about 100,000 and decreasing; and for northern fur seals the level is about 800,000 and currently stable after decreasing sharply in the 1960s and 1970s.

Walruses are distributed throughout Arctic seas from western Alaska, Canada and Greenland to northern Eurasia. The world population is about 250,000 animals and may be slightly decreasing.

Except for the tropical Hawaiian monk seal, phocid seals are distributed throughout the world's temperate, subpolar and polar waters. The **pagophilic** (or "ice loving") seals are the most numerous of all pinnipeds and occur near both poles. Antarctic seals include the Weddell seal, leopard seal, Ross seal and crabeater seal (the most abundant pinniped in the world, with a population that may exceed 15 million animals). Arctic pagophilic seals include the ribbon seal, bearded seal, spotted seal, harp seal, hooded seal and ringed seal (the most abundant pinniped in the Northern Hemisphere at about four million animals). Subpolar and temperate species include gray seals, harbor seals, elephant seals, Caspian seals (in Caspian Sea only), Baikal seals (in Lake Baikal, Soviet Union, only) and Mediterranean monk seals.

San Miguel Island is unique among all locations where pinnipeds are found in that six species are known to breed on or use the island during the year. Northern elephant seals, harbor seals, California sea lions and northern fur seals all breed there. Northern sea lions (which bred there up until about 1982) and Guadalupe fur seals frequently haul out on the island to rest.

Sea otters, once almost extirpated throughout their range by commercial hunters, inhabit water ranging from the Kuril to Commander islands in the U.S.S.R. far east, and throughout most of the Aleutian Islands and Gulf of Alaska, where the current population exceeds 200,000 animals. Small colonies also can be found in central California (about 1,400 animals), northern Washington State (about 100 to 150 animals) and British Columbia (about 350 animals).

General Biology

Form and Function

All marine mammals exhibit **convergent evolution,** or the tendency of unrelated animals to acquire similar adaptive structures in their streamlined, fusiform shape. Such a form reduces drag while swimming through the water, which in turn facilitates the capture of prey and lowers the amount of energy required for propulsion. Cetaceans are the most advanced of the marine mammals in this regard. Their cervical vertebrae are shortened and, in some cases, fused, eliminating the neck, and their bodies taper from chest to tail. The tail bears a pair of horizontal flukes, and all traces of the hind limbs are internal. The forelimbs are shortened, stiffened and modified into flippers used for steering and balance; only the shoulder joint is mobile, and it is just beneath the body surface. In most cases, the rostrum is elongated, as in beaked whales and some dolphins, and the nostrils have "migrated" to the top of the head to facilitate breathing (odontocetes have only one external nostril and mysticetes have two). The nostril or "blowhole" is often crescent-shaped, closed by a fibrous plug and opened by muscular effort. The blowhole and nasal passage are connected directly to the lungs by the larynx, allowing cetaceans to swallow and breathe at the same time. Whales have no vocal cords, but produce sounds of varying frequencies in their nasal cavities. These sounds are used as a form of communication and echo location for detecting and capturing prey. Sounds are received by the lower jaw and transmitted to the middle ear by a long body of specialized fat in the jaw. There are no ear pinnae (ear flaps) or other appendages extending outward from the body and there is no hair, except for a few bristles around the mouth and head in some species. The sexual organs are internal.

Pinnipeds exhibit the fusiform shape common to other marine mammals but not to the extreme of cetaceans. Because pinnipeds reproduce on land or ice, they have appendages modified for terrestrial locomotion and for propulsion in the water. In otariids, the front flippers are large and used for propulsion, whereas in the phocids and odobenids the rear flippers serve this function. The otariids and phocids have a small tail that fits into the gap between the rear flippers and helps with streamlining. The ear pinnae are present but reduced in the otariids and not evident at all in the phocids and odobenids. The teats are retracted and the penis is withdrawn within the body surface. Unlike cetaceans, most of the skin is covered by hair, which helps to regulate body temperature and serves as a protective covering. However, unlike their terrestrial relatives, pinniped hair does not have the attached erector muscle and remains flat against the skin. There is no noticeable telescoping of the skull; the nostrils remain anterior and, as in cetaceans, are closed in the relaxed position to prohibit intrusion of water while diving.

Feeding

The division of cetaceans into the mysticetes and odontocetes is based on their different feeding modes. The mysticetes are "strainers" and use their baleen to filter food items from the sea; the odontocetes are "graspers" and capture fishes or cephalopods (squid and octopus) with their teeth. The teeth are used for grasping prey only; no chewing occurs, although in some cases prey may be torn into smaller pieces before swallowing. Because the teeth are used principally for grasping and holding prey, they all have a similar shape (called **homodont**), usually conical or pegged-shaped.

There are two basic feeding methods in the mysticetes: those that "skim" the surface with their mouth open (like bowhead and right whales) and those that gulp or take in large volumes of water and then force it out through the baleen to extract food (like fin whales). The skimmers typically have long, narrow baleen plates while the gulpers have short, broad plates. Baleen plates extend from the upper jaw and are arranged in parallel rows perpendicular to the long axis of the head. The plates are flattened and are longest near the back of the mouth; they grow in length with age. The interior edge of each plate is fringed and appears as "hairs" that become intertwined to form a sort of strainer (the term mysticete is from the Greek mystax, or mustache). The mysticetes feed by taking in large quantities of food-laden water, then using the tongue to force the water back out past the fringed "hairs" of the baleen, where the prey are entangled. The prey are then swept off the baleen by the tongue, deposited in the throat and swallowed. Unlike other mysticetes, gray whales leave a record of their foraging activities by producing depressions in the bottom sediments while extracting benthic invertebrates.

The feeding mode of pinnipeds is similar to the odontocetes, with few exceptions. However, some teeth are **heterodont** (that is, they have different shapes and function) and are grouped into incisors (2 to 3 in front), canines (one on either side of the incisors), premolars and molars (grouped together as "post-canines"). As with the cetaceans, little chewing occurs, so the molars and premolars in many species have tended to develop into the conical, homodont shape.

The diet of marine mammals includes a large variety of animal forms, from small zooplankton to large fishes and other marine mammals (Figures 17.5 and 17.6). The diet varies for each species by season and location, and in some species by age and sex. For instance, as female northern fur seals migrate from the Pribilof Islands in the Bering Sea to southern California, the food items change. In the Bering Sea they feed principally on walleye pollock, capelin (a smelt), Atka mackerel and squid; in the Gulf of Alaska they feed on sand lance, Pacific herring and capelin; and off California they feed principally on anchovy, squid and capelin. The diet of northern sea lions, which do not migrate, changes by season depending on the availability of prey. Those that feed near Kodiak Island from January to March consume octopus, pollock and Pacific cod, and from July to September they eat capelin, salmon and pollock.

The amount of food eaten by marine mammals is staggering. The primary reason for this is that these animals continuously lose body heat by conduction to the water. To compensate for heat loss, most marine mammals increase heat production by elevating their metabolic rate. The increased metabolic rate requires the animals to consume large volumes of food. This is especially true for small forms, such as the sea otter, which may consume up to 25% of their body weight in food per day! It is not true for large animals, like the baleen whales, because of the surface area to volume relationship (surface area is proportional to the two-thirds power of the volume), which reduces the relative area from which heat is lost with increasing size. The feeding rate for most

Food relationships of toothed marine mammals in the subarctic Pacific.

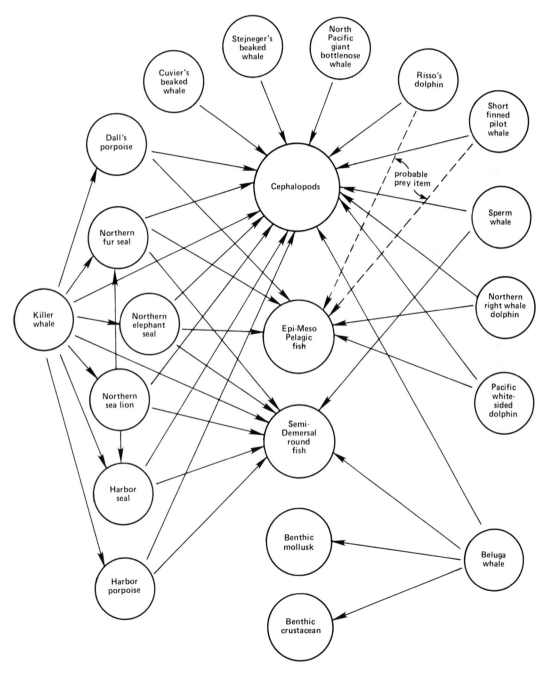

Figure 17.5. Diagram showing generalized food relationships of toothed marine mammals and their principal prey categories in the subarctic Pacific Ocean region.

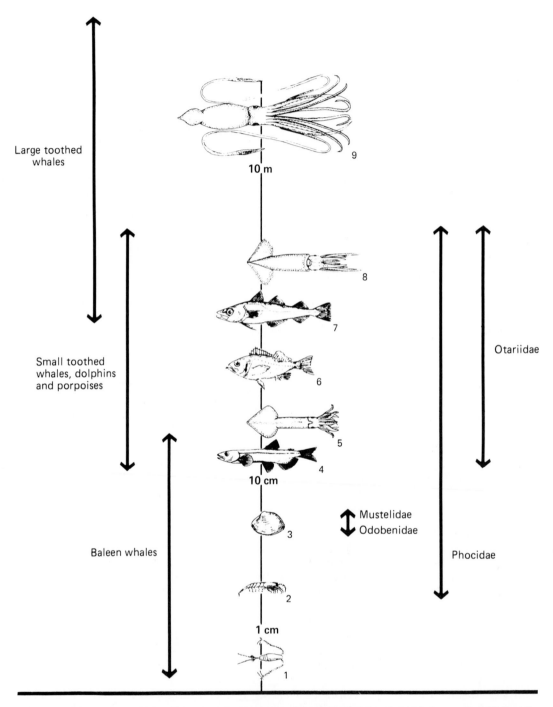

Figure 17.6. Diagram depicting the generalized size ranges of prey for marine mammals in the subarctic Pacific region and Bering Sea. Prey size increases on a logarithmic scale. (1) Calanoid copepods; (2) euphausiids; (3) benthic invertebrates; (4) shallow and midwater fishes; (5) market squid; (6) bottom fishes; (7) cods (principally juveniles); (8) bottom and midwater squids; and (9) giant squids.

marine mammals approximates 6% to 10% of their body weight per day. The estimate of the total biomass of food consumed by female northern fur seals when in the Bering Sea from July to October is over 300,000 metric tons. Northern sea lions in the Gulf of Alaska may consume over 500,000 metric tons of food a year (in contrast, the commercial total allowable catch [TAC] of groundfish was about 200,000 metric tons in 1989).

Physiology

The difficulties of living in the marine environment have resulted in the evolution of complex physiological adaptations among marine mammals that separate them from their terrestrial counterparts. Only those pertaining to diving (and thus feeding) and temperature regulation are discussed in this section.

There are two principal problems for a diving mammal: carrying adequate amounts of oxygen with it during the dive, because lungs are poorly suited for breathing underwater; and coping with gases under pressure at depth. In considering the first problem, we note that marine mammals have considerable breathhold capacities. Some species remain underwater over an hour (most dives, however, are considerably shorter). Even though they have relatively small lungs, marine mammals have high oxygen stores in their copious blood and muscle tissues. The muscles contain a molecule similar to hemoglobin, called **myoglobin**, that holds onto oxygen until it is needed for metabolism. Also, the muscle fibers are able to tolerate high levels of lactic acid, a molecule produced during strenuous activity. During breaths, gases are efficiently exchanged thanks to an oblique diaphragm that accounts for the high volume of air exchange and reduced dead-air space. During a dive, the heart slows down (called **bradycardia**), and the amount of blood leaving the heart is reduced. There is a shunting system in the vessels to redistribute the blood away from nonvital organs to vital ones, such as the heart and brain. Recent evidence indicates that the core body temperature may drop by 2°C.

The second problem pertaining to diving is coping with dissolved gases under pressure. For every 32 feet of vertical distance an animal dives, the pressure increases by one atmosphere (based on atmospheric pressure at sea level, 15 lbs/in^2 or 760 mm of mercury). When an animal dives, progressively more gases dissolve in the body fluids. These gases come out of solution during ascent and form gas bubbles that may cause pain in the joints (commonly called the bends). To obviate these problems, the lungs of marine mammals collapse at depth, forcing the air to areas where it will not be absorbed, such as the bronchioles, trachea, and in cetaceans, the air sacs located just below the blow hole. Also, pinnipeds generally exhale before diving to reduce the amount of gas absorbed into the blood.

Temperature regulation in the aquatic environment is another major problem confronting marine mammals. Marine mammals, like their terrestrial counterparts, have a constant body temperature (i.e., they are **homeothermic**). However, marine mammals continually lose heat to the surrounding water more rapidly and efficiently than land mammals. Consequently, a variety of methods have evolved to combat lowering body temperature through heat loss. In man and other land mammals, 10% to 20% of body heat may be lost during exhalation. Marine mammals breathe less often, thus conserving heat. Some cetaceans and the elephant seals reduce expired air temperatures through **countercurrent heat exchange**. This is a method for conserving body heat by using warm, core blood to warm cool blood as it returns to the core from the extremities. The lowering of expired air temperature also condenses moisture in the air, facilitating conservation of body water. Most marine mammals are **selectively homeothermic**; that is, they allow some parts of the body to become as cool as ambient temperature by shunting blood away from exposed surfaces. Some of the pagophilic seals may lay on the ice without melting it by maintaining a core temperature of 100°F (38°C) and a skin temperature near or below freezing. Other methods to combat cold include a well-developed blubber layer (in some animals the blubber may represent 58% of body weight), thick

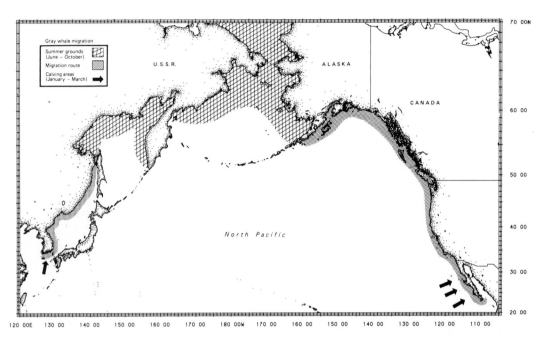

Figure 17.7. Migratory routes, feeding grounds, and calving areas of Pacific gray whales. Note that the area used during migration is usually within 3 miles of the coastline along the United States and is exaggerated here for illustrative purposes.

skin, insulative fur (not all pinniped fur serves as an insulator; some merely protects the skin from abrasion), increased metabolic rate, behavioral modifications and countercurrent heat exchange.

Migration

Migration can be defined as the mass movement of animals to and from feeding or reproductive areas. Many marine mammal species migrate; however, others do not. For instance, northern sea lions do not migrate but disperse widely after the reproductive season, with animals seemingly going in all directions. After reproducing at preferred rookeries, they do not congregate in any one area, nor do they seem to prefer any location over another. Some sea lions from the Kodiak Island area, Alaska, are known to travel northeast into Prince William Sound; others have gone to Southeast Alaska; yet others have gone west toward Unimak Pass and the Aleutian Islands. Those species that do migrate move predictably to and from the same areas year after year. Examples include a cetacean, the California gray whale, and a pinniped, the northern fur seal. Both species migrate from northern to southern latitudes at about the same time each year.

There are two types of causality regarding migration (applicable to other biological phenomena as well). The first is the **proximate** or immediate cause, such as the need to give birth or to gain better feeding areas. The second is the **ultimate** cause, which is a combination of complex factors that has resulted in natural selection favoring the observed event. These types of causality will be discussed further later in this section.

Gray whales in the eastern Pacific Ocean travel more than 10,000 miles (16,000 km) during their annual migration from summer feeding grounds in the Bering, Chukchi and Beaufort seas to winter calving grounds in Baja California—the longest migration of any mammal (Figure 17.7). The migration begins in October as the arctic seas begin to freeze. The whales, fat after spending the summer feeding in shallow water on benthic organisms, swim south across the Bering Sea

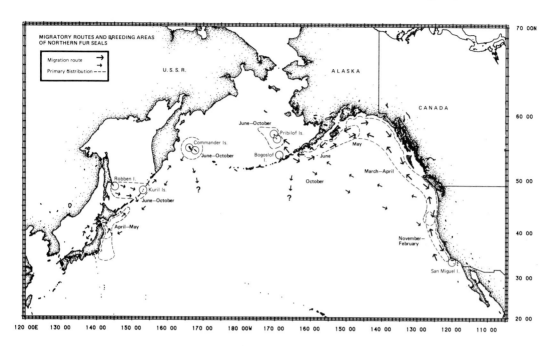

Figure 17.8. Breeding islands (within circles), generalized distribution around the breeding islands and migratory routes of northern fur seals in the North Pacific Ocean and adjacent seas. Not shown, for purposes of clarity, is the general oceanic distribution north of the subarctic boundary throughout the North Pacific Ocean (see York, in Croxall and Gentry, 1987). (Modified from Gentry and Looyman, 1986.)

through Unimak Pass and into the Gulf of Alaska. The first wave of animals to migrate south along the coast consists primarily of pregnant females, which arrive in the calving lagoons in December through mid-January. Next come adult females that have recently ovulated and have weaned their calves born 2 years before. Adult males and immature females follow, and last are the immature males. The reverse migration begins with newly pregnant females moving north from February to April, followed by adult males about 2 weeks later. Next are the females that have failed to conceive and the immature animals of both sexes. In May, females with newborn calves migrate north. Gray whales probably mate successfully during the southward migration, but copulation is common in the lagoons during winter and elsewhere at all times of the year. Pregnancy lasts about 13 months, and a female gives birth every 2 to 3 years. Calving occurs mostly in the lagoons.

Northern fur seals migrate at about the same time as gray whales, but their cycle is reversed (Figure 17.8). Fur seals give birth during the summer in the Bering Sea, when gray whales are there to feed, and migrate south during fall and winter to escape the cold and use offshore feeding grounds in California and Oregon waters. Typically, fur seals of all ages and both sexes leave the Pribilof Islands in October and November. Females and some young males migrate as far south as southern California and are found along the continental slope and shelf in pursuit of prey; they remain pelagic until they return to the Pribilof Islands in spring and summer to breed; they do not touch land from the time they leave the Pribilof Islands until their return. Most adult males winter in the North Pacific Ocean south of the Aleutian Islands and eastward into the Gulf of Alaska.

The causes for these extensive migratory patterns are complex. The proximate causes are certainly related to reproduction and feeding for both species. The ultimate causes can be related to the general areas where each species evolved and to temperature regulation. For gray whales, which probably evolved in warm tropical lagoons or temperate waters, the migration is from the birth lagoons (tropics) north to feeding grounds (arctic) and return. For fur seals, which evolved in cool subarctic waters of the North Pacific, the migration is from the rookery islands (subarctic) south to feeding grounds (temperate) and return. Both species leave the arctic environment during fall to escape the onset of winter sea ice and lowering water temperatures. For fur seals, the primary problem is with temperature regulation in cold water, especially for the small females and new pups; males are able to remain in the cool subarctic waters because of their larger size. Gray whales, which feed on benthic invertebrates, must leave because their feeding locations become covered with ice, and natural selection has continued to favor those animals that migrate to warm, protected bays to the south.

Reproduction

Most fishes reproduce by laying great numbers of eggs at tremendous metabolic expense, and the success of the species depends on the survival of a small number of those eggs. Most mammals, however, produce relatively few young, but the probability of their survival is much higher. Mammals are born at an advanced level of development and are nourished after birth by energy-rich milk from the mother. Parental care lasts until the young are capable of caring for themselves. Marine mammals are no exception to this general mammalian pattern; however, because of the need to maintain body heat, marine mammals must attain a critical size before birth or before entering the water after birth.

For instance, the smallest marine mammals, sea otters, weigh only about 4.5 lbs (2 kg) at birth. Sea otters have very dense fur (adults have over 650,000 hair fibers per square inch!), which helps protect them from the cold. In addition, the pup usually lies on the mother's stomach as she floats on her back; the pup only comes in contact with the water during the mother's feeding dive, and then the pup floats high, minimizing its contact with the water. Once the pups grow enough and increase their surface area and weight, they develop more independence. The smallest pinniped at birth is the ringed seal, which weighs about 9 lbs (4 kg). Ringed seals give birth on ice in a unique, snow-covered birth cave, or lair, where the pup nurses on energy-rich milk for 2 weeks, tripling its weight, before venturing out of the lair. The pup does not enter the water until it attains the critical size that allows it to remain in the water without severe loss of body heat. At the other extreme are the baleen whales, which give birth to calves that may weigh over a ton and that generally have a developed blubber layer at birth. However, even though these animals are very large at birth, most cetacean species give birth in warm temperate or tropical waters, perhaps to reduce the burden of heat regulation for the newborn calves.

Calves must also gain as much weight as possible in the shortest amount of time. Thus, the female's milk is very rich to aid in weight gain, and the metabolic costs of lactation place a heavy burden on the mother. During the lactation period, she may eat 1.5 times as much food as a nonlactating female. Consequently, marine mammals typically give birth to only one calf at a time; twins are unusual and are rarely reared successfully.

The amount of time a seal nurses her pup varies with species. Female hooded seals of the North Atlantic Ocean lactate for the shortest time, nursing the pup for only 4 days before leaving it and migrating north. During the nursing time, the pup increases in size four to five times the birth size. Northern fur seals nurse their single pups 3 to 4 months before migrating, and northern sea lions may nurse their pups for a year or more. Cetaceans nurse their calves from about 6 months to 1 or 2 years, depending on the species.

All seals and sea lions exhibit a phenomenon called **delayed implantation**, which is a delay in the attachment of the fertilized egg to the uterus with resultant delay in fetal development. The delay is such that in most species the fertilized eggs begin development in synchrony, resulting in the birth of pups at about the same time each year. This timing also facilitates mating since males aggregate on the beaches to mate after the females have given birth. Delayed implantation is unrecorded among cetaceans.

Marine Mammal and Fishery Interactions

Man and marine mammals have interacted for as long as man has lived near the sea. In the Bering Sea, the interactions can be divided into four periods: (1) the subsistence period; (2) the northern fur seal period (which included the harvest of sea otters for their fur); (3) the whaling period; and (4) the commercial fishing period (Table 17.2). Unlike the other three periods, which are characterized by the exploitation of marine mammals for subsistence or commercial gain, the last period is characterized by the interrelationship between man's commercial fishing activities and marine mammals and the effect that each has on the other. The level of interaction between the two has been described as:

direct or operational interactions, where

- marine mammals cause damage to a fisherman's gear and/or catch;
- marine mammals are injured or killed as a result of contact with fishing gear or fishermen; and

indirect or biological interactions, where

- predation by marine mammals reduces the quantity of a target species that is available to a fishery;
- harvests by a fishery reduce the amount of prey available to marine mammals; and
- marine mammals function as hosts for parasites that may reduce the marketability of commercially caught fishes.

Table 17.2. The four periods of interaction between man and marine mammals in the Bering Sea including the duration and type of interaction.

Name of Period	Duration	Type of Interaction
Subsistence	ca. 28,000 years ago to present	Subsistence hunting of marine mammals by natives.
Northern fur seal	1786 to 1984	Commercial harvest of fur seals (and sea otters to 1911).
Whaling	1845 to ca. 1914	Commercial harvest of whales and walrus
Commercial fishing	early 1900s to present	Interactions between marine mammals and fisheries

A list of marine mammal and fishery interactions in Alaska is presented in Table 17.3. The fisheries involved with these interactions include those where the target species is also important in the diet of a marine mammal. However, many interactions occur in fisheries where the marine mammal is incidentally caught by fishing nets targeting on fishes unrelated to the diet of the marine mammal. An example is the incidental catch of porpoise in tuna purse seine fisheries.

Direct Interactions

There are numerous instances where a marine mammal causes damage to gear or catch in a commercial fishery. For instance, Alaskan killer whales that pull sablefish off hooks during longline retrieval have become so proficient that they cost fishermen over $2,000 a day in lost fish. Harbor seals and California sea lions remove or damage salmon caught in gill nets or on trolling gear; this also may result in significant economic loss to the fishermen. It is principally during these acts that seals and sea lions are shot and killed by fishermen. Virtually every location in the world where pinnipeds and commercial fisheries overlap includes documented cases of gear or catch damage by pinnipeds. Aside from the killer whale/sablefish interaction just described, there are relatively few instances of cetaceans causing such damage. Cetaceans do damage gill nets and, in some cases, entire nets are lost as they are towed away by whales. The whales also probably swim through nets, leaving holes. However, there is little documentation of this because no one can distinguish the cause of holes in nets after the fact.

Table 17.3. Marine mammal interactions with commercial fisheries in Alaska and the type of damage caused by each.

Fishery	Principal Marine Mammal	Type of Interaction
Groundfish trawl net	Northern sea lion Northern fur seal Harbor seal	Incidental take; minor catch loss and gear damage
Salmon gill net	Northern sea lion Harbor seal Spotted seal Dall's and harbor porpoise Cetaceans Sea otter	Catch loss, gear damage Catch loss, some gear damage Minor catch loss and gear damage Incidental catch Infrequent gear damage Minor gear damage
Salmon troll	Northern sea lion Northern fur seal	Minor catch loss and gear damage Minor catch loss and gear damage
Salmon purse seine	Northern sea lion	Catch loss, gear damage
Halibut longline	Northern sea lion Northern fur seal	Catch loss, gear damage Minor catch loss and gear damage
Sablefish longline	Northern sea lion Killer whales	Catch loss, gear damage Catch loss, gear damage
Crab pot	Northern sea lion	Minor gear damage

Marine mammals are often killed or injured as a result of contact with fishing gear or catch. As mentioned, they are frequently shot by fishermen, but they are also caught in the gear incidental to fishing operations. In the purse seine fishery for yellowfin tuna (*Thunnus albacares*), thousands of dolphins are killed each year. Before the late 1950s, almost all tuna were caught by hook and line. With the advent of synthetic net materials and the power block used for hauling nets in the late 1950s and 1960s (see Chapter 15), however, purse seines became efficient tools for harvesting schoolfish.

Fishermen had long known of the association of certain pelagic dolphins with yellowfin tuna; whether it is a symbiotic relationship for better food acquisition or for some other reason, schools of yellowfin tuna move in association with large schools of several species of dolphins. Fishermen learned to set their purse seines around the dolphins so as to also encircle the tuna. After pursing the net, they would gradually bring it in (''dry it up'') and brail in the fish. During this procedure, many dolphins became entangled in the mesh and drowned or were killed in other ways. Sometimes tens or hundreds of animals were killed in a single set. Eventually, fishermen developed a procedure called ''backing down'' (Chapter 15) in which they could run the ship in reverse and pull the net backwards, literally sliding the net out from under the dolphins that tend to gather or ''raft'' at the far end of the pursed net while keeping the deeper swimming tuna in the net. Nevertheless, there were still occasional sets involving high mortality depending on the expertise of the skipper.

Passage of the Marine Mammal Protection Act in 1972 (discussed later in this chapter) required the fishing industry to reduce the number of dolphins killed and resulted in an annual quota of about 20,000 dolphin deaths in the U.S. tuna purse seine fishery. The U.S. industry has made dramatic improvements in reducing the number of dolphins dying in the fishery; unregulated tuna fleets of foreign nations now present a far more serious source of mortality. In 1987, 70 foreign vessels from eight nations fished for tuna in the eastern tropical Pacific and were estimated to kill over 100,000 dolphins. While the foreign fleet is about twice the size of the U.S. tuna fleet, the foreign fleet's estimated dolphin mortality rates are about four times as high. Resolving this problem will be complex with regard to concern for the marine mammals involved, economy of the fishery, consumer rights and foreign trade.

Another direct interaction is the incidental catch of marine mammals in commercial trawl fisheries. Northern sea lions (and other species) are commonly caught in trawl fisheries in Alaska and have long been blamed for damaging fishing gear and caught fish, often becoming entangled in nets themselves. They have been caught incidentally in foreign commercial trawl fisheries in the Bering Sea and Gulf of Alaska since about 1954, when those fisheries developed. During the period 1978–1981, northern sea lions were the predominant marine mammal species taken incidental to commercial trawl fishing in Alaskan waters. Presently, the population of northern sea lions in Alaska is declining rapidly (it is one-half the level it was 20 years ago), and one suggested reason for the decline is the synergistic effect of fisheries. The factors in fisheries that would contribute to the decline include incidental take in nets, intentional kill (shooting), harassment, removal of prey resources (the commercial species being fished) important to the animal's well-being and other factors. The synergistic effect is the working together of two or more of these factors, resulting in a greater impact on the population than could be achieved by the factors separately.

Indirect Interactions

In most cases, determining the severity of indirect interactions between marine mammals and fisheries is difficult. Fishermen have long claimed that the removal of predatory marine mammals from fishing areas would increase their catch, but there is currently no scientific basis, in most cases, to predict whether the reduction of marine mammals would have a positive or negative effect on fish catch. Fishermen in Finland claim that the dramatic decline in the cod fishery there was a result of the appearance of seals in the area. But available data are not adequate to determine if their claim is true or if hydrographic changes caused cod numbers to decline. Because of complex effects of other species in the ecosystem, it does not necessarily follow that if the number of seals is reduced more fishes would become available to man. Other natural predators, such as fishes and sea birds, may increase their take of prey formerly eaten by seals. Similarly, there are no conclusive data to predict the impact of fish removal by a fishery on a marine mammal stock, although commercial catch has been blamed for the decline in abundance of southern elephant seals and other pinnipeds.

Another form of indirect damage caused by marine mammals to fisheries, and one which has been considered a serious economic threat, is the role of seals as the host of a nematode worm, commonly referred to as the cod worm. This worm lives as an adult in the stomachs of gray seals and harbor seals, principally in the Northeast Atlantic. Worm eggs are expelled into the water with the seal's feces, and the hatched eggs are consumed by benthic crustaceans, which are in turn eaten by cod and other fishes. The larvae encyst in the mesenteries and muscles of the fish until eaten by the seal, thus completing the cycle. The presence of the larvae in the cod, although not a serious human health threat, reduces the marketability of the fish, thereby causing economic hardship on the fishermen. A similar problem may have developed recently with herring worms. There have been numerous attempts, some successful, to force European governments to cull seals to reduce the number of hosts and thereby reduce the incidence of cod worm infestation in the fish.

Use of Marine Mammals

Subsistence Uses

Man has made use of marine mammal for centuries, beginning with subsistence use by aboriginal peoples and gradually involving commercial harvesting employing sophisticated technology for meat, fur, oil and other products. Stone Age peoples of northern coastal Europe have used beached whales or killed gray seals, harp seals and ringed seals for food, fuel, clothing and implements. The pinnipeds were a convenient source of prey in that they came ashore to rest or breed and were fairly easy to kill, particularly the pups. In the North Pacific region, early aboriginal cultures (e.g., Aleut, Koryak, Kamtchadal, Eskimo [Inuit] and Chukchi) developed some elaborate methods to capture marine mammals for food, clothing and implements. The particular uses of pinnipeds by these peoples can be approximated by describing their use by Eskimos prior to the arrival of western whalers.

Seals, principally ringed seals, were killed by Eskimos on land and ice or from small boats with harpoons or clubs. Once killed, the animal's meat was eaten raw or dried and, in some situations, cooked. The blubber was also eaten, but its most important use was as fuel for lanterns and for heat. The Eskimos used skins and fur as clothing, rope and as the walls of their boats—the small one-person **kayak** or larger **umiak,** which held six or more hunters. The intestines and other body parts were used for clothing, ropes and boots; stomach walls for oil pouches; and in some cultures, the whiskers (**vibrissae**) were used as hat decorations, pipe cleaners and as ties. The ivory of walrus was used for implements, such as harpoon heads (**naulang**). Virtually the entire animal was used. In today's Eskimo society, however, and since the introduction of high-powered rifles, outboard motors and snow machines, there is little incentive to use the animals to the extent their ancestors did, and the Eskimo's dependence on them has declined.

Commercial Uses

The principal product of whaling was the rendered oil. The oil from right whales and sperm whales was particularly prized for its lubrication properties and for illumination. Most oil came from the blubber (which contains about 60% oil), but the meat, some organs and bone were also sources. The use of oil for illumination declined when petroleum took its place, but the discovery in the 20th century of hydrogenation, or the hardening process for oil, allowed it to be used in soaps and margarine. Fin whale oil was preferred for this process. Glycerin, a byproduct of oil hydrogenation (but not just whale oil), was used in industry and medicine. Other by-products of whale oil were used in the manufacture of numerous articles, including varnishes, linoleum, printing ink and dynamite.

Some whale meat was used as food, depending on the species, while some was rendered to oil. As with all animals, some whale meat tastes stronger or better than others. Baleen whale meat was generally preferred for consumption and is either eaten raw or cooked after being quick-frozen. Today whale meat is eaten primarily in Japan and is classified and named depending on its color and location on the animal.

The meat, bone and blood may be ground into a meal or powder and used for animal food and fertilizer. Other products include a tanned leather from the skin, gelatin, vitamin A from the liver, insulin from the pancreas and ambergris (once used in the manufacture of perfume) from the large intestines of sperm whales. In modern whaling, the baleen has no real commercial value, but in earlier days it was used as supporters for women's corsets, ribs of umbrellas, whip handles and an assortment of products that today use spring steel.

Pinnipeds have been commercially harvested primarily for their furs. The United States harvested northern fur seals on the Pribilof Islands until 1984 (Figure 17.9). Male seals between 2 and 5 years of age were killed and their furs processed for use in clothing. Annual harvest averages were as follows: 71,500 between 1940 and 1956; 40,000 between 1957 and 1959; 36,000 in 1960; 82,000 in 1961; and ~27,000 between 1972 and 1984. The male harvest was supplemented with a female harvest between 1956 and 1968, when over 250,000 females (average of 20,000 per year) were killed in an effort to increase pregnancy rate and female survival. It was predicted that this, in turn, would produce a sustained annual harvest of 55,000 to 60,000 males and 10,000 females on the assumption that a reduction in female densities would produce an increase in female productivity. Predicted results did not occur and the population abundance declined. The carcasses of males and females were often rendered into pet foods. White-coated harp seal pups from Canada, Greenland and the Soviet Union were harvested for their pelts. The pups were dispatched with clubs while lying on the ice. Fur seals in the Southern Hemisphere have also been harvested extensively for furs.

With the demise of commercial whaling in the 1980s, the commercial use of whales and other marine mammals has turned to a more benign industry. In the United States and elsewhere, whale and seal watching is becoming a popular activity for many people who otherwise may have never had the opportunity to observe marine mammals in the wild. In some situations, the number of boats following or observing whales has become so numerous that the government has had to regulate vessels to protect the whales from well-intentioned observers.

Exploitation and International Treaties

The early exploration and settlement of the North Pacific region was largely based on the search for wealth from the fur, animal oil and ivory that could be derived from the abundant marine mammals living in this area. One of the early Russian-financed expeditions into this region, led by the Danish explorer Vitus Bering during 1741–1742, returned with accounts of an abundance of valuable marine mammals, principally sea otters and fur seals. Bering died while shipwrecked on the Commander Islands during his second expedition to North America, and members of his crew, including the naturalist George Wilhelm Steller, survived by eating individuals of a remnant population of the now extinct Steller sea cow.

The Russians hunted sea otters and fur seals in an ever-increasing eastward direction, across the Aleutian Islands and into North America. They established trading posts and domination over local aboriginal cultures as they went. In the resultant flood of exploitation, fur seals and walrus were drastically reduced, the Steller sea cow was hunted to extinction and sea otters were extirpated from most areas of their range.

The pattern of exploitation of other pinniped species was similar around the world. Southern fur seals were harvested by Europeans and Americans intensively from the late 1700s to the early 1800s. Yankee whalers, forced to give up their fishery in the western North Atlantic because of the war with Great Britain, undertook sealing voyages to the subantarctic islands of South Georgia, South Orkney, Kerguelen, etc. Skins were traded in China for other goods, which in turn were sold in Europe for hard currency. As fur seal populations were reduced in the southern oceans, and sea otters and fur seals were reduced in the North Pacific, the sealers turned to elephant seals for oil. Many hundreds of thousands of animals were indiscriminately slaughtered and both the northern and southern elephant seals were almost exterminated. The northern elephant seal was so drastically reduced that only a single herd of about 100 animals remained on Guadalupe Island off Mexico by 1890. Since then there has been no commercial exploitation of northern elephant seals, and the herd

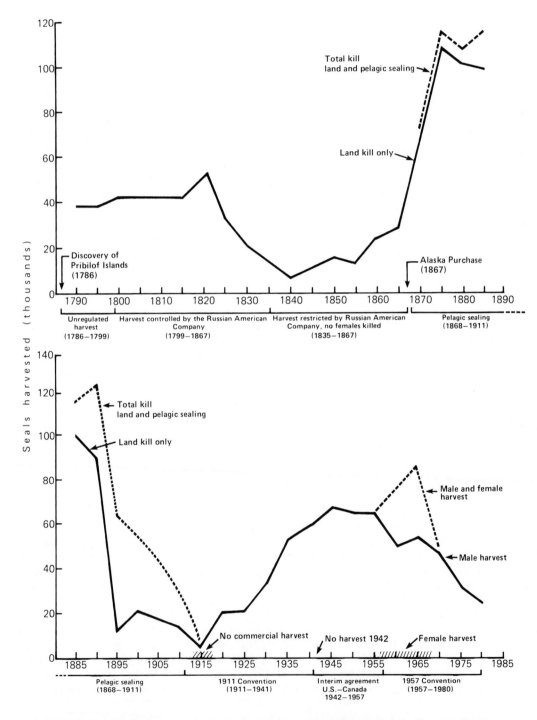

Figure 17.9. A history of the commercial harvest of northern fur seals on the Pribilof Islands, Alaska, 1786–1984. Data are 5-year averages. (From York, in Croxall and Gentry, 1987.)

has expanded to a population estimated at about 100,000 animals in 1987, while the southern elephant seal has recovered to a level of about 600,000 individuals.

The Fur Seal Treaty

By the 1830s, the Russians began leaving the west coast of North America as the quantities of furs diminished and the sealers could no longer maintain the expense of supporting those colonies. By the late 1800s, sea otter and walrus populations had been harvested to such a degree that it was no longer profitable to hunt them. The northern fur seal population, having partially recovered during the last decades of Russian control of Alaska and the Pribilof Islands, was subjected to a new round of exploitation from pelagic sealing in combination with extremely poor harvest practices on the rookeries (Figure 17.9). Between 1867 (when the United States acquired Alaska and the Pribilof Islands from Russia) and 1892, fur seals were harvested by private companies under a lease from the U.S. government, resulting in declining populations. In 1892, the United States and Great Britain entered into an agreement to encourage studies of the fur seal, to study the practice of pelagic sealing and to submit matters under dispute to a tribunal of arbitration. In 1893, the tribunal upheld the practice of pelagic sealing, and the overharvesting continued unabated with an average annual kill of over 70,000 animals from all sources for the period 1870 to 1897. In 1897, the United States attempted to unilaterally prohibit pelagic sealing under domestic law, but Great Britain and other countries continued the practice for the next 14 years.

In 1910, a new U.S. law was passed (which is no longer in effect) that continued the domestic ban on pelagic sealing, terminated the leasing system of fur seal management and harvest and banned the killing of females and pups on land. The federal government took over direct control of the fur seal harvest and herd management. Thus, domestic legislation and a determined public helped sway international opinion (much as it does today) to abolish pelagic sealing and pave the way for a multinational treaty in 1911, when the northern fur seal population was at a historical low point of 200,000 to 300,000 animals.

The exploitation of northern fur seal and sea otter populations thus led to the first international treaty wholly dedicated to the management and conservation of wildlife resources. In 1911, the Pacific rim nations of Russia, Japan, Great Britain (for Canada) and the United States signed into law the Treaty for the Preservation and Protection of Fur Seals and Sea Otters. This landmark treaty placed a total ban on pelagic sealing, vested responsibility for future seal harvests in the nation having jurisdiction over the rookeries and established a *quid pro quo* of sharing the skins from such harvests with the other nations not governing the rookeries. The treaty also established a certification system to prevent illegal commerce in seal skins and a nominal enforcement policy where ships suspected of pelagic sealing could be seized.

The Fur Seal Treaty worked exceptionally well and became a model of successful international wildlife management. The fur seal population increased rapidly at a rate of 6% to 8% over the next 3 decades, while at the same time allowing a modest harvest. From 1912 through 1917, only subsistence harvesting was allowed (although the skins were sold for fur processing). Commercial harvesting resumed in 1918 with a take of just over 34,000 animals; it continued at reasonable and controlled levels until 1940. In that year Japan, having tried in 1926 and 1936 to revise the treaty and increase the harvest level, gave notice of abrogation of the treaty. The treaty lapsed in 1941, and no efforts were made to develop any substitute until the post-war years. In 1957, the Interim Convention on the Conservation of the North Pacific Fur Seals was signed by the same four nations as the 1911 treaty. While the new agreement was pointedly designed to manage populations to the level that would "provide the greatest harvest year after year," the Interim Convention also established a scientific research program to attain the stated goal of maximum sustainable productivity of the fur seal resources.

The Interim Convention remained in effect, providing effective international stewardship over northern fur seal resources through several renewals and extensions. The Convention was officially disbanded in 1988 after the United States decided not to ratify the 1984 extension agreement—bitter irony considering the efforts originally expended to conclude this landmark treaty 77 years earlier. It is also true, however, that conditions have changed markedly in the last 30 years—sufficiently, in the view of animal protectionists, to preclude the need for a convention expressing a goal of obtaining maximum yield year after year. The Pribilof herd has suffered a decline of over 50% since the mid-1950s, initially from a reduction in herd size by deliberate take of females from 1956 to 1968. The herd declined further in the 1970s, perhaps from entanglement in marine debris and possibly other causes. As a result, protectionists lobbied against ratification, contending that the reduced population level cannot support a commercial harvest and that there are other means available for preventing resumption of pelagic sealing. The last commercial harvest of fur seals in the United States was in 1984; however, Aleut natives living on the Pribilof Islands continue to take about 2,000 fur seals a year for subsistence.

The International Whaling Convention

Whaling was practiced by the earliest coastal peoples, but generally only when animals beached or stranded themselves. Europeans began whaling from ships before the New World was colonized, but they rarely traveled great distances from shore. By the 17th century, however, when America was being settled, adventurous whalers began harvesting whales farther and farther from home. British, Dutch, Basque and German whalers were taking Greenland right whales from near Spitsbergen and the east side of Greenland. At the same time, Yankee whalers were taking the black right whale, possibly the last remnants of the Atlantic stock of gray whales and occasionally humpback whales.

Right whales were clearly preferred because of their thick blubber, long baleen, slow swimming speed and tendency to float when killed. Consequently, local populations were severely reduced before the mid-1700s. In New England, exploitation shifted to humpback whales, and after the War for Independence, whalers expanded their operations into the South Atlantic and eventually Pacific oceans. Whalers initially pursued the same species—right, humpback and sperm whales—but also began to take gray whales by the 1840s. Subsequently, an estimated 10,800 gray whales were taken along the California and Baja California coasts between 1846 and 1874 by over 60 ships and processed at 11 shore stations.

Changes in the design of whaling ships and gear allowed development of a truly pelagic fishery that led to increased catches. Changes included development of the explosive harpoon in 1868, steam vessels replacing sailing ships and the advent of the stern slipway in 1925, enabling factory ships to bring whales on board for processing. Such modifications led to further depletion of the gray whale population by the Japanese and Soviets on the feeding grounds in the Bering and Chukchi seas in the 1930s and raised doubts about the continued existence of the species.

Other whales succumbed even more quickly. American whalers, perhaps following gray whales or other species into their summer feeding grounds in the Bering Sea, took their first bowhead whales there in 1848. Within 50 years, the population was severely depleted. In the first 20 years of the bowhead fishery, an estimated 60% of the stock was removed, and over the 65- to 70-year history of the fishery at least 18,650 bowheads were killed.

As dramatic and obvious as these incidents of overexploitation were, the first international effort at whale conservation did not occur until the Convention for the Regulation of Whaling was convened in 1931, 20 years after the signing of the Fur Seal Treaty in 1911. Developed under the auspices of the League of Nations, the whaling convention, like the League, was doomed to failure because a number of key whaling nations (Japan, Germany and the Soviet Union) never joined and were not bound by its mandates. However, the convention established several important conservation principles, particularly protection of females with calves, suckling calves or immature whales and all right whales. Unfortunately, it did not inhibit overexploitation of other whales because it did not impose catch quotas and it basically allowed the whaling companies to establish economically dictated production targets.

With increasing concern over the future of exploited whale stocks and, recognizing the inability of the 1931 convention to provide any meaningful controls, the International Whaling Conference was convened in London in 1937. The main problem continued to be that whaling companies actually controlled the kill levels and, amid threats by nations to withdraw from the conference, very little was accomplished. Nothing was done to reduce Antarctic blue whale catches (the average kill of blue whales from 1928–29 to 1939–40 was 16,000 whales a year), nor to try to place rational limits on the take of any species other than right whales.

Another conference in 1938 seemed to accomplish some regulation by banning the take of humpback whales. It established season and size limits, defined five Antarctic Statistical Areas and created a sanctuary from the Shetland Islands to the eastern Ross Sea. However, again no catch limits were set, and still another meeting in 1939 actually lost ground because the members made concessions to Japan to get that nation's cooperation with existing season and size limits and to ratify the agreement. All efforts at regulation were interrupted with the advent of World War II. Some whaling continued, but most ships were converted to military use.

In 1944, 1945 and 1946, the Allied Powers met in London to establish regulations on post-war whaling. The 1944 conference included a proposed catch limit of 16,000 blue whale units (1 BWU = 110 barrels of oil, and 1 blue whale = 2 fin whales, or 2 1/2 humpback whales, or 6 sei whales); this was roughly one-half the number of whales taken during the excessive 1937–38 season. The 1945 conference established a requirement that weekly catch reports be sent to the International Committee for Whaling Statistics for timely monitoring and to ensure that catch limits were not exceeded.

The 1946 conference resulted in the signing of the International Convention for the Regulation of Whaling. This momentous agreement finally established the principle of catch limitations by quotas (which were unfortunately based upon the BWU). It also included all the major whaling nations of the world, even Japan and the Soviet Union, two achievements that were sorely lacking in earlier agreements. In addition, the convention established the International Whaling Commission (IWC), which could designate protected stocks or species; fix open and closed seasons and areas; specify size, sex and catch limits; and specify methods of whaling. On the negative side, however, there was neither an effective means of enforcement nor an observer program to document infractions. The economic needs of many nations rebuilding after the war also overshadowed the conservation desires written into the convention. The continued use of the BWU exemplified this, because by setting quotas in blue whale units, the IWC was establishing an economic quota. By not relating the quota to any species' sustainable population, the IWC guaranteed overharvesting from the beginning. Thus, as the stock of blue whales declined in the Antarctic, fin whales and then sei whales became the target species. As the total Antarctic catch declined in numbers as well as oil production, pelagic whaling fleets shifted their emphasis to the Pacific and elsewhere.

The last period of American whaling coincided with the shift in modern whaling to the North Pacific. As described earlier, the initial phase of Yankee whaling concentrated on local stocks of right, humpback and sperm whales in the late 1700s and early 1800s. The second phase was represented by the buildup of distant-water whaling in the mid to end of the 1800s. This period included the exploitation of gray and bowhead whales in the North Pacific Ocean and Bering and Chukchi seas. Most whaling by Americans had ceased by the 1920s except for a few shore stations that continued to operate in California and Baja California into the 1930s. The 1946 Convention brought full protection to gray and bowhead whales, but as the uses of whale products increased and Antarctic whale production declined, the industry began to shift to the North Pacific for other baleen and sperm whales. Three pelagic whaling fleets (factory ships with catcher boats) were active from 1954 to 1961, then jumped to seven fleets by 1963 and, by 1967, the number of fleets in the North Pacific surpassed the Antarctic as the principal whaling region in the world.

The United States no longer took whales pelagically, but in 1956 the Del Monte Fishing Company opened a new whaling station in San Francisco Bay at Richmond, California. It was joined by a second shore station nearby in 1958, and a small company began operations in Astoria, Oregon, in 1961. The latter took only a few whales and closed in 1965, and the second Richmond station closed in 1966. The Del Monte Company, however, remained in operation through 1970, taking 73 whales that year, but finally closed permanently in 1971 because its license was not renewed. During the period 1956 to 1970, the two California shore stations took a total of over 3,200 whales, including over 1,000 fin, 300 sei and 800 sperm whales; 800 humpbacks and 20 blue whales before they came under protection in 1966; and 300 gray whales under scientific permits. Thus, American commercial whaling, which pioneered so many distant-water fisheries in the 18th and 19th centuries, quietly slipped into the pages of history in 1971 with the closing of the last shore station.

During the 1950s, the scientific community expressed increasing concern about overharvesting owing to excessively high quotas. By 1959, the Antarctic whaling nations failed to reach agreement on national quotas for the 1959–60 season, and Norway and the Netherlands withdrew from the commission. The whaling nations adopted voluntary catch levels which totalled more than the previous IWC quotas (17,500 BWU instead of the 14,500 of 1957–58). To get the two nations back into the IWC in 1960, the IWC agreed to suspend catch limits for the 1960–61 and 1961–62 seasons. The two nations rejoined the IWC and agreed to a quota for 1962–63 of 15,000 BWU. The actual take for that season, however, fell short of the quota (11,306). This tended to confirm the earlier warnings of scientists that quotas were set too high. In 1963, the IWC had another crisis after receiving a report from an independent committee of scientists ("Committee of Three"), appointed in 1960 to assess baleen whale numbers in the Antarctic. The committee strongly recommended banning blue and humpback whale catches and eliminating the BWU, advocated setting quotas by species and proposed radically reducing the number of fin whales taken from 28,000 down to a maximum of 7,000. As a result, the IWC instituted a ban on humpback whale catches and a partial ban on blue whale take. However, they retained the BWU—and though the quota was reduced to 10,000 BWU, it was still much too high. When the actual take totalled only 8,429 BWU, substantially below the quota, it became obvious that whale stocks were in serious jeopardy. Still, the whaling nations could not accept the overwhelming scientific evidence and refused to lower the quota as recommended. In 1964, no agreement was reached and no official quota was set, but a voluntary limit of 8,000 BWU was agreed upon by the three remaining pelagic whaling nations of Japan, Norway and the Soviet Union (the Netherlands and United Kingdom had given up pelagic whaling the previous year). When the catch again fell far short of the voluntary limit, confirming the warnings of the scientific committee, serious doubts were raised about the future of the IWC. In an extraordinary meeting of the commission in 1965, member nations made the first real efforts to base quotas on the best available scientific information. They agreed on a step-wise reduction over 3 years (1965–66 through 1967–68 seasons) down to 3,200 BWU. The commission still could not deal with species quotas or an objective observer program, but with the sharply lower quotas, the pressure was reduced for the time being. In 1968, a new crisis threatened when estimates of sustainable yield were revised downwards on the basis of a re-evaluation of the methods for ageing whales and other adjustments affecting stock assessments. Fortuitously, Norway temporarily halted pelagic whaling for a 2-year period, allowing reduction of the total quota without imposing a reduction in the national quota of Japan and the Soviet Union. Quotas remained at 2,700 BWU for 1969–70 and 1970–71, but then were again lowered to 2,300 for 1971–72.

In 1972, spurred by a growing conservation movement around the world, a United Nations sponsored "Conference on the Human Environment" was held in Stockholm, Sweden. Among a series of major recommendations from the conference was a near-unanimous recommendation for strengthening the IWC and a 10-year moratorium on commercial whaling. At the next meeting the IWC did not adopt the moratorium, but it eliminated the BWU, adopted catch limits by species and established an international observer scheme. In subsequent years, the commission placed increasing emphasis on recommendations from its scientific committee, resulting in adoption of the "new management regime" in the mid-1970s (which established "protection" and "management" of stocks) and new criteria for determining the status of harvested stocks. The commission gradually improved its operations and increased the conservation focus of its efforts. In 1982, the IWC finally adopted an indefinite moratorium on commercial whaling which went into force and effect in 1986. Today, the IWC is wrestling with the problems of "subsistence" or "aboriginal" whaling, how this is defined, how to establish appropriate quotas for such needs and the kinds of scientific research that can best contribute to the required reassessment of whale stocks to begin in 1990.

We have described the Fur Seal Treaty and Whaling Convention in considerable detail to show how the patterns of exploitation developed and how the resources were seriously depleted before an international conservation agreement could be obtained. In the case of the IWC, attainment of a viable conservation agreement was an evolutionary process that took about 30 years before effective management and conservation could be developed; by then, many of the stocks had been overharvested to the point of being commercially and almost biologically extinct. These examples also illustrate the power of public opinion, both domestically and world-wide. American public opinion played a major role in creating a fur seal agreement and, 70 years later, it played a major role in that treaty's demise. Similarly, the mobilization of U.S. and international protectionist opinion was instrumental in achieving the necessary changes in the IWC to turn it into a conservation-oriented organization. As discussed below, much if not all of our domestic legislation and international treaties are the result of public agitation in response to various kinds of overexploitation.

The Convention for the Conservation of Antarctic Seals

The Convention for the Conservation of Antarctic Seals was agreed to in June 1972 because high seas (including pack ice) seal populations were not covered under the original Antarctic Treaty of 1959. The seal treaty applies to and sets a zero quota for southern elephant seals, Ross seals and southern fur seals, and sets catch limits on leopard seals (12,000 annually), Weddell seals (5,000 annually) and crabeater seals (175,000 annually). The treaty applies to the seas south of 60°S latitude. It also establishes three seal reserves, closed seasons and sealing zones for those species that may be harvested and requires timely reporting of any take to the Scientific Committee on Antarctic Research (SCAR). Few Antarctic seals have been taken since the treaty has been in place.

Convention for the Conservation of Antarctic Marine Living Resources

In 1982, the Convention for the Conservation of Antarctic Marine Living Resources (CCAMLR) was concluded among the Antarctic nations. This treaty also applies to the area south of 60°S latitude. It establishes an ecosystem concept for managing living marine resources in a multinational treaty and provides for management of fishes, krill, birds and mammals. It encourages coordinated research programs by member states. Finally, it establishes a goal of the "greatest net annual increment" as the lowest level of a range of population size. This level is similar to the "maximum net productivity" level of the Marine Mammal Protection Act of 1972 and basically seeks to maintain a population size that will produce the maximum number of surviving animals after subtracting deaths from births.

Convention on International Trade in Endangered Species of Wild Fauna and Flora

Concluded in 1973, the Convention on International Trade in Endangered Species of Wild Fauna and Flora (CITES) established regulations on international import and export of endangered animal and plant species and products. The treaty established three categories based on population status with appendices that listed the species in each; for example, Appendix I lists those species most endangered and for which no trade is allowed except for scientific purposes or perpetuation of the species. Appendix II lists those species that may be "threatened" with becoming endangered but for which limited trade might be allowed under strictly controlled conditions. Appendix III contains the "introduction from the sea" provision, where control must be exercised by other parties to help control movement of a species being protected by one party when it is not listed on Appendices I or II.

The signatory parties meet every 2 years to add or subtract species from the appendices and to consider other aspects of protecting endangered species. Although enforcing the CITES regulations varies tremendously between nations, the treaty has helped control what was once wide-open traffic in these plants and animals.

U.S.-U.S.S.R. Environmental Protection Agreement

The U.S.-U.S.S.R. Environmental Protection Agreement was signed in 1972 and contains 11 "areas" or sections devoted to joint research on air pollution, water pollution and a variety of other environmental aspects. Area V, Protection of Nature and the Organization of Preserves, is composed of eight projects including Project 6—Marine Mammals, which promotes cooperative research on the biology, ecology and population dynamics of marine mammals of concern to both countries. In its first 15 years, Project 6 achieved over 15 joint research cruises in the Chukchi and Bering seas and the North Pacific, at least 14 joint field studies, 12 project meetings, consultations and special workshops, and more than 12 laboratory research exchanges.

International Convention for the High Seas Fisheries of the North Pacific Ocean

The International Convention for the High Seas Fisheries of the North Pacific Ocean and its resultant Commission (International North Pacific Fisheries Commission, or INPFC) were originally created in 1952 to conduct research and management to ensure the maximum sustained productivity of the fishery resources of the North Pacific Ocean and to encourage their conservation. Marine mammal amendments were added in 1978 to create a framework for conducting joint research on the biology and incidental take of marine mammals (primarily Dall's porpoise) in the Japanese high-seas gillnet fishery for salmon. By cooperating in joint research with U.S. scientists, the Japanese were issued a permit for incidental take of Dall's porpoise in their salmon gill nets while fishing in the U.S. Exclusive Economic Zone (EEZ; the old term was fisheries conservation zone or FCZ). However, objections to the incidental take of Dall's porpoise and other marine mammals led to a lawsuit in 1987 that resulted in an injunction against issuance of the incidental take permit. Without the permit, the Japanese were unable to fish within the U.S. EEZ, resulting in problems of salmon fishery management within the INPFC. The lawsuit also terminated a major research program that contributed valuable information on Dall's porpoise biology.

Polar Bear Treaty

In 1973, the governments of Canada, Denmark, Norway, the Soviet Union and the United States signed an agreement for the protection of polar bears throughout the Arctic. The Polar Bear Treaty provides for strict control of take by member nations and prohibits the use of aircraft or large motorized vessels in hunting bears. Since one of the major problems with this species is a lack of information on life history, movements and behavior, the treaty seeks to promote or encourage a multinational research program. It also calls for member nations to take action to protect the polar bear's ecosystem.

Domestic Legislation

Prior to 1969, there were only a few domestic laws that addressed management needs or provided protection for marine mammals. Most of these were passed to implement international treaties. For example, a law to implement the 1911 fur seal treaty between Russia, Great Britain, Japan and the United States was passed in August 1912. Earlier, another domestic act related to the fur seal treaty was passed in May 1910 just prior to development of the treaty. The 1910 act was passed "to protect the seal fisheries of Alaska, and for other purposes," and coincided with the expiration of the last Pribilof Island commercial sealing lease. It provided interim protection for the Pribilof seal herd until the United States could achieve international agreement for a treaty. As mentioned earlier, the fur seal treaty was abrogated by Japan in 1941; consequently, the 1912 implementing legislation was superseded by another domestic fur seal act in 1944 that implemented the 1942 provisional agreement with Canada. The provisional agreement remained in effect until the more comprehensive multilateral interim agreement of 1957, and the 1944 domestic act remained in effect until 1966, when the current fur seal act was passed (see below). The only other major legislation directed at marine mammals, passed in 1949, implemented the new International Whaling Convention.

The Whaling Convention Act of 1949

The Whaling Convention Act of 1949 implements the Whaling Convention of 1946. It provides for the appointment of a U.S. Commissioner and Deputy Commissioner, establishes the prohibitions of the convention as unlawful acts under U.S. law and establishes licensing procedures for taking whales. The act also provides for scientific research permits and enforcement requirements, including a system of fines and penalties.

The Fur Seal Act of 1966

The Fur Seal Act of 1966 implements the Interim Convention of 1957 and prohibits any taking or possession of northern fur seals, except by government regulation or for subsistence by Indians, Aleuts and Eskimos. The act requires the Secretary of Commerce to implement the scientific research requirements of the convention and establishes a U.S. Commissioner and Deputy Commissioner to the Fur Seal Commission. The act also establishes continued administration of the Pribilof Islands by the federal government, along with the necessary community services.

The act also protects sea otters by extending the same basic protection to them as afforded to fur seals. The act provides for enforcement and penalty provisions in the same manner as for fur seals.

Although the Fur Seal Treaty was formally disbanded in 1988, the Fur Seal Act remains in force and currently provides, along with the Marine Mammal Protection Act of 1972, the management framework for the federal government to deal with fur seal issues.

The Endangered Species Conservation Act of 1969 (ESCA)

In 1969, the Endangered Species Conservation Act was passed, expanding upon the original Endangered Species Preservation Act of 1966. The 1966 act applied only to "native species" of fishes and wildlife, and it had so many deficiencies that it allowed little more than enabling the Secretary of Interior to acquire needed lands to meet the goal of protecting endangered species and their habitat.

The 1969 act expanded upon the international aspects of protecting endangered species and authorized the Secretary of Interior to develop a list of species or subspecies of fishes and wildlife "threatened with worldwide extinction," and to prohibit their importation into the United States, with some exceptions. The act was passed in response to public opinion opposed to the excesses in pelagic whaling and the deaths of dolphins caught incidental to commercial tuna purse seine fishing. One of the principal results was the inclusion in the first endangered species list of eight species of great whales (blue, fin, sei, humpback, right, gray, bowhead and sperm whales). Several of those species were listed because of the whaling nations' tendency to overharvesting. These species still appear on that list, although the gray whale appears to have recovered to pre-exploitation population levels, and several other species are found in numbers not usually associated with endangered status (i.e., fin, sei and sperm whales).

The 1969 ESCA required the Secretaries of Interior and State to seek the convening of an international ministerial meeting to conclude a binding international convention on the conservation of endangered species. The result was the CITES described previously. While that convention deals specifically with controlling trade of endangered species, it nevertheless had important implications and influence on domestic legislation, in particular in recognizing varying degrees of vulnerability (CITES Appendices I-III) and in providing regulatory measures designed to reflect that vulnerability.

The National Environmental Policy Act of 1970

The passage of the National Environmental Policy Act of 1970 (NEPA) established the principle that the environment should not be degraded and that it should be restored where previously damaged. This act created the "environmental impact statement" (EIS) and required preparation of an EIS when major federal actions might significantly affect the human environment. The clear intent is to force federal agencies to consider environmental impacts of proposed actions before it commits itself to them. Therefore, whenever federal funds are being used for a development project, or a federal permit is required (as in Army Corps of Engineers dredging permits), any federal agency with oversight or permitting authority must prepare an EIS. During preparation of the EIS, the agency must consult with any other federal agency with special expertise or jurisdiction pertaining to the expected environmental impacts. The EIS and consultation comments are available to the public and remain a part of the record throughout the review process. Thus, the public can request litigation over an agency's interpretation of the environmental impacts and over the agency's recommended actions regarding the perceived environmental impacts, if any. Just such a case was brought against a comprehensive research permit granted by the National Marine Fisheries Service (NMFS) under the Marine Mammal Protection Act to allow research and assessment of killer whale populations and a small live take for studies on propagation in captivity. The resultant court decision in 1986 nullified the NMFS permit and established the precedent that extensive and controversial marine mammal research permits may require an EIS.

The Pelly Amendment of 1971 to the Fishermen's Protective Act of 1967

The Pelly Amendment of 1971 gives the President of the United States authority to prohibit importing fish products from nations conducting fishing operations in such a way as to diminish the effectiveness of any multilateral fishery conservation treaty or agreement in which the United States is a participant. Under the Pelly Amendment, the Secretary of Commerce must "certify" to the President when such deleterious fishing operations are being conducted by foreign nations, and the Secretary of the Treasury is responsible for enforcement. The President may choose to apply any level of enforcement from no embargo to a total embargo of fishery products from an offending nation.

The Pelly Amendment was originally applied only to fishery agreements, but in 1978 the Fishermen's Protective Act was further amended to provide protection to wildlife and to give certifying authority to the Secretary of Interior.

While the Pelly Amendment is a powerful weapon for influencing the behavior of foreign nations, it is a two-edged sword and must be used with caution, lest it spark an economic war. In 17 years of existence, it has been used only twice in the context of influencing whaling conservation policies; the most recent instance was in 1987–88, when Japan was certified for conducting scientific research whaling for minke and sperm whales in the Antarctic in opposition to specific recommendations by the IWC.

Marine Mammal Protection Act of 1972

A large part of the growing environmental consciousness of the late 1960s focused on the exploitation of marine mammals. Not only was the U.S. public concerned about the depletion of large whale populations, it was opposed to the incidental kill of hundreds of thousands of dolphins in the yellowfin tuna purse seine fishery and the harvest of white-coat harp seal pups in the northwest Atlantic. It was public outcry for protection of these groups of animals that resulted in the Marine Mammal Protection Act of 1972 (MMPA).

The MMPA split federal jurisdiction over species by giving responsibility for whales, seals and sea lions to the Department of Commerce and walrus, sea otter, polar bear, manatees and dugongs to the Department of Interior. The fundamental element of the MMPA was that it established a moratorium on the taking of marine mammals, with a few notable exceptions, such as providing a permit system to allow incidental take in commercial fisheries and for scientific research and public display. It established a ban on importing marine mammals and marine mammal products, with certain exceptions. The MMPA established the three-member Marine Mammal Commission, supported by a nine-member Committee of Scientific Advisors that was charged with oversight responsibilities in reviewing government regulations and the entire permitting process. The commissioners also are involved in shaping U.S. policy in international treaties that deal with marine mammals. The MMPA includes an exemption from the moratorium for Aleuts, Indians and Eskimos in Alaska for subsistence take and use of marine mammals in arts and handcrafts. It also allows for designating species as "depleted" which, until 1988, prohibited incidental take in commercial fisheries. However, under the 1988 amendments to the MMPA, some incidental take of depleted populations is allowed for 5 years; during this time, the government must determine levels of incidental take and the impact of the take on affected marine mammal populations, and it must develop a permanent management regime for incidental take.

During the 16-year lifetime of the MMPA, about 300 research permits have been granted, along with a like number of public display permits. The agencies have established extensive regulations governing general permits for incidental take in commercial fisheries and have supported minimal research programs in response to the requirement of the act to monitor the status of marine mammal species.

The principal failing, thus far, has been the inability of the agencies and the Marine Mammal Commission to develop a workable system to enable return of marine mammal management to the states. Only one state, Alaska, has made a meaningful attempt to secure return of management, but its efforts foundered on the problems of determining population sizes and status relative to optimum sustainable population levels as required in the act and, more recently, the problems of designing a management plan that could equally address native peoples and other Alaskans in the context of the basic waiver under the act.

Nevertheless, the MMPA is a landmark piece of legislation and, even though foreign peoples sometimes have great difficulty in understanding the extreme concern that the United States has for marine mammals, the concepts are becoming more accepted throughout the world and more entrenched in American public opinion.

The Endangered Species Act of 1973

The Endangered Species Act of 1973 (ESA) was built upon the ESCA of 1969 and corrected many of the shortcomings of the earlier act, as well as enhancing its regulatory capabilities. The ESA retains the split jurisdiction over marine mammals as established in the MMPA.

The ESA provides for the conservation of endangered and threatened species of fish, wildlife and plants. It is not restricted entirely to U.S. species but can be applied to foreign species if there is substantial U.S. involvement in habitat use or other factors. For example, the United States listed a Mexican fish in 1979, the totoaba, because it was impacted by destruction of its estuarine breeding areas resulting from reduced Colorado River water flowing into the Gulf of California. The ESA established two lists of endangered and threatened species, to be reviewed and updated every 5 years. Briefly, the criteria for listing species are:

- the present or threatened destruction, modification or curtailment of its habitat or range;
- overutilization for commercial, sporting, scientific or educational purposes;
- disease or predation;
- the inadequacy of existing regulatory mechanisms; or
- other natural or man-made factors affecting its continued existence.

The act also contains a key provision for designating **"critical habitat"** of listed species to provide added protection for the species' environment. It requires the development and implementation of "recovery plans" to ensure a listed species' survival and conservation. At the present time, there are recovery plans for several species of marine mammals, including Hawaiian monk seals, humpback whales and right whales. One disadvantage of the act is it applies only to federal actions or federally funded actions, except when a cooperative agreement has been established with affected states. Section 7 of the act establishes a consultation process between the project agency and the Secretaries of Commerce or Interior, as appropriate. This process embodies the key provisions of the National Environmental Policy Act in that a biological assessment must be prepared and an EIS if any effects are foreseen.

During the ESA's existence, it has contributed significantly to the preservation of populations of endangered or threatened species. With regard to marine mammals, it has provided a framework within which to identify and monitor actions that may impact these animals—a prime example is the extensive and continuing consultation process, including mitigating measures required of the Minerals Management Service over all aspects of outer continental shelf energy development. Because of the existence in coastal waters of several listed species, including some whales and sea otters in California, all phases of development from initial leasing to production are scrutinized by involved agencies and the public.

Except for a few lingering problems such as inadequate funds to implement recovery programs and lack of review of proposals to list species and designate critical habitat in a timely manner, the ESA must be considered a highly successful landmark legislation.

The Magnuson Fishery Conservation and Management Act of 1976

The Magnuson Fishery Conservation and Management Act of 1976 (MFCMA) is another landmark piece of fisheries legislation in that it provides for conservation and management of fish stocks within an exclusive fishery zone that is expanded from the old 12 miles (19 km) to 200 miles (320 km). The MFCMA also established eight Regional Fishery Management Councils that prepare fishery management plans for applicable species and deal with seasonal restriction, allowable catch, etc. An early amendment to the MMPA made its provisions applicable under the MFCMA. A prime example of the implications of this is the requirement that foreign fishermen obtain permits if they might take marine mammals incidental to their fishing operations within the U.S. EEZ.

The Packwood Amendment to the MFCMA provided some additional enforcement leverage to the Pelly Amendment discussed earlier in that the Secretary of Commerce may certify a foreign nation under the MFCMA for diminishing the effectiveness of a conservation agreement. If a nation is certified, the Secretary is required to reduce that nation's allowable catch within the EEZ by 50%. Unfortunately, this provides continually decreasing influence on foreign nations as U.S. domestic fisheries use a larger portion of available catch.

The Whale Conservation and Protection Study Act of 1976

The most recent and, perhaps, least significant domestic legislation dealing with marine mammals is the Whale Conservation and Protection Study Act. Passed by Congress in 1976 to augment research on the ecology, habitat requirements and population dynamics of all whales found in waters under U.S. jurisdiction, the Act was never implemented to its intended level. No funding was obtained specific to this act and, consequently, neither an increase in whale research nor the bilateral agreements with Canada and Mexico required under the Act were achieved.

Finally, it is interesting to note that most of the significant domestic legislation, such as the ESCA, NEPA, the Pelly Amendment, the MMPA, ESA and MFCMA and several of the important international treaties that affect marine mammals (Antarctic Seal Treaty, CITES, U.S.-U.S.S.R. Environmental Agreement and the Polar Bear Treaty) were all achieved during a relatively short period from 1969 to 1976 (Table 17.4). In contrast, during the last 12 years, relatively few domestic laws and international treaties dealing wholly or in part with marine mammals have been developed. The most significant of these were the amendments to the INPFC in 1978, CCAMLR in 1982 and recent agreements designed in part to control the discarding of potentially entangling debris in the oceans. While there are undoubtedly a number of factors that contributed to that situation, one inescapable conclusion is that the environmental advocates of the 1980s are much less successful in achieving lasting statutory goals than the generation of the 1960s and early 1970s, and that the executive and legislative branches of government during the productive period of 1969–76 were apparently more environmentally conscious than in recent years.

Table 17.4. Chronology of marine mammal exploitation and protection.

Before 1600	Subsistence hunting by coastal peoples on local stocks—worldwide.
1600–1750	Beginning of "distant water" take of right, bowhead, humpback, and sperm whales by Basque, Dutch, English and American whalers.
1750–1840	Development of large-scale commercial exploitation of pinnipeds, including southern ocean fur seals and northern fur seals and sea otters. Continued development of distant-water whaling; Americans move into Pacific Ocean and elsewhere.
1840–1900	Sealers shift to elephant seals after commercial extinction of northern and southern fur seals. Whaling increases in North Pacific (gray whales are primary target, followed by bowheads).
1868	Explosive harpoon developed, forever changing the efficiency of taking whales. United States begins leasing system for Pribilof Island fur seal harvest; other countries pursue pelagic sealing.
1900–1918	Northern fur seal population at lowest level; recovery begins under new laws.
1910	First U.S. law for marine mammals passes, a fur seal act prohibiting killing of females and pups on land and terminating the leasing system of fur seal management and harvest.
1911	First multilateral marine mammal treaty—the "Treaty for the Preservation and Protection of Fur Seals and Sea Otters"—prohibiting pelagic sealing and incorporating protection of females and pups.
1912	U.S. law implements the 1911 fur seal treaty.
1918	U.S. resumes limited commercial harvest of fur seals; population begins to increase.

Table 17.4.—*Continued.*

1919–1944	Antarctic whaling shifts from shorebased to pelagic.
1925	Stern slipway invented allowing for truly pelagic whaling; annual take increases by five- to tenfold.
1931	First International Whaling Convention; protection for right whales and females with calves, but no catch quotas or other means of limiting take; several key whaling nations refuse to join.
1937–1939	A series of international whaling conferences to institute conservation principles into a new convention.
1941	Japan abrogates 1911 fur seal treaty.
1942	United States and Canada establish a provisional agreement to continue fur seal protection.
1944	New U.S. fur seal act passes with all the protective features of the earlier acts and implements the 1942 provisional agreement with Canada.
1946–1971	Beginning of the end of marine mammal exploitation.
1946	International Whaling Convention incorporates whaling quotas—but only as "blue whale unit" (BWU) equivalents; protects gray and right whales.
1949	U.S. Whaling Convention Act implements the 1946 Convention. Whaling nations ignore warnings of scientists on depletion of stocks; Antarctic quotas can not be met by end of 1950s.
1957	Interim Convention on Conservation of North Pacific Fur Seals signed by same four nations as in 1911; treaty embodies scientific research and goal of sustained yield of fur seals.
1960s	Whaling emphasis shifts to North Pacific and surpasses Antarctic by 1967.
1963	Report on status of whale stocks by "Committee of Three" proposes ban on take of blue and humpback whales and elimination of BWU.
1965	IWC agrees to ban blue and humpback whale take and reduces Antarctic quotas over 3 years.
1966	New domestic fur seal act implements 1957 interim convention and incorporates protection and scientific principles.
1969	Endangered Species Conservation Act passes; 8 species of great whales listed as endangered.
1970	National Environmental Policy Act (NEPA) establishes protection for marine mammal environment.
1971	Last U.S. whaling station goes out of business in California.
1972 to present	The period of protection—has the pendulum swung too far?
1972	U.N. sponsored "Conference on the Human Environment" recommends a 10-year moratorium on commercial whaling. Subsequently, IWC eliminates the BWU, sets whale quotas by species, establishes an observer scheme and emphasizes scientific recommendations; whale quotas gradually reduced or eliminated for most stocks of large whales in the Antarctic and North Pacific. U.S. Marine Mammal Protection Act provides protection to all marine mammals under U.S. jurisdiction.

Table 17.4.—*Continued.*

1972—con't.	Antarctic Treaty nations sign Convention on Conservation of Antarctic Seals providing protection to 3 species and sustainable quotas on 3 others. U.S.-U.S.S.R. Environmental Protection Agreement promotes cooperative research on marine mammals.
1973	Convention on International Trade in Endangered Species of Wild Fauna and Flora establishes international controls on the trade of endangered species, and ancillary protection for many marine mammals under similarity of appearance provisions. The Polar Bear Treaty establishes a multinational management and research effort for polar bears. U.S. Endangered Species Act of 1973 builds upon the 1969 law and expands habitat protection and incorporates a consultation process among federal agencies under NEPA provisions.
1974 to present	Some marine mammal populations increase under MMPA, leading to increasing fishery interaction problems.
1976	Magnuson Fishery Conservation and Management Act extends jurisdiction of the MMPA throughout the U.S. Exclusive Economic Zone. Whale Conservation and Protection Study Act attempts to bolster domestic and international whale conservation (ineffective).
1978	Fishermen's Protective Act extends provisions of the Pelly Amendment to wildlife, putting teeth into enforcement of IWC recommendations. Annexes to the International Convention for the High Seas Fisheries of the North Pacific Ocean provide a framework to deal with incidental take of marine mammals, mainly Dall's porpoise, in high seas fisheries.
1980s	Entanglement in marine debris implicated as one source of mortality of marine mammals, sea birds and sea turtles.
1982	IWC agrees to a moratorium on commercial whaling. Antarctic nations agree to a Convention for the Conservation of Antarctic Marine Living Resources, establishing the concept of ecosystem management of natural resources.
Mid-1980s	Treaties, domestic regulations and educational efforts begin to produce results in reduction of marine debris.
1986	IWC moratorium on commercial whaling goes into effect.
1988	U.S. fishing industry and environmental community helped draft amendments to MMPA that will allow limited take of depleted species.

Additional Reading

Bonner, W. N. 1982. Seals and Man. University of Washington Press, Seattle. 170 p.

Breiwick, J. M. and H. W. Braham, eds. 1984. The status of endangered whales. Special section, Marine Fisheries Review 46(4).

Croxall, J. P. and R. L. Gentry, eds. 1987. Status, biology, and ecology of fur seals. Proceedings of an international symposium and workshop, Cambridge, England, 23–27 April 1984. U.S. Dep. Commerce, NOAA Technical Report NMFS 51. 212 p.

Environmental Law Institute. 1977. The Evolution of National Wildlife Law. Prepared for the President's Council on Environmental Quality, Washington, D.C., U.S. Govern. Printing Office. 485 p.

Fay, F. H. 1982. Ecology and biology of the Pacific walrus, *Odobenus rosmarus divergens* Illiger. U.S. Fish and Wildl. Serv., North American Fauna, 74:1–279.

Gaskin, D. E. 1982. The Ecology of Whales and Dolphins. Heinemann, Exeter, New Hampshire. 459 p.

Gentry, R. L. and G. L. Kooyman, eds. 1986. Fur seals: Maternal Strategies on Land and at Sea. Princeton Univ. Press, Princeton, New Jersey. 291 p.

Haley, D., ed. 1986. Marine Mammals of Eastern North Pacific and Arctic Waters. Pacific Search Press, Seattle, WA. 295 p.

Harrison, R. J. and J. E. King. 1965. Marine Mammals. Hutchinson University Library Press, London and New York. 192 p.

Henderson, D. A. 1972. Men and Whales at Scammon's Lagoon. Dawson's Book Shop, Los Angeles, CA. 313 p.

Kenyon, K. W. 1969. The sea otter in the eastern Pacific Ocean. U.S. Fish and Wildl. Serv., North American Fauna 68:1–352.

King, J. E. 1983. Seals of the World. Comstock Publishing Associates, Cornell University Press, Ithaca, New York. 240 p.

The Library of Congress, Congressional Research Service. 1974. Treaties and other International Agreements on Fisheries, Oceanographic Resources, and Wildlife to which the United States is Party. U.S. Govern. Printing Office, Washington, D.C. 968 p.

Macdonald, D., ed. 1984. The Encyclopedia of Mammals. Facts on File Publication, New York. 895 p.

Rice, D. W. and A. A. Wolman. 1971. The life history and ecology of the gray whale (*Eschrichtius robustus*). Amer. Soc. Mammalogists, Spec. Publ. No. 3:1–142.

Scammon, C. M. 1968. The Marine Mammals of the North-western Coast of North America, Described and Illustrated, Together with an Account of the American Whale-fishery. Dover Publications Inc., New York. 319 p.

Scheffer, V. B., C. H. Fiscus and E. I. Todd. 1984. A history of scientific study and management of the Alaska fur seal (*Callorhinus ursinus*), 1786–1964. NOAA Tech. Rep. NMFS SSRF-780:1–70.

Schevill, W. E., ed. 1974. The Whale Problem: A Status Report. Harvard University Press, Cambridge, MA. 419 p.

Tillman, M. F. and G. P. Donovan, eds. 1983. Historical whaling records, including the proceedings of the international workshop on historical whaling records, Sharon, Mass. Sept. 1977. Repts. Internat. Whaling Comm. Spec. Issue No. 5. Cambridge. p. 1–269.

VanBlaricom, G. R. and J. A. Estes, eds. 1988. The Community Ecology of Sea Otters. Ecological Studies 65, Springer-Verlag, New York. 247 p.

18

Resource Uses in Conflict

Robert R. Stickney and Frederick G. Johnson

Introduction

Assume for the moment that your livelihood depends upon income from a fishery. You might be a charter boat operator for a recreational fishery, the manager of a fish processing plant or a halibut longliner, but as time goes on, you find that your income and livelihood are in jeopardy. We have just arrived at a very common problem—what happens if there are not enough fish for everyone who wants them? As fisheries develop, whether they are commercial or recreational in nature, this problem tends to emerge, and its solution generally begins with **allocation**.

Allocation is the "A" word in fisheries, and it deals with the decisions of fishery managers (who have the responsibility to establish social policy regarding fisheries) to determine who gets to catch fish. Another way of looking at the allocation issue is that it dictates who does not get to catch fish. The allocation of fishery resources in high demand almost guarantees that conflicts will greet its outcome, no matter how the allocation is distributed across the users of the resource. That, however, is only the beginning of the conflicts that might arise.

Let us also assume that the ecological habitat which supports the fishery resource in question supports human interests other than fisheries. This is generally the case with nearshore marine or inland fisheries, in that the same habitats in which fishes or shellfishes reside also serve as water sources, transportation corridors, electric power generation sites, municipal sewage discharge sites, logging or agricultural areas or a host of other alternate uses. To the extent that these alternate uses adversely impact fish production, we essentially have to "allocate" a formerly harvestable portion of our fishery stocks to cover these non-fishery induced losses. Perhaps this is easier to understand if we return to a surplus production model or spawner/recruit model as outlined in Chapter 13.

These models quantify the excess productivity of a fishery stock to turn that excess productivity into a renewable harvest. If we refer specifically to a spawner/recruit model, the model predicts that if we allow a certain number of adult salmon to travel upstream to spawn (that is the **escapement**), we then expect a certain number of their returning progeny (the **harvestable surplus**) to be available for harvest, while still leaving the escapement needed for future returns. The fact of the matter is that if various human activities degrade the habitats these fish occupy at different times in their lives enough to adversely affect the survival of the fish, all of these additional losses to the resource must be subtracted from the harvestable surplus and not from the escapement. In that sense, by using fishery habitats for other purposes, we are allocating a portion of our fishery resources to the alternate uses.

In this chapter, we examine several major ways in which uses of natural resources present conflicts. We will first discuss examples of conflicts related to fisheries themselves, followed by examples of how alternate uses of land and water conflict with fisheries.

Fishery Conflicts

As implied above, many fishery conflicts are related to patterns of allocation of a fishery resource among fishermen. A wide variety of fishing user groups may be involved, depending upon the fishery and its location. Several of these user groups should be mentioned and the generic conflicts between them noted.

The recreational versus commercial dichotomy is a classic and recurrent source of discord within fisheries. Often the same individual will take part in both kinds of fishing, but that does little to take the heat out of the conflict between the two groups. Fishery managers must protect the interests of both groups, but striking the appropriate balance of allocation between them is analogous to walking a tightrope.

Another dichotomy that is increasingly emerging as an issue in governmental jurisdictions concerns the extension of the treaty rights of native peoples to modern fishing situations. Across the United States and Canada, indigenous cultures and tribes are exerting, for good reason, the conditions of agreements they signed long ago that secured their rights to harvest fishes and shellfishes as they had done for centuries. Fishery managers and governments are compelled to uphold the stipulations of these treaties when it comes to allocation between tribal and non-tribal user groups.

Treaties between native tribes and modern governments essentially reflect two different nations allocating resources in the same geographic area. International treaties, on the other hand, concern agreements between nations in different geographic areas. Highly migratory species such as salmon, tuna, whales and many others cross the jurisdictions of many coastal nations in the process of their migrations. Nations that wish to share the harvest of these stocks must reach agreement among themselves on how to do so while protecting the resources at the same time. In large part, the history of fishery management reflects the history of these international negotiations.

There are many other allocation-related conflicts that fishery managers must deal with, including high-technology commercial interests versus subsistence or low-technology artisanal interests, which is a particularly acute issue in developing countries. Perhaps one of the more perplexing and ecologically complicated issues, however, deals with user groups that target the same species, but at different times in the life cycle, or those that target the normal prey of commercially important species. The herring fishery in the North Pacific is a good example of both cases. Multiple fisheries for Pacific herring (*Clupea harengus*) involve capture of the whole fish for use as food, the whole mature fish as a source of roe, or the roe alone after it is deposited on seaweed. The same areas (Alaska, British Columbia and Washington State) support important salmon and halibut fisheries, and the herring are important forage fishes for the salmon and halibut. A perplexing problem faces those responsible for allocating the herring resource: How many herring should be left for each of the herring harvesters, and to the ecosystem that supports the other fisheries?

Incidental Catches

Another source of friction between fisheries concerns incidental catches, which are also known as bycatches. An extreme example of a bycatch problem concerns shrimp trawlers in the Gulf of Mexico, where the target organisms (*Penaeus* spp.) are often only a relatively small portion of the total weight of the catch. The remainder consists of many species of fishes that might be caught by other fishermen, both sport and commercial, such as croakers, seatrouts, porgies, sea robins and spot. In the northwest Gulf of Mexico, for example, the targeted shrimp represent an average of just 7% of the trawl catch, and the rest ends up primarily as waste. This problem is inherent in the **selectivity** of trawl gear, which in general is quite low (for the target species as opposed to non-target species).

Another bycatch problem with the Gulf of Mexico shrimp fishery involves the protection of sea turtles, which may also be taken in the trawl nets. In order to minimize the turtle bycatch, special trawl net configurations called turtle exclusion devices (TEDs) have been designed and used, occasionally over the objections of commercial shrimpers. TEDs, net mesh restrictions, changes in mesh orientation and escape vents are all measures intended to reduce bycatches that may be imposed upon commercial fishermen at the discretion of fishery managers. In many cases, requirements for these modifications are resisted by the fishermen because they are expensive and because they may reduce catches of desired species. Often, however, commercial trawlers welcome design changes that increase the selectivity of their gear if the changes reduce the amount of time it takes the fishermen to sort the catch. Shrimp trawlers on the Pacific Coast, for instance, were greatly assisted by changes in net design that excluded fishes from the catch of small shrimp because the fish have to be manually removed before the shrimp undergo mechanical processing.

Perhaps the most publicized and emotionally charged bycatch problem relates to the kills of porpoises and dolphins by tuna seiners. This topic is covered in the previous chapter, and has been an important issue since passage of the U.S. Marine Mammal Protection Act in 1972. The Inter-American Tropical Tuna Commission is currently grappling with the mammal bycatch issue on an international level.

Another bycatch issue now in the international limelight is incidental catches of marine mammals, birds and salmon by the red squid driftnet fleet. Driftnets are nylon monofilament nets that are fished in international waters of the North Pacific, primarily by Japanese, South Korean and Taiwanese fishermen. Each vessel lays out a series of long driftnets that measures 30 miles, and on a given day as much as 30,000 miles of driftnets is deployed in the North Pacific. The target species are large squid and smaller species of tuna, but many animals that encounter the nearly invisible nets become entangled and die. Salmon fishermen in Alaska are concerned that the driftnets may intercept enough salmon of certain species to curtail their own catches. Environmentalists are concerned about the incidental kills of marine mammals and birds, and have described driftnet fishing as the "biological strip mining" of the sea.

Ghost Fishing

Ghost fishing refers to the capacity of gear that has been lost or abandoned to continue to capture fishes or shellfishes. Lost driftnets, gill nets, sections of netting material and lost traps for shrimp, crab, lobster or fish present the most serious potential for ghost fishing. There are two major factors that exacerbate this problem. One has to do with the persistence of nylon polypropylene netting and certain other components of fishing gear in the marine environment. Since these substances are not biodegradable, the losses of valuable species that become entangled or trapped by a piece of gear can continue for years. The second factor is that when one animal becomes entangled, it often acts as bait to attract others, which also become entangled, and the killing cycle can continue in this manner.

Solutions to the ghost fishing problem include conscientiously attending gear so that it does not become lost in the first place, properly disposing of gear that is worn out, and incorporating biodegradable substances. Biodegradable fasteners for the tops of crab pots are being considered as a feature that would disintegrate and allow the crabs to escape from a lost pot. Biodegradable plastics or other synthetic fibers that retain their strength, non-visibility and durability over a fishing season would go a long way towards reducing the ghost fishing of lost or sunken gill nets and driftnets.

Nuisance Species

Some of the species that we wish to protect from incidental harm from fishing activities, such as birds and marine mammals, can also be serious nuisances in certain fishing situations. Black cod longliners, for example, find that killer whales eagerly strip black cod from longlines as the lines are being retrieved. Diving birds are a common problem for net-pen fish farmers and aquaculturists raising mussels and other shellfishes by suspended line techniques. Gill netters, and sometimes even trollers and recreational fishermen lose salmon to marauding sea lions. All of these examples, in addition to many not mentioned here, complicate the issue of protecting some species while at the same time harvesting others.

Aquaculture Controversies

In Chapter 16, we saw how net-pen salmon rearing in Washington has been met with opposition by upland view property owners. Initial opposition to net-pen salmon farming came in the form of "visual pollution" and has spread to concerns about environmental contamination from wastes escaping from the net-pens, the release of chemicals and antibiotics into the water, escape of exotic strains and species, and the potential transmission of diseases. In some instances, opposition has come from commercial fishermen who look at aquacultured salmon as direct competition. Opposition to aquaculture has spread to upland commercial operations in some instances and is not restricted to Washington. Net-pen aquaculture has also become controversial in Maine and British Columbia. In addition, concerns about further expansion of the trout farming industry in the Hagerman Valley of Idaho have been raised because of fears that the Snake River is being negatively impacted by fish farming wastes.

In the southern United States, where catfish farming dominates the aquaculture industry, there has been little, if any, opposition to aquaculture development. Fish farms are viewed as desirable enterprises that provide jobs and enhance local economies. Permitting regulations, which are very stringent with respect to coastal aquaculture in many states, tend to be very loose with respect to inland fish farms, particularly in the south, where 30 years of experience in fish farming has been gained.

Much of the difference between regions that accept and support aquaculture development and those wherein aquaculture is actively opposed can be viewed in terms of human population density and land use practices. The Puget Sound area of Washington, for example, is home to a large percentage of the people in the state. Land values are high, the water receives a great deal of recreational use, and there is an actively pursued environmental ethic that aims to retain the quality of life. New industries, particularly those which carry any hint that they might negatively impact environmental quality, are viewed in an extremely negative fashion by some groups of people. The situation in places like the Hagerman Valley of Idaho, however, is quite different with respect to human population density. Unlike the Puget Sound region, no large concentrations of people live along the Snake River. Use of the area for recreation is high, and people are concerned that further expansion of the trout industry could cause environmental degradation that would impair sport fishing and other water-related activities.

Salmon and trout farms largely exist in non-agricultural areas, while catfish farms developed as a result of agricultural diversification. Many catfish farmers practiced animal husbandry and grew row crops before entering the catfish business. In fact, it is not uncommon to see catfish production as one component in a multi-faceted farming activity today. Large catfish farms, such as in Mississippi and Arkansas, were constructed in areas where such crops as cotton, rice and tobacco once dominated. Land use was agricultural in the past and continues to be agricultural today. Catfish farming is generally viewed positively by that segment of the local public that is not actively involved in the industry.

In Japan and Norway, net-pen culture of various species is being practiced with a high level of public support. One reason for the difference between net-pen enterprises in those two countries and in the United States is the fact that seafood production is critical to the populations of Norway and Japan. The United States, a major net importer of fish and fishery products, does not need to raise fish, either from an economic standpoint or to provide food for its citizens. In Norway, a country where only a small fraction of the land can be used for agriculture, the economy depends heavily on the sea (large deposits of North Sea oil are the other major source of income). In Japan, the bulk of the animal protein in the human diet is derived from seafoods. Japanese fishing boats traverse the world, and aquaculture has been developed for a large variety of species. Net-pens that are eyesores to many in Puget Sound are viewed positively in Japan because they provide much-needed food. There is no doubt that Japan pushed net-pen culture to its limit in some areas. Pollution from

Figure 18.1. Nori culture nets in a Japanese bay. Nets and the poles from which they are suspended cover many acres of water and are viewed positively by the Japanese because they provide a source of much-needed food.

net-pens had some severe impacts on local environments, but regulations on the numbers of net-pens have greatly improved the situation in recent years. Today, net-pens, commercial fisheries and various other water-related activities co-exist in harmony throughout Japan. The same situation will occur in certain parts of the United States coastal zone only when people perceive that we have a basic need for more fish.

Somewhat less controversy surrounds the production of oysters, clams and other species that can be reared in conjunction with the bottom. As long as oysters are grown on intertidal or subtidal bottom areas, there is often public acceptance in the United States. When oyster culturists attempt to hang their animals from rafts and longlines (hanging culture), which have floats at the surface and ropes criss-crossing the water, major opposition can be anticipated. Obstruction of navigation and detraction from views are among the criticisms. This contrasts with Spain and Japan, where string and raft culture are employed extensively. Japan has large areas devoted not only to hanging culture of invertebrates, but also to the production of seaweeds such as nori (*Porphyra* spp.). During the fall and winter, large areas in certain bays are dotted with thousands of poles, which support the nets upon which nori is grown (Figure 18.1). Aquaculture is seen as a highly valued enterprise with positive economic impact on local communities and as being within the best interest of the nation. In the United States, the economic impact of oyster production benefits a small percentage of the population, and the perceived negative impacts of developing the industry are often viewed as outweighing the public benefits.

Exotic Species Introductions

When a species or organism is moved from a location where it is indigenous to one where it is not native, the movement is known as an exotic introduction. Under this definition, it is not necessary to move a plant or animal from one nation or continent to another. For example, walleye (*Stizostedion vitreum*) are native to various northern and northeastern states but can currently be found in the west and in various southern and midwestern states outside its native range. Its occurrence in all those regions is the result of exotic introduction.

Many exotic introductions have been highly beneficial to humans. Cattle, chickens, wheat, soybeans and, in fact, the majority of the types of livestock and grains produced in the United States were introduced from other regions of the world (corn and turkeys are two notable exceptions since they represent native species). Some fishes that were introduced to North America have been well received. Examples include brown trout (*Salmo trutta*) and goldfish (*Carassius auratus*), which are not generally seen as a nuisance. Other species of fishes are not universally appreciated. Many people are not pleased with the presence of common carp (*Cyprinus carpio*), and there are mixed feelings about the presence of tilapia (*Oreochromis* spp. and *Tilapia zillii*), which have been introduced as aquaculture species to control aquatic weeds and to provide forage for sportfish. In some cases, tilapia have displaced more desirable native species, which can certainly lead to controversy.

Perhaps the most controversial fish introduction of this century involves the grass carp or white amur (*Ctenopharyngodon idella*). Introduced from the Amur River in Asia in the 1960s, grass carp were first evaluated in the laboratory for their value in controlling aquatic weeds. The fish is a strict herbivore except during the early stages of its life, when it consumes plankton. After a few years of study, the state of Arkansas was so pleased with the fish that it was legalized for introduction and stocked as a biological weed control agent.

Many biologists insisted that the fish could not successfully reproduce in the United States because its spawning requirements were very precise, and appropriate spawning conditions did not exist in the United States. However, within a few years, grass carp fry were found in the Mississippi River, and have also been reported from other rivers. The jury is still out on the survivability of grass carp in the United States, however, as few, if any juveniles have been reported in areas where spawning has apparently occurred. The grass carp fry may not be able to effectively avoid predators and thus may disappear. In any case, fear that grass carp would become a problem by eating all of the aquatic vegetation available or that it would displace more desirable native fishes prompted over 30 states to ban the introduction of grass carp into their waters.

During the 1970s, sterile hybrids became available from a commercial fish farm in Arkansas. The hybrid between grass carp and a distant relative produced sterile fish that could still be counted on to consume aquatic vegetation. More recently, triploid grass carp (fish with three sets of chromosomes produced by exposing fertilized eggs to certain types of chemicals) have become available. Triploids are also sterile and are being legalized in a number of states that outlawed the normal grass carp.

Another fish that caused quite a stir, at least in Florida, is the walking catfish (*Clarias batrachus*), which was also introduced from Asia. Although brought into the United States as an aquarium fish, the species is widely raised in Asia and elsewhere for human food. Walking catfish can live in water of extremely poor quality because they have the ability to remove oxygen directly from the air so long as their gills are wet. They are also able to move about on land by "elbowing" their way along with their pectoral spines (thus the term "walking catfish").

Walking catfish released into Florida ponds, for whatever reasons, often hiked off to other locations. Rumors arose to the effect that walking catfish were prone to sneaking up on and consuming dogs. Perhaps they could even eat small children. Walking catfish are relatively small and are not aggressive animals, and it is likely that there was some confusion between the fish and alligators in the minds of some individuals. In any case, escape of the walking catfish led to near panic, the passage of strict laws and attempts to eliminate the fish from the state of Florida. Georgia was not too excited about being invaded by the supposedly dog-eating fish either. The panic finally died down and walking catfish (undoubtedly still present in Florida) faded from the limelight.

Legitimate concerns about moving fishes from one area to another exist in the United States. The Lacey Act makes it a federal offense to move a fish from one area where the animal is legal to another where it is also legal, if the fish passes through any jurisdiction (local, state or federal) which does not allow the fish. Stiff fines and jail sentences can await violators of this act. Besides the introduction of harmful species, fish transportation also raises the concern that exotic diseases might be imported.

Exotic invertebrates and plants have not received as much attention as fishes, but there have been some which have caused severe problems. Notable are the Chinese snail (*Corbicula* spp.) which can be found in prodigious numbers in many streams in the eastern United States and has displaced more desirable native species, and the hydrilla (*Hydrilla* spp.), which was introduced from Africa as an aquarium plant. Hydrilla has been known to grow in water from the shoreline to depths reaching 30 feet (10 m) in reservoirs and can be so thick that many fish cannot penetrate it. This prevents centrarchids (bass and sunfishes) from spawning because they cannot get to the spawning beds that occur in shallow water. Thus, whole species of fishes can eventually be eliminated from some reservoirs. Grass carp were introduced, in part, to control hydrilla and other noxious aquatic weeds. Introduction to the Pacific Coast of the United States of the Pacific oyster (*Crassostrea gigas*) from Japan also brought with it the oyster drill, a gastropod predator of young bivalves, and a predatory flatworm. These two species have become extremely troublesome to west coast oyster growers.

The above examples sufficiently demonstrate that uncontrolled exotic introductions can lead to serious problems. While controls on exotic introductions have been imposed in the United States, they were not put into place before some significant problem releases had already occurred. Additional mistakes with respect to exotic introductions can be avoided if sufficient testing is undertaken under controlled conditions before the organisms are released into the natural environment.

Land and Water Use Conflicts

Human uses of water and land for purposes that are unrelated to fisheries almost always affect fishery habitats located in the vicinity of these developments, and they often affect fisheries adversely. Sometimes, effects on fisheries are detectable at considerable distances from the sites of the alternate activities, as is the case with certain kinds of water pollution. We shall consider here a number of water and land uses that can impact fisheries.

Dredging and Filling

The transport of many goods and commodities takes place on the water in ships and barges. In the United States, networks of navigable waterways have been established on major rivers such as the Ohio, Mississippi and Missouri. In addition, navigable waterways along the coasts provide barges and shallow-draft vessels an opportunity to transit long distances without being required to move offshore into unprotected waters. Along the eastern seaboard is the Atlantic Intracoastal Waterway, while along the Gulf coast is the Gulf Intracoastal Waterway. Barge traffic can move from New England to Florida along the former and from Florida to the South Texas coast along the latter.

Maintaining the minimum depths of water in the inland and coastal waterways, as well as the passages into and within the harbors of this nation and various others around the world involves the periodic removal of sediments. This is accomplished by dredging. Several types of dredges have been developed, with a common one being the hydraulic dredge (Figure 18.2). Major ports that ac-

Figure 18.2. A hydraulic dredge.

commodate large ships maintain channels of 35 feet (about 12 m) or more deep, while the intracoastal waterways and riverine barge channels are maintained at a depth of at least 12 feet (4 m).

In some instances, dredging for channel maintenance may be required yearly, though in most instances several years pass between dredging episodes. The material that is removed from the channel (known as dredge material or dredge spoil) must be disposed of in some manner. Historically, this has been done by piling it on marshes and upland areas. In some cases, islands of dredge material have been formed. Maintenance dredging of public waterways is under the supervision of the U.S. Army Corps of Engineers, which has developed spoil disposal sites adjacent to the various waterways. As private development along waterways has occurred, dredging by the private sector to create access for pleasure boats to waterfront homes and marinas has become a common practice.

For many years the practice of disposing dredge material on marshlands, in the water adjacent to the dredged channel or in other types of wetland areas was not questioned. With rising environmental awareness by the public and the development of environmental protection legislation at the state and federal levels, disposal of dredge material became a subject of some concern. Some dredge materials, particularly those removed from harbors, are contaminated with toxic chemicals. Their safe disposal presents a problem, no matter where they are deposited. In some cases, barging dredge material offshore for disposal is a satisfactory solution. In others, the expense and potential for marine environmental contamination make that solution impractical or unacceptable.

Studies aimed at making beneficial use of dredge material have been conducted. Apparently, it is possible to make serviceable bricks out of at least some dredge materials. Beaches and salt marshes have also been created. The latter uses seem more appropriate than brick manufacture since there are tens of millions of cubic yards of dredge material to be disposed of annually—it is not clear that brick manufacturing would be able to keep up with the supply of raw material!

There has been no final resolution to the problem of dredge material disposal. In general, upland and offshore disposal have increased as a means of protecting sensitive intertidal and near-shore areas that were once the recipients of the bulk of the material.

Trace Metals

One of the primary uses that industrialized societies have for aquatic habitats involves using them as receiving waters for the discharge of wastes.

Virtually every kind of chemical that exists on earth can be found in water. Many are naturally occurring and beneficial, and indeed may be required by living aquatic organisms for proper growth and metabolism. Others are man-made and may be harmless, toxic or of unknown significance. Among the hundreds of thousands or millions of chemicals that can be found in water, trace metals have been of particular interest to scientists. Trace metals are elements that may be found dissolved in the water as ions and may also be found in the sediments. In many instances, equilibrium reactions occur between the water and sediments; removal of the trace metal from the water leads to its replacement in the sediment, and vice versa. Thus, the aggregate concentrations in the two media remain fairly constant, as no outside inputs are added.

Some trace metals that we are all familiar with include mercury, lead, zinc, iron, nickel and copper, though there are many others. Some of the more exotic ones include cadmium, cobalt and molybdenum. Some trace metals are required by organisms for normal metabolism. Zinc occurs in certain enzyme systems and those systems will not function properly if zinc is absent. Iron is found in the chemical hemoglobin, which occurs in the blood of vertebrates and is necessary for life because it carries oxygen to the tissues. In certain invertebrates, a similar compound, called hemocyanin, employs copper instead of iron.

Many metals have no known function in biological systems, though they may accumulate in tissues and eventually lead to toxicity. Overexposure to mercury in man, for example, can lead to severe nervous system disorders. Mercury accumulates in fat tissue in humans and is not readily eliminated. Thus, long-term exposure leads to ever-increasing levels in the body, which is the reason why concern was expressed during the 1970s about what were felt to be excessive levels of mercury in certain fishes, such as swordfish. A particularly serious example of mercury accumulation in the food chain took place in Minamata Bay, Japan. Many people who ate fish from the bay contracted what is now known as "Minamata disease," a very serious set of health problems even long after they ingested the fish.

Warnings about consuming excessive amounts of swordfish and tuna were put out by the U.S. government. As we have learned more about natural background levels of mercury in seafoods and other types of food, the "mercury problem" seems to have moderated. There remain legitimate concerns about high levels of mercury in aquatic organisms living near certain types of manufacturing plants that utilize the metal and may subsequently lose large amounts of it to the water. Portions of rivers that have received inputs of mercury and other trace metals have been closed to fishing because the resident fish populations have dangerously high levels of mercury within their tissues.

Some toxic trace metals can mimic required metals in the diet. The classic example is cadmium, which is chemically similar to zinc. Cadmium will compete with zinc and take the place of zinc in the molecular structure of enzymes. Those enzymes do not work correctly when cadmium has replaced zinc, so the organism becomes unable to perform certain biochemical activities that are required for proper metabolism. As long as the ratio of zinc to cadmium is high (that is, there is relatively more zinc present than cadmium), there will be few instances of cadmium toxicity. When the ratio becomes reduced (as may occur around metal plating factories where large amounts of cadmium are released), sufficient cadmium levels may get into fish to impair the enzyme systems and thus impair fish performance and even limit survival.

Trace metals are usually measured in parts per million, parts per billion, or even parts per trillion! Modern analytical techniques allow us to measure those concentrations with high precision. If you were to take two water samples, one well away from areas where ships had passed, and one from the shipping lanes, it would be possible for an analytical chemist to differentiate between the two on the basis of the levels of trace metals present. Ships constantly lose metals to the water from their hulls, and the increased concentration of those metals in regions where shipping is heavy is readily measured in modern laboratories.

Most of the wastewater in the United States is subjected to sewage treatment, which removes solids and may reduce nutrient levels substantially. Trace metals are commonly unaffected by passage through sewage treatment plants, however. Their removal from water is very expensive and, in many instances, no attempt is made to accomplish that task.

The trace metal story is not entirely one of gloom and doom. As indicated above, trace metals do come into equilibrium with the sediments, which serve as a so-called "sink" for these metals. Once the metals move sufficiently deep in the sediments to be undisturbed by burrowing animals, they are effectively lost to the biological system since deposition of new sediments at the top of the sediment-water interface is continuous. Contaminants can, however, be mobilized once again when an area is dredged or otherwise disturbed. Contaminated dredge material has been a significant problem in some instances.

While we should not be complacent about the levels of trace metals in our aquatic environments, and should, in fact, be greatly concerned with respect to areas known to be heavily contaminated, the general situation is that nature handles trace metal inputs quite effectively. Normal chemical processes keep the levels of most trace metals in water within acceptable limits. When man perturbs the system by dumping concentrated trace metals as a result of carelessness or ignorance of the problem, areas of concern are created. The U.S. Environmental Protection Agency and various state environmental agencies are responsible for monitoring the waters of the United States and identifying areas where contamination has occurred or is occurring. Protection of these waters from excessive contamination was mandated by the Clean Water Act in 1972. Presently, efforts are underway to clean up some of the more highly contaminated areas.

Petroleum and Halogenated Hydrocarbons

While trace metals are not biodegradable, many hydrocarbon contaminants are biodegradable to some extent in aquatic ecosystems. Petroleum hydrocarbons include the thousands of different compounds present in crude oil that contain hydrogen and carbon in their molecular structure. Halogenated hydrocarbons include one or more of the halogens, such as chlorine or bromine, in addition to hydrogen and carbon. As a rule, halogenated hydrocarbons are broken down much more slowly in aquatic environments than non-halogenated hydrocarbons. Certain bacteria and other microorganisms are able to utilize hydrocarbons as a food source and break them down into smaller compounds biologically. Scientists are using genetic engineering to enable such microorganisms to break down various contaminants more effectively, although the use of genetically altered organisms in the ecosystem is controversial. In addition to biodegradation, these compounds may also be broken down by physical means, such as photo-oxidation under the influence of sunlight. Other modes of removal from contact with aquatic organisms include evaporation to the atmosphere and deep burial within sediments.

Petroleum products are transported over and under aquatic habitats more often and in greater volumes than any other toxic substances. Petroleum hydrocarbons are hydrophobic compounds that have limited solubility in water. They also have a lower density than water in most cases, and therefore they float as a "slick" at the water surface. The most soluble fractions of petroleum are also the most toxic, and this class of compounds is called aromatic hydrocarbons. **Aromatic hydrocar-**

bons have one or more benzene rings (six-carbon cyclic structures) in their molecular framework. Non-aromatic hydrocarbons present in crude oil are called **aliphatic** or **paraffinic hydrocarbons**. In addition, crude oil has an assortment of sulfur-containing compounds and trace metals.

As was seen in the spill of crude oil in Prince William Sound, Alaska, in 1989, introduction of these compounds into natural ecosystems can bring about widespread mortalities of animals and plants and can pose a serious threat to fisheries. It takes years of observations and research before the impact of large oil spills on aquatic environments can be fully understood. Short-term impacts include **acute toxicity**, which results in rapid death of organisms by poisoning or coating with oil. Long-term impacts accrue from **chronic toxicity**, which reflects more subtle effects such as reproductive failure, reduction in growth, cancerous lesions and disease, or changes in physiology and behavior. The full spectrum of ecological consequences following spillage of a toxic substance is very difficult to quantify. In very general terms, acute toxicity of petroleum hydrocarbons to aquatic organisms occurs at levels in the range of parts per million, while chronic toxicity may occur at hydrocarbon concentrations in the parts-per-billion range.

Oil at the surface can be recovered by pumping or adhesion to absorbent booms. Part of the surface oil evaporates to the atmosphere and other fractions dissolve in the water. The latter two cases involve primarily aromatic hydrocarbons having relatively low molecular weights. Petroleum fractions that remain often become whipped by wave action into an oil-in-water emulsion sometimes described as "chocolate mousse." Oil that comes into contact with sediment particles generally adheres to the particles and may be deposited at the bottom, where it can remain for many years. Applying dispersants to a surface slick to dissipate the oil maximizes the amount of toxic substances that reach aquatic organisms, and thereby maximizes the resulting toxic effects. The only factor that is minimized by the application of dispersants to oil spilled in aquatic habitats is the cost of cleaning up the oil.

Halogenated hydrocarbons include insecticides, herbicides, polychlorinated biphenyls (PCBs) and a host of compounds used in industry and agriculture. Some of these compounds, such as PCBs and the now banned DDT, are very persistent in aquatic environments and also accumulate in aquatic food chains. At the level of accumulation in the tissues of aquatic organisms, they may not be high enough to exert toxic effects on the organisms but are still of concern to humans who consume those organisms. If a fish, for example, takes up a small amount of petroleum from the water after an oil spill, its flesh becomes "tainted" and the petroleum fractions can often be detected by consumers as an off-flavor. PCBs and other halogenated compounds are not generally detectable by taste when they occur in fish and need to be analyzed chemically. These compounds are of concern in seafood because some of them are thought to increase the risk of developing cancer.

Sewage

Some chlorinated hydrocarbons are formed when sewage effluents are chlorinated to kill bacteria before release into aquatic environments. These include chlorophenol, but are not of major concern because they are produced in small amounts. Of greater concern regarding sewage discharges are nutrient and organic enrichment, pathogenic microorganisms and other toxicants.

Primary sewage treatment involves little more than allowing particles to settle and form sludge, which is disposed of separately from the liquid effluent. Secondary treatment includes a process in which bacteria break down many of the organic compounds in sewage into smaller compounds and nutrients. The nutrients enhance plant growth in the receiving waters. While many of the bacteria are then killed by chlorination, pathogenic viruses are often not killed. Concern over the spread of human diseases such as typhoid, hepatitis and cholera through ingestion of filter-feeding shellfish taken from sewage discharge areas has resulted in many of these areas being closed to shellfish harvesting. Fecal coliform bacteria are monitored in these areas as an indicator of the likelihood that human pathogens from sewage may be present.

Organic compounds serve as food substrates for microorganisms in the receiving waters, and the metabolic activity of these microorganisms may use up the available dissolved oxygen in the water if the discharge of organic compounds is too high. The enhanced demand for oxygen associated with waste discharges is called **biological oxygen demand** (BOD). BOD and phytoplankton blooms resulting from nutrient enrichment combine to reduce water quality in areas receiving high doses of sewage in proportion to their rates of dilution and water replacement ("flushing" rates).

Finally, as mentioned earlier, toxic contaminants such as heavy metals may be present in sewage effluents. Other industrial wastes, including solvents, PCBs, aromatic hydrocarbons and waste oil, may also be present, particularly in urban discharges. These contaminants are generally not removed by sewage treatment.

Power Plants

Another major alternate use of water involves the generation of electricity. Water is used to generate electricity directly in hydroelectric power plants, or to provide cooling in thermal (nuclear and fossil fuel) plants.

Hydroelectric power generation takes advantage of the potential energy inherent in water held at high elevation. When the water is allowed to flow to a lower elevation, some of this energy is translated into the motion of turbine blades, which results in the production of electricity through a generator. For the water to be held at elevation, a dam must be constructed that backs up the water to a given elevation, or **head**, above the **turbine intakes** at its base. Hydroelectric dams are constructed in rivers that have steeply sloping margins, and the water impoundment turns the riverine habitat into a reservoir.

As one might imagine, this bears significant ecological consequences. The Columbia River system in the Pacific Northwest is probably the most obvious case of large-scale hydroelectric development and its ecological consequences. Beginning with construction of the Bonneville Dam in the early 1930s, construction of hydroelectric installations on the Columbia and Snake river systems continued through the 1970s, leaving a total of over fifty dams in place. These dams have been devastating to some of the Columbia River's magnificent salmon and steelhead trout runs, despite considerable effort to protect the fishery resources. Mitigative measures include the construction of fish ladders and other passage facilities for returning adult salmonids, traveling screens and spillways to divert downstream migrants from turbine intakes, and hatchery releases to make up for fish that are destroyed. Despite these costly efforts, the rebuilding of the Columbia River fisheries has been slow.

There are also nuclear power plants on the Columbia River and other river systems and marine coastlines around the world. Nuclear plants circulate large volumes of water in their cooling towers, where the water is used to condense steam before it is recycled into the reactor core. After the cooling water has passed through the series of heat-exchanging condensing tubes in the cooling tower, it is released into the aquatic environment at a slightly higher temperature. The cooling water intakes are screened to prevent entry of fishes and debris into the cooling towers, but small fishes may be killed if they are pinned against the screens by the velocity of the inflowing water. This problem is called **impingement**; in contrast, **entrainment** is when a fish is carried by the water *into* the intake structure.

Besides any effects of temperature increases and impingement brought about by nuclear plants, there is the possibility of chemical effects due to chlorine. Condensing tubes develop a film of microorganisms on surfaces that are exposed to the cooling water, since the cooling water is drawn from the environment and is not sterile. In order to maintain effective heat exchange, these microorganisms are periodically killed by chlorination. The chlorine and its resulting chemical by-products are then discharged into the receiving waters along with the cooling tower "blowdown."

In addition to hydroelectric and nuclear power plants, there are power plants that burn fossil fuels, namely coal and natural gas. The fossil fuel is used to fire huge boilers that produce steam. The steam turns turbines that, in turn, produce electricity. After the steam passes by the blades of the turbine, it passes through pipes which are exposed to water, usually from a lake or river. The water cools the pipes through which the steam is passing, causing the steam to cool and condense back to liquid water. The water in the pipes is then recycled back to the boiler so the process can be repeated. The water in the river or lake that is used to cool the condensor pipes becomes heated somewhat in the process. In most instances, the change in temperature is kept to no more than about 10°F, but that is still significant to the biology of the receiving waters.

Aquaculturists have looked at the waste heat from power plants as a resource. By exposing aquaculture species to the increased temperature, growth may be enhanced. Or, more commonly, the growing season can be extended by using the heated effluent from a power plant to maintain summer water temperatures to which the culture animals are exposed long after the normal water temperature has begun to decrease in the fall. Several additional weeks of growth may be provided. In the deep South, where winters are generally mild, year-round growing seasons may be provided through the proper use of heated effluents.

There are also problems with waste heat. Exotic tropical fish species, which might not survive during winter under normal conditions, may find sufficiently high temperatures available to support their year-round survival in the thermal effluent of a power plant. This has occurred in the case of tilapia in Texas. A population of 6 million tilapia were present in a power plant lake of about 700 acres. The fish were all crowded into a discharge canal that received the warmwater from the power plant. One day during winter in the late 1970s, the power plant was shut down for repair and all 6 million fish died within a few hours. Needless to say, there was a stinking mess to be cleaned up.

Fossil fuel power plants have been identified as one of the main sources of acid rain in the United States and Canada. Certain gases (sulfur dioxide and nitrous oxides) that are released to the atmosphere through the smokestacks of power plants form acids when they are mixed with water. The gases rise up into the clouds, the acids are formed when the gases mix with water vapor, and the resulting rainfall has a low pH. The problem (discussed in Chapter 11) is of major concern, particularly in the northeastern United States and in southeastern Canada.

The release of carbon dioxide as a consequence of burning fossil fuels has also been implicated in the apparent trend toward global warming. One theory states that increased levels of carbon dioxide in the atmosphere keep radiant energy from the sun from escaping back into space as efficiently as when lower levels of the gas are present. The trapped radiant energy leads to an overall warming of our planet through a phenomenon known as the **greenhouse effect**. Scientists continue to argue about whether global warming is, in fact, occurring, and if it is, whether it is due largely to man's activities or to natural fluctuations in the earth's climate. In any case, it does appear that a warming trend is underway, and the recent severe droughts and increases in the occurrence of the El Niño phenomenon (see Chapter 10) are attributed to this warming trend. An increase in the average temperature of only 1 degree will cause a significant rise in sea level (because of the melting of polar ice), leading to coastal erosion and the inundation of coastal lands. At least some of this problem may be related to the heavy use of fossil fuels for electric generation, not to mention their use in motor vehicles.

A simplistic solution would be to shut down fossil fuel power plants, do away with motorized vehicles, and thereby stop changing the composition of the atmosphere. While this solution appeals to certain segments of the population, the reality is that our national, and even global economy is based upon the use of fossil fuels to produce the power we need to produce our crops, manufacture the goods that we use, provide us the means to get to and from work and support the various other activities in which humans engage. The fact is that civilizations that do not have large numbers of private automobiles or even rural electrification are often more polluted than highly developed nations. The smog created from hundreds of charcoal or wood cooking and heating fires, the water pollution caused by the wastes of draft animals and the untreated sewage of nations that are over-populated and too poor to provide proper sanitation add significantly to both global warming and water pollution.

The reality of the situation is that the earth has just too many people on it. Until mankind comes to grip with that problem and finds a way in which to control the population of humans on the planet, their levels of consumption or both, the problems that we are currently facing will continue to increase.

Forestry and Agriculture

The use of land to grow trees and crops presents more potential interactions with fisheries, in both beneficial and adverse ways. Logging practices can harm riparian and riverine habitats by producing excessive siltation, loss of vegetative cover, changes in food chains and physical blockage of streams. Siltation is probably the most serious problem associated with logging. Herbicides applied in connection with forestry practices may also find their way into streams and rivers. Among fisheries, the salmonid fisheries are probably the most sensitive to forestry practices.

Fertilizers applied in connection with either forestry or agricultural practices may either help or harm fisheries by causing increases in primary production of receiving waters. Farmers also need large volumes of water to irrigate their fields, and much of this water evaporates on the fields before it returns to a river or reservoir. The water that does return to an aquatic habitat is generally warmer than the water that was withdrawn, and may carry herbicides and pesticides in addition to the fertilizers.

Water quality in many parts of North America has been affected by agriculture. In many regions, forests have been cleared and prairies plowed under to make way for agricultural crops. As forests are removed, the amount of sunlight reaching streams increases, leading in some instances to changes in the temperature regime and a consequent alteration in the sustainable fauna. Perhaps more dramatically, runoff from agricultural land has carried large quantities of silt and clay to the waters of our continent, causing increased turbidity and more rapid sedimentation.

During the present century, **biocides** (herbicides and pesticides) have been developed and heavily applied to agricultural crops. Some of these chemical compounds are relatively short-lived, lasting only a few days. Others, such as the now banned DDT, persist in the environment for many years, even decades. Many of the biocides that have been developed and put into use are directly toxic to fishes and other forms of aquatic life. Herbicides will kill phytoplankton, filamentous algae and rooted aquatic plants as easily as they kill terrestrial weeds. Pesticides designed to kill terrestrial insects have the same effect on aquatic insects and crayfish, and if present in sufficient concentrations, they can kill fish.

Biocides can enter the aquatic environment through rain runoff from agricultural lands, or improper application of the biocides may cause them to be sprayed directly onto streams, lakes and estuaries. Aerial spraying (crop dusting) is often a source of contamination when wind drift carries a biocide away from the target crop and out over an adjacent water body.

By practicing good soil conservation and applying chemicals properly, foresters and farmers can reduce the amount of soil, biocides and nutrients lost from their land. It is in the farmers' best interest to reduce the amount of soil, biocides and nutrients lost from their land; by practicing good soil conservation and applying chemicals properly, farmers can minimize such losses. It is also in the best interest of the aquatic environment to keep runoff of materials from forest and agricultural land to a minimum.

Conclusions

User conflicts have become one of the major issues of our time. There are few absolutes since environmental degradation in the mind of one person represents the livelihood of another. While everyone should be interested in maintaining healthy aquatic environments for future generations, we cannot ignore the fact that many human activities require water and their curtailment would have economic and social consequences that are intolerable in modern society. Our goal should be to promulgate the wise use of our natural resources. This means protection of extremely sensitive environments from the encroachments of mankind and the wise stewardship of those resources that man requires for survival and advancement. The boundaries between the extremes are ill-defined, and therein lies much of the controversy.

19

Inland Fisheries Management

Christopher C. Kohler and Gilbert B. Pauley

Introduction

Diversity of Freshwater Fisheries

Inland fisheries management is the art and science of producing annual crops of freshwater and anadromous fishes for either recreational (sport) or commercial uses. Most of inland fisheries management is directed at maintaining or increasing the recreational aspect of the fishery resource. At the turn of the century, inland fishing took place primarily on natural lakes, small streams and rivers. The increase in the number of man-made impoundments over the last few decades has caused a boom in freshwater fishing. This increase in fishing opportunities has been coupled with numerous advances in sport fishing equipment such as nylon line, fiberglass rods, spinning reels, depth sounders and electrical trolling motors. Freshwater sport fishing is now considered to be one of the most popular of leisure pursuits (Figure 19.1). Indeed, it was estimated by the Sport Fishing Institute that in 1985 the American public expended 27 billion dollars on boats, fishing paraphernalia, food, lodging and other expenses associated with recreational fishing! It can be expected that an increased burden upon management agencies to provide fishing opportunities for an expanding and sometimes demanding public will exist for the foreseeable future.

The **fishery manager**, whose job it is to exercise wise stewardship over the fishery resource, must have a knowledge of the biology of the different species of fishes within the systems being managed as well as an understanding of the complex interrelationships that exist between species, their environment and people (Figure 19.2). For example, the interrelationships between timber, fish and wildlife in stream drainages are inseparable. Inland fishes are found in highly diverse environments (see also Chapter 12)—from small farm ponds with only a few or even a single species of gamefish to the Great Lakes with both resident and anadromous fishes. They are found in small, rapidly moving headwater streams often with fish occupying highly specialized niches or large, slow moving navigable rivers with highly diversified numbers of fishes and complex interrelationships. They can be in pristine high mountain lakes and include only a very few species or in weed-choked lowland lakes in which many species or just a few co-exist. With respect to temperature gradients, they may be in warm waters with fishes belonging to the catfish (Ictaluridae), bass (Percichthyidae) and sunfish (Centrarchidae) groups, cool waters with fishes such as perch (Percidae) and pike (Esocidae), or cold waters in which salmonid groups (Salmonidae) are the primary sport fishes of interest. The species of fishes present in these various inland waters, as well as the quantity and quality (size and edibility), are determined by the complex interrelationships of environmental variables such as water quality; availability of food, cover, spawning and nursery habitat; and species composition.

Multiple Resource Use

Inland waters have many uses other than providing fishing opportunities. The waters may be used for hydroelectric power, drinking water, agricultural irrigation, commercial navigation, recrea-

312

Figure 19.1. Fishing and other leisure pursuits focusing on freshwater resources are expanding at a rapid pace throughout North America.

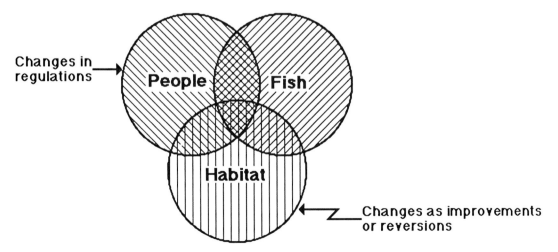

Figure 19.2. Resource components and interactions that fishery biologists need to address in the management of fish populations for recreational use. Alterations not involving the fish directly are shown.

Figure 19.3. Reservoirs are often created to control flooding or some other non-fisheries use. The impounded waters also offer excellent opportunities for fishing, boating and swimming. Excellent fisheries often develop in the tailwaters below the dam.

tional boating, swimming, etc. In many cases, fishing may have a lower priority than one or more of the other uses of the resource. Accordingly, fisheries managers must operate within certain restrictions that sometimes ensure poor fishing. For example, a reservoir built primarily for flood control (Figure 19.3) will experience periodic water level fluctuations that sometimes literally leave fish spawning nests high and dry.

Fish, like the waters within which they reside, are also subject to multiple users. An extreme example would be in waters where court rulings of treaty rights allow certain Indian tribes to take a percentage of certain harvestable fish stocks using commercial gear while non-Indian fishermen may catch the remainder using only sport fishing tackle. Many other commercial and recreational fisheries also coincide. Another example would be some fishermen seeking panfish while others are after trophy-size species such as largemouth bass or northern pike.

The larger the number of multiple users of a resource, the more complex the management of that fishery will become with an increasingly convoluted set of regulations. It is the responsibility of the fishery manager to ensure that each user group gets a fair share of the available fishery resource. In some cases, fish management strategies need to be evaluated for costs and benefits to help determine where management emphasis should be placed.

Management Philosophy and Strategies

Inland fisheries are managed for the same reasons as their marine counterparts—to manipulate habitat, biota and human users to produce sustained benefits for mankind. Inland fisheries management followed the same trend as in the marine environment in that it was initially directed toward **maximum sustained yields (MSY)**, in which the goal was to achieve maximum catches year after year. MSY as a management goal has been replaced by a more pragmatic approach, **optimum yield (OY)**, in which numerous socioeconomic factors are taken into consideration to obtain a "quality" fishery. Harvest rates under an OSY regime are always less than would occur where MSY is the goal.

Inland fisheries management differs from marine management in that the former is almost totally devoted to the recreational aspects, though some important inland commercial fisheries do exist, and that the issue of ownership, or more appropriately, stewardship, has not been as hotly contested. State, provincial and federal governments are the stewards in that they hold the aquatic resources in trust for the general use of their citizens throughout North America. Political boundaries are much easier to define for inland waters than for oceans. Nevertheless, many cases exist with respect to inland fisheries where it is still not clear under whose jurisdiction the actual management befalls. For example, the Mississippi River is bordered by several states, is largely maintained navigable by the U.S. Army Corps of Engineers, and has wetlands managed in part by the U.S. Fish and Wildlife Service. Thus, a number of state and federal agencies interact, overlap and may even have conflicting goals, as they all pursue the management of the river's resources. With respect to the Great Lakes, additional complications exist because the boundaries of two countries meet. Often co-management is a solution, such as exists with the salmon and steelhead runs in the Columbia River, where three states, two federal agencies and several Indian tribes all have management responsibilities for these fishes. The point to remember is that all agencies work for the common good, but they each use their own definition for having achieved it.

Inland fisheries management also differs from marine management in the ways in which various fisheries can be manipulated. Habitat modification can only be achieved to a limited extent in the marine environment but can often be accomplished with relative ease in freshwaters (note: the Great Lakes would be similar to the oceans in this regard). Such modifications are commonly referred to as habitat improvements. However, in many cases the so-called improvements are in actuality attempts by managers to revert habitat to their more natural states. Such manipulations would be more accurately termed habitat reversions. Examples of habitat improvements/reversions will be described in sections that follow.

Because inland fisheries are largely recreational, it should come as no surprise that much of their management is directed at the angler. Such management is achieved via regulations. **Regulations** (e.g., **length limits, creel** or **catch limits, restricted seasons, area closures, gear restrictions**, etc.) are designed to manage the resource through the user. Regulations have historically been imposed to protect aquatic populations from overexploitation. However, in modern fishery management, regulations are often used as tools to enhance as well as protect aquatic populations. Thus, regulations might be enacted in order to adjust the size composition of fish stocks, to manipulate predator-prey population sizes, or even to control undesirable species. Some specific examples are provided in subsequent sections dealing with particular freshwater systems (streams, ponds, lakes, etc.).

Stocking is a major tool used in inland fisheries management. Stocking programs can be categorized as (1) introductory, (2) maintenance, (3) put-grow-and-take and (4) put-and-take. **Introductory stocking** refers to placing species in systems where they did not previously exist. An example of introductory stocking is the placing of a lacustrine (lake-inhabiting) species in a newly impounded reservoir or adding a new sport fish into the mix of an existing fishery. Introductory stockings are an important fisheries management tool provided the major ecological ramifications of such stockings are fully considered. **Maintenance stocking** is implemented to supplement natural recruitment (addition of new members to population) of target species. It is an enhancement procedure used when spawning habitat is limited and/or the species is experiencing high levels of exploitation. When recruitment completely fails (spawning habitat has become totally degraded or never existed in the first place), the population must be maintained by regularly stocking subadults and allowing them to grow to a harvestable size; this is **put-grow-and-take** stocking. When fish of harvestable size are stocked with the intent of having them soon caught by the angler, the management strategy is called **put-and-take** stocking.

Poor fishing is usually caused by overpopulation and stunting of desirable fish species or an over abundance of undesirable fish species and a concurrent shortage of acceptable fish. To determine what caused these problems and what solutions are available, the population must be evaluated. The overall quality of a fish population can be estimated and compared to other fish populations by a **condition factor**, or **relative weight index,** which correlates the surface or length to the volume or weight of the fish. **Proportional Stock Density (PSD)** is another method of assessing the overall quality of a fish population. PSD is an index of population balance that reflects the size structure acceptable to the angling public for a given fish species in a particular water body. Condition factor and PSD are two of the best ways to evaluate a fish population. Other methods include estimating *standing crop* (the poundage of a given species or complex of species present in a body of water at a specific moment in time), **carrying capacity** (the maximum poundage of a given species of fish that a specific habitat will support during a stated interval of time), **productivity** (the biomass that a body of water will provide naturally within a given time) and growth rate of species of concern. Several models are available that assist managers in predicting annual sport fish **yield** in terms of totals for either single or mixed fish species. The best known of these models is the **morphoedaphic index** for lakes and reservoirs, which is simply the total dissolved solids in mg/liter divided by mean depth in meters. With information obtained from these various methods, fishery managers can balance the desires of the fishing public and the objectives of the management agencies.

Role of Agencies

As previously mentioned, a number of governmental agencies are involved in the management of inland fisheries. In the past, fishery needs were addressed by short-term problem solving methods. In other words, fishery management was **reactive** in that actions were not taken until a specific need, usually a problem, presented itself. Even the measures taken were not always well-conceived, with the approaches being largely "trial-and-error." Modern fisheries management is based on a more rational approach for decision-making, with specific strategies designed to meet certain goals—it is pro-active. For example, in June 1988, the U.S. Fish and Wildlife Service published the "National Recreational Fisheries Policy." The policy sets the framework for decision-making and should guide much of fisheries management at the federal level. Similarly, a number of states devise intricate strategic plans for the management of their aquatic resources.

As in all other walks of life, governmental agencies are in partnership with the public sector to meet the needs and desires of their constituents. Thus, management is a dynamic process in which those needs and desires are met under the prevailing political and financial constraints. To summarize, governmental agencies (1) define strategies based on scientific knowledge and public pressures (which sometimes are in conflict), (2) implement management procedures to meet objectives, (3) evaluate effectiveness of management through monitoring programs, and (4) are held accountable by their constituents. As long as the public sector is fully integrated into the process, aquatic resources generally are managed in a wise and efficient manner. The above statement requires a major qualifier to be accurate—that is—the public sector must be comprised of informed individuals. In other words, the public sector needs to be cognizant of the ecological and financial realities which drive much of management activities. Therefore, educating the public is as important as a management tool as are habitat improvements/reversions, setting regulations and stocking fish. Fisheries management is about managing people as much as it is about managing aquatic populations and their habitats.

Figure 19.4. Smallmouth bass with individually numbered floy tag.

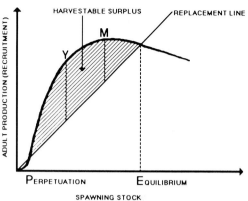

Figure 19.5. Spawner/recruit curve or surplus yield model showing the optimal yield area of harvestable or surplus fish. M is the maximum number of fish produced, while Y is the optimum yield.

Fisheries Management in Flowing Waters

Small Streams

Because of their predominantly cool water and velocity, small northern or high elevation streams tend to be managed for some type of salmonid in most regions. Either they are managed for resident fishes or, in some cases, for anadromous fishes. In some areas, streams are dammed to create small impoundments which may then be managed as ponds for other fish species if the temperature rises. Small streams are very susceptible to alterations of their environment. Habitat losses have occurred from such alterations as channelization for erosion and flood control, siltation, agricultural run-offs and municipal discharges. Therefore, much of the management of small streams is aimed at altering various parts of the habitat in an attempt to revitalize old habitat or to create new habitat for spawning and rearing purposes—gravel restoration, riffle and pool creation, velocity and flow alterations, and both instream and riparian cover additions. Resident fishes may be managed as native or hatchery populations. Small streams or sections of them lend themselves nicely to quality fisheries either through "catch and release" fishing or catch, size and gear restrictions.

Large Rivers

Depending upon their size, velocity, geographic location and degree of eutrophication, large rivers support different populations of fishes, which may include a wide variety of both anadromous and resident gamefish. Rivers, like small streams are subject to environmental alterations, but it is more difficult to effect habitat improvements or reversions in rivers. Therefore, existing habitat must be preserved because suitable rearing and spawning are only found in a small percentage of the available water. Effective methods of sampling fish populations in flowing waters (streams and rivers) are electroshocking, seines and **creel census** (sampling the angler's catch). **Punch cards** are often used to monitor catch statistics, but care must be taken in evaluating them because there is a bias toward returns by successful anglers. Tags (Figure 19.4), fin clips and dye marking can be used to evaluate populations and fish movements. In systems where anadromous fishes are found, it is not unusual for them to take management priority. Such choices are generally mandated by the fishing public.

Fish in rivers may be managed as quality fish or as put-and-take hatchery fish. Length limits are imposed for two reasons: first to get the fish to an "acceptable size" for the angler to catch or second to ensure that the fish will spawn at least once. This is true in other systems as well. "Acceptable" minimum size varies between the different species of gamefish and in different geographic areas for the same species of gamefish. Many anadromous fishes are managed on the basis of harvestable numbers that are available after designating a specified number of adult fish as spawners to perpetuate the run (Figure 19.5). This is often called a **spawner-recruit curve** or **surplus yield model**. Rivers are often managed by segments with different regulations for various sections of river.

Fisheries Management in Slow-Moving Reservoir Waters

Reservoirs are man-made impoundments, large in size relative to ponds, where water is retained but eventually passes through (Figure 19.3). Modified riverine systems in which navigational pools are formed by locks and dams are not classified as reservoirs because they have storage ratios less than 0.01 (through-flow is 100 times greater than average annual pool volume).

Reservoirs generally are not manageable in the same manner as rivers, natural lakes or ponds, though they have characteristics of all three. However, if one thinks of a reservoir as a series of small lakes with a slow moving river winding through, then various aspects of the reservoir can be manipulated and therefore managed. Manipulation of water levels can be done to favor certain fish species and depress others. This will drive small forage fish out of the coves and protected areas into the open water where desirable gamefish can more easily utilize them. Water manipulation, mechanical harvesting and chemical treatment with herbicides can reduce the amount of aquatic macrophytes present. Such plants can interfere with spawning ground access by gamefish and hide small forage fish from predacious gamefish, allowing them to increase in numbers and upset the predator-prey balance. The protection provided by too much vegetation may also result in over-population and stunting of panfishes. However, some fish species, such as pike (*Esox lucius*), pickerel (*E. americanus* and *E. niger*) and muskellunge (*E. masquinongy*), require vegetation for spawning so that total elimination of vegetation is never a management goal. Moreover, it has been shown that predacious gamefish such as largemouth bass (*Micropterus salmoides*) have a parabolic relationship to the density of aquatic plants with optimum quantity and quality occurring between 20% to 40% of maximum weed cover while insectivorous fishes show a linearly increasing relationship with an increasing density of plants (Figure 19.6).

Certain species of fish not ordinarily found in reservoirs can be introduced to occupy a certain niche and to contribute directly to the catch of quality fish. For example, white bass (*Morone chrysops*) and striped bass (*M. saxatilis*) have been widely introduced to occupy the open water niche of reservoirs. A hybrid cross of these two fish, which grows to 8 lbs in only a little over 2 years, has been used increasingly in recent years, especially in smaller reservoirs and lakes. Smallmouth bass (*Micropterus dolomieui*) are often introduced to occupy the shoreline area of rocky reservoirs to utilize existing crayfish populations. Other fish such as threadfin shad are sometimes introduced to serve as forage. Scrap or rough fish sometimes can be controlled by introducing predators, such as largemouth bass, smallmouth bass and tiger muskellunge (a hybrid of muskellunge and northern pike). The tiger musky is a favorite for this purpose because it can be easily reared in the hatchery and can grow to 28 inches in 2 years. Where coves and specific areas can be isolated from the main reservoir, rotenone can be used to eliminate undesirable fishes prior to planting with desired fish (rotenone kills fish by interfering with their oxygen uptake).

The success of the angler can be controlled by gear restrictions, increasing or restricting the seasons, or altering the catch and size limits. Electrofishing (Figure 19.7) and creel surveys are very useful techniques to assess the effects of size limit regulations. In **oligotrophic reservoirs**, gamefish can be increased by adding artificial structures in selected parts such as shallow arms or

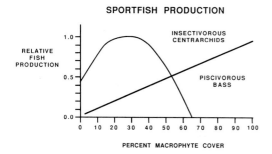

Figure 19.6. Relative production of piscivorous largemouth bass and insectivorous centrarchids are a function of the aquatic macrophyte cover and biomass.

Figure 19.7. Electroshocking unit as set up for a boat for utilization in ponds, lakes and reservoirs.

bays, which can either improve the spawning and rearing area, concentrate bait or forage species, or provide protection for adult predator fishes.

Depending upon their stage of development (see Chapter 12), reservoirs will contain a wide variety of fish species. Most large reservoirs possess mixed species fisheries. In general, cool and clear **oligotrophic reservoirs** contain trout or salmonid fisheries, while highly productive warm water **eutrophic reservoirs** contain largemouth bass, sunfish (*Lepomis* spp.) and other centrarchids. **Mesotrophic reservoirs**, which are intermediate in development to the other two types, often contain walleye (*Stizostedion vitreum*), sauger (*S. canadense*) and smallmouth bass. Since large reservoirs often have developmental characteristics that resemble two of these types of waters (and even all three in certain instances), it is not unusual to find fishes preferring each of these stages within a single reservoir. Likewise, a deep reservoir might have warmwater fishes in its **epilimnion** and coldwater fishes in its **hypolimnion** (provided it contains a suitable content of oxygen). This is called a **two-story fishery**.

Fisheries Management in Standing Water Systems

Ponds

Ponds are bodies of water that are smaller than lakes, are not formed by glaciers or rivers and are often artificially constructed to some degree. Because they are generally very small, ponds can be manipulated very easily with respect to both the fish populations and their environment. For example, some ponds may be drained while others may be poisoned to completely kill all fish, plants and invertebrates present. A subsequent stocking of desired fishes at appropriate densities may then be done. Population balance of gamefish and their prey usually is a major management problem. The balance and structure of fish populations will change depending upon the amount of natural mortality and fishing pressure that occurs. **Underfished populations** generally are characterized by large numbers of stunted fish, with high survival of many year-classes and few individual fish of desirable size to the anglers. **Overfished populations** have an overabundance of very young fish and those year-classes below the legal minimum size limit, with most fish caught and retained at or very near the minimum size limit. A number of mathematic relationships using weight measurements have been in common usage for assessing balance. Balance in the population often can be maintained by planned cropping through variations of catch and size limits. A size limit is more critical than a catch limit, because a size limit operates on every fish that is caught, while a catch

Figure 19.8. Gill nets are often used by fisheries biologists to sample fish populations. Nets with larger mesh sizes are also used where legal by commercial fishermen.

limit does not affect the harvest until an angler reaches that limit. In some situations, both a minimum and maximum **slot length limit** might be required.

Generally, trout do not do well in small ponds because of the high temperatures that can occur and perch are not advised because of their enormous reproductive capacity and propensity toward stunting. Therefore, management should focus on some balanced fish populations, which can most often be accomplished by use of largemouth bass and some type of sunfish. Crappies (*Pomoxis* spp.) are usually not a good choice for reasons similar to the yellow perch (*Perca flavescens*) and, additionally, they will compete directly with the bass. Bluegill sunfish (*Lepomis macrochirus*), which primarily eat insects, and redear sunfish (*L. microlophus*), which eat insects and snails, are good management choices because they provide forage for the bass, do not compete directly with the bass and will grow to sizes that will offer sporting quality. Hybrids of certain sunfish are gaining popularity because specific crosses can yield nearly all-male populations, thereby reducing stunting problems. The hybrids exhibit **heterosis** (hybrid vigor) and are excellent sport fishes. Catfish may in some instances be utilized, but bullhead should be avoided for the same reasons as perch and crappie. Population assessment can be performed in a pond by either electroshocking or net capture (various types of nets are available such as gill (Figure 19.8), fyke and trap nets with different uses). Shoreline seines (Figure 19.9) are also used for assessing reproductive success. Some of the fishes mentioned above are not legal in certain regions.

Lakes and Large Impoundments

Lakes are natural bodies of water created by glaciers or rivers and are usually larger than a pond, although some high mountain lakes are very small. Population structure, density and age-and-growth of fish species of interest are important factors in proper lake management. Population structure can be ascertained in

Figure 19.9. Shoreline seines are an effective gear to sample reproductive success of fishes in farm ponds and shallow coves of reservoirs.

Figure 19.10. Fisheries biologists use a variety of gears to sample aquatic biota and limnological aspects of the lake.

several ways but the proportional stock density (PSD) is extremely useful for bass and sunfish. Some method of determining the catch-per-unit-effort (CPUE) would be useful because catch rates for gamefish are directly related to fish densities. Angler needs can be projected from the CPUE in some cases. Lakes near urban areas usually encounter the heaviest fishing pressure. Fish composition and limnological data needed for lake analysis can be obtained in several ways, such as electroshocking, net capture and creel census to gather fish data (Figure 19.10). If a lake is small, it can be treated like a pond in which the manager attempts to keep a balanced population structure with a single apex predator. On the other hand, if the lake is large, it should be managed more like a reservoir with more than one apex predator, which often will occupy several niches.

The types of fishes present in lakes will depend upon the developmental stage of the lake (see reservoirs above). Management may be based on keeping the natural fish populations in balance, it may rely on supplements of hatchery fish (with subsequent mixing and dilution of the gene pool), or it may depend entirely upon hatchery fish in a put-and-take fishery. The first would be more easily accomplished in lakes that are distant or isolated from population centers, while the later is often a necessity in lakes in or near urban areas. Lowland lakes, which are usually more productive than high mountain lakes, are managed more as put-and-take lakes with hatchery plantings occurring just before opening day, while the high lakes distant from urban areas and open only for short periods after the ice melts are managed for native populations. These native populations may be supplemented by hatchery fish at irregular intervals over a period of years. Care must be taken with the high mountain lake supplemental stockings because some fishes can reproduce rapidly and grow slowly because of the very cold water temperatures. Large populations of stunted fish are thereby produced, such as brook trout (*Salvelinus fontinalis*).

The Great Lakes

In order to understand the management of the Great Lakes' fisheries, it is necessary to have a historical perspective as to the major changes that have occurred since European colonization. The top predators of the Great Lakes originally consisted of lake trout (*S. namaycush*), burbot (*Lota lota*) and numerous species of coregonids. The Great Lakes contained many unique species or sub-species that are now extinct such as the blue pike, Michigan grayling and several of the coregonids. The Great Lakes have suffered nearly every insult that mankind can impose. First, fish populations were heavily overfished, with the largest individuals of the largest species being the first removed—the fishing up process witnessed in nearly every new fishery. Following the construction of the Welland and Erie Canals, a wave of invasions of exotic biota occurred. The most deleterious of these was the sea lamprey (*Petromyzon marinus*), which coincidentally has the same species and size preferences as humans and thus accelerated the declines of the top predators. The alewife (*Alosa pseudoharengus*), which likely also made its way to the upper lakes via man-made canals, found a situation very much to its liking and, in the absence of the major predators, experienced a population explosion. Nowhere was this more evident than in Lake Michigan, where it was estimated that by the mid-1960s alewife comprised nearly 80% of the lake's biomass. Pollution and other forms of habitat degradation have also run rampant in the lakes.

The introduction of Pacific salmon (*Oncorhynchus* spp.) coupled with development of chemical control measures for lampreys turned a very bleak situation into an almost overnight success story. Ironically, the Pacific salmon fishery has become so popular and so many salmon have been stocked into the lakes that the once over-abundant forage fishes have not been able to keep pace. Overall, the management of the Great Lakes has encompassed both the worst and best that such efforts have to offer. In the latter regard, within the past 25 years, various forms of management have resulted in (1) reduction of overexploitation of stocks, (2) control of sea lamprey populations, (3) introduction of a highly popular Pacific salmon fishery, (4) some recovery of native fish stocks, (5) a decline in density of alewife and rainbow smelt (*Osmerus mordax*) populations largely due to salmonid predation and (6) recognition and attempts to curb the problems of pollution and habitat destruction. This last aspect of fisheries management in the Great Lakes is not amenable to any quick-fix solutions.

Stresses and Messes

Good fisheries management requires good environmental management—not just of the aquatic system but also all of its surrounding watershed and atmosphere. Thus, pollution and habitat degradation are two of the greatest challenges facing the modern fisheries manager. Yet these are problems whose causes and remedies are often not within the purview of the manager or his agency. Nevertheless, such problems as acid rain, hazardous wastes, agricultural runoff, erosion, channelization, gravel removal, etc., will have to be dealt with by fisheries managers and environmental agencies for years to come. The sentiment expressed by the Sport Fishing Institute's motto that "the quality of fishing reflects the quality of life" will need to become ingrained within the public at large if the financial resources are ever to be made available to stem the tide of environmental insults. Fisheries managers can only do so much as the first line of defense.

Unlike pollution and habitat degradation, a second problem facing many inland fisheries, overfishing, is well within the realm of control of the fisheries manager. It has been estimated that one out of four Americans participates in the angling experience, obviously some more than others. With increasing population growth, longevity, affluence and leisure time, it is clear that fishing pressures will continually expand. The Sport Fishing Institute estimates that between the years 1985 and 2025, angling will increase by about 40%. Abating pollution and habitat degradation, expanding habitat improvement and reversion measures, implementing stocking programs in their various forms and setting realistic and enforceable regulations will all be required to meet the future demands for quality angling experiences.

Because fisheries management is as much art as science, mistakes have and will continue to be made. How well fisheries managers learn from the mistakes of the past and those currently being made will govern the pace of evolution from art to science. Such "exercises in fish management futility" as stocking salmonids in a California lake for a decade despite a predictable summer oxygen deficit in its hypolimnion can no longer be tolerated. Introducing species outside their native range without forethought to the potential ecological ramifications of such actions must end. Cautious test stocking and the use of sterile fish when possible should become standard practice for evaluating efficacy and impact of fish in new environments. The increasing demands being placed on aquatic resources in all their various forms leaves little room in the future for "trial-and-error" fisheries management.

The Challenge Ahead

The preceding sections briefly described the basic concepts and strategies employed in the management of freshwater fisheries. A number of the challenges being faced were also presented. How well fisheries managers meet these challenges will ultimately define what inland fisheries can and will be. One can only speculate where inland fisheries management might be heading, but it will likely fall within one of the following three scenarios, which entail attitudes of pessimism, optimism, or pragmatism. The pessimistic scenario is one where the problems facing our inland fisheries are considered to be so overwhelming that the only hope fisheries managers have to provide fishing opportunities to a burgeoning angling population is to rely almost totally on hatchery-produced fish. If this scenario proves to become the heart and soul of fisheries management, then most inland systems will be "artificial" in nature and will require continual refurbishment by managers—who would become "masters" rather than stewards. Once the initial intricacies of such controlled management have been worked-out, the actual management of these systems might become quite simple. On the other end of the spectrum—the optimistic scenario—one could foresee that all challenges to the integrity of inland fisheries will be met head-on, the public environmental conscience will be heightened, resulting in sufficient financial resources being made available to the appropriate state and federal agencies, and our scientific knowledge base will expand at a pace that will keep mankind one step ahead of man-induced catastrophes. In this scenario, inland fisheries will return to their more natural states, allowing them to sustain themselves under the watchful eyes of their stewards. Such a fisheries utopia is likely to be relegated to the imaginations of idealists, minorities in every society, and thus a more pragmatic scenario appears to be more within reach. In this scenario, managers will use all the tools available within existing financial and societal constraints to protect, enhance and, where necessary, to overtly manipulate aquatic resources to optimize the multiple uses of inland fisheries resources. Such blending of artificial and natural fisheries will likely be the modern approach to modern-day fisheries problems.

In summary, inland fisheries encompass a diversity of systems, a diversity of uses and a diversity of abuses. Fisheries managers are embarking into an era in which there are many problems to be faced—but also an era in which they are prepared better than ever to find pragmatic, fiscally responsible and lasting solutions.

Additional Reading

Bennett, G.W. 1970. Management of Lakes and Ponds. Van Nostrand Reinhold Company, New York. 375 p.

Clarke, R.N., D.R. Gibbons and G.B. Pauley. 1985. Influences of recreation. *In* W.R. Meehan (Tech. ed.), Influence of Forest and Rangeland Management on Anadromous Fish Habitat in Western North America. U.S. Forest Service, Pacific Northwest Experiment Stations. Portland, OR. Gen. Tech. Rep. PNW-178. 31 p.

Cohen, F.G. 1986. Treaties on Trial. University of Washington Press. 229 p.

Cooke, G.D., E.B. Welch, S.A. Peterson and P.R. Newroth. 1986. Lake and Reservoir Restoration. Butterworth Publishers, Boston. 392 p.

Hall, G.E. and M.J. Van Den Avyle (eds.). 1986. Reservoir Fisheries Management Strategies for the 80's. Reservoir Committee, Southern Division American Fisheries Society, Bethesda, MD. 327 p.

Lackey, R.T. and L.A. Nielsen (eds.). 1980. Fisheries Management. John Wiley & Sons. New York. 422 p.

Meehan, W.R., T.R. Merrell and T.A. Hanley (eds.). 1984. Fish and Wildlife Relationships in Old-Growth Forests. American Institute of Fishery Research Biologists. Moorehead City, NC. 425 p.

Nielsen, L.A. and D.L. Johnson (eds.). 1983. Fisheries Techniques. American Fisheries Society. Bethesda, MD. 468 p.

Salo, E.O. and T.W. Cundy (eds.). 1987. Streamside management: forestry and fishery interactions. Institute of Forest Resources, University of Washington Contribution No. 57. 471 p.

Sternberg, D. 1987. Freshwater Gamefish of North America. The Hunting & Fishing Library. Prentice Hall Press. New York. 160 p.

Stroud, R.H. (ed). 1986. Fish Culture in Fisheries Management. American Fisheries Society. Bethesda, MD. 481 p.

II

Modern Sport Fishing Techniques

20

Marine Fishing:
Salmon

Charles White

Introduction

Saltwater sport fishing is a very popular and growing recreational activity throughout the United States and Canada. According to recent statistics from British Columbia, sportsmen take approximately 4% of the total salmon catch each year. Commercial nets take almost 100% of the sockeye, chum, and pink salmon. While sport anglers take a significant portion of the chinooks and coho, especially in nearshore waters of Canada, overall numbers still favor the commercial sector in all salmon species. The sport catch reflects a somewhat higher percentage in Washington, Oregon, and California, but is still small when compared to the commercial catch.

However, in economic terms, the picture is entirely different. Sport-caught fish, especially salmon, have a much greater value per fish than commercially caught fish. Recent surveys in British Columbia show that recreational fisheries contribute $50/lb in economic activity versus $3/lb for the commercial catch. On the U.S. side, the statistics are even more dramatic. One survey in Washington showed sport fishermen had an economic impact 29 times as great as their commercial counterparts.

Approximately 50 million sport fishing licenses are issued in the United States and Canada annually. Saltwater anglers in Alaska, British Columbia, Washington, Oregon, and California contribute about 2 million toward that total.

Background

Saltwater sport fishing on the west coast is continually evolving and becoming more sophisticated. (Native Indians and early settlers on the west coast used a piece of fish flesh impaled on bone hooks and dragged on a heavy line behind a canoe or row boat.) Some of the earliest artificial lures were spoons and jigs made from shell or hammered metal, and carved wooden plugs. As the number of anglers increased, local entrepreneurs began making various lures, and heavy duty rods, reels, and line were introduced from the east coast of the United States, England, and Australia.

In the past 25 years, there has been an explosion of new equipment ranging from more sophisticated outboard power to electronic fish-finding devices. Some of the mechanical and informational aids available to modern saltwater sport fishermen include:

- Specialty boats with light-weight, efficient power plants—These new boats, ranging from 12 feet to more than 40 feet long, greatly extend the anglers' range, allowing them to move quickly from one area to another (Figure 20.1). Boat trailers further extend this range, allowing even small boats to get to remote locations for launching near the "hot spots."
- Depth sounders and fish finders—These electronic echo-reflection devices have become the "underwater eyes" for today's angler. Using sophisticated display screens, they not only accurately locate the bottom and indicate its composition (mud, rock, weeds), they

Figure 20.1. Today's fishing boats are fast and comfortable. They often come equipped with rod holders, live wells, depth finders and other amenities that the modern fisherman requires.

show schools of baitfish and often the salmon or bottomfish themselves. Avid anglers have become quite attached to their fishing accessories. (I once tracked down my electronics technician over a holiday weekend to repair my unit when it quit unexpectedly!)

- Better rods, reels, and line—Fishing rods have evolved from tree branches and cane poles to fabricated combinations of light-weight fiberglass and graphite, and some with Kevlar, which is so tough and resilient that bulletproof vests are made from it!

  Reels are available with single-action spools, with gears to multiply the winding action, and with friction drags (slipping clutches) that allow the fish to run without the angler releasing the reel handle. Spinning rods and spinning reels have revolutionized casting techniques. The line peels off the spool in an almost frictionless motion, allowing casts of 100 to 200 feet or more from the shore, boat or pier.

  Fishing lines have changed from the old cutty-hunk braided cotton thread to monofilament nylon, which is less visible to the fish. Nylon will also stretch up to 30%, providing a shock absorber to keep the fish from breaking the line or tearing out the hooks. For those who want to eliminate stretch in order to control the lure and the fish more tightly, non-stretch Dacron has many advantages.

- Downriggers—Increasing pressure from both commercial and sport fishermen have led to fishing for salmon in deeper water in many areas. Trolling in deep water requires some techniques that are unnecessary in shallow water. Trolling a lure on light tackle deep enough to catch the fish is difficult. Downriggers are designed to allow sport fishermen to overcome this problem.

  A downrigger is a large gurdy wheel with a wire line extending through a pulley over the side of the boat and a lead weight on the end (Figure 20.2). The fishing line is attached

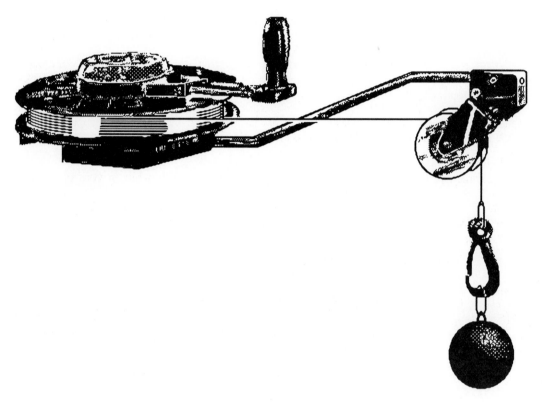

Figure 20.2. A typical downrigger with cannonball weight attached to the hook at the end of the line.

to a release clip on the wire line near the lead. The lead weight drags the fishing line to the proper depth, where the release clip trips open when the fish strikes. This allows the angler to play the fish directly on a light rod, without the interference of the weight.

Downriggers also allow precision depth control. Sophisticated counters measure the length of wire line from the boat to the lead weight.

- Instructional books, magazines, films, videos, and seminars—Sport fishing is such an inexact science that most anglers are hungry for any new fact or "tip" to improve their knowledge and their fishing success. (I have written nine books on fishing and marine life which have sold more then 400,000 copies in the Pacific Northwest. The top seller, "How to Catch Salmon—Basic Fundamentals," has sold over 130,000 copies. My film shows and seminars have brought close to one half million anglers to a theater or classroom for the latest in fishing lore. I have developed a remotely controlled underwater TV camera that for the first time showed the actual strike in close-up slow motion, and allowed instant replay!)

Know Your Prey

Knowing the habits and feeding characteristics of the target species is vital to consistent success. Chinook salmon always seem to be near some reference point, like a steeply shelving reef, a kelp bed, the shore line or near the bottom itself in deeper water. Coho, pink, sockeye and chum salmon are more commonly found in open water. Some fish prefer specific depths. Migrating sockeye are often found at 50 feet while pinks are usually found in somewhat shallower depths. Coho can run right on the surface and sometimes very deep (200 feet or more). Migrating chinooks often hug the shoreline at about 30–50 feet, but feeding chinook are caught by sportsmen at 200 feet or more offshore and over 500 feet deep by some commercial trollers

There is an annual pattern of fish behavior just as with anything in nature. Like the crocuses and snowdrops that poke through the snow each winter, or the regular return of the swallows to Capistrano, the underwater world is, in some respects, as predictable as the dates on the calendar.

Herring and anchovies move in a regular cycle of feeding and spawning, as do shrimp and zooplankton. The predatory fish follow this same pattern to be near the food supply, and then respond to their ripening eggs and milt (sperm) by moving toward their spawning grounds.

Keep a Diary

A simple and very effective method of learning more about fish habits in your own fishing area is to keep a diary, listing pertinent information from each trip. You can build a personal record of fish habits and preferences in your own favorite fishing locations. Most "high-liner" fishermen keep such a record and find it to be one of their most valuable fishing tools.

Some important facts to record include:

- *Weather conditions*—Sun, clouds, rain, wind speed, and wind direction can affect your fishing. Salmon, especially chinook, tend to go deeper on bright days, and wind can significantly alter trolling speed and direction. Even color selection for lures differs when clouds or rain alter the amount and quality of light reflected from your lure.
- *Tidal conditions*—Most fishermen know that fish feed in a cyclical pattern, with one or two heavy feeding periods per day, each lasting about half an hour to 45 minutes. This exciting time is called "the bite" in angler's lingo. When I started salmon fishing almost 40 years ago, I thought it was superstitious hocus-pocus when the old timers told me that some tides were "good tides" and others were "bad tides." Then I realized that tides do have a sometimes dramatic effect on fishing success.

    In many areas, the "bite" is most likely to occur during the hour before or the hour after slack tide. Slack tide occurs during the tide change; that is, the transition time during which the tide changes from ebb to flood or vice versa. This is the time when the water moves slowly or not at all, making it easy for the baitfish to catch plankton or euphausid shrimps. And when the baitfish are out feeding, the predatory gamefish are on the prowl to find them!

    The well-known Solunar Tables by John Alden Knight show the times that fish and other forms of animal life tend to be more active and feeding. There are two major solunar times and two minor ones each 24-hour period. In studying these tables, I find that the major solunar times when I fish correspond roughly with the time of low slack tide and the minor ones correspond with high slack tides.

    In narrow passes where strong tides create boiling rapids and whirlpools, mid-tide is the most productive time for fishing. During this time, the plankton and baitfish are swept by the bigger and stronger tides into quiet back eddies, followed by the salmon. The fish become concentrated in relatively small pockets, where they are targeted by moochers, jiggers or trollers (see Methods of Sport Fishing, this chapter).

    In some large bays and inlets, the tidal currents are minimal. The area just fills up like a bathtub, with no pronounced currents or back eddies. These areas usually provide the best fishing in early morning or late evening.
- *Color and size of lure*—Baitfish and shrimp take on various colors and shades in an attempt to camouflage themselves (to match water color or bottom color). Changes in water color often result from concentrations of various types of plankton. Water temperature, salinity, and the mud and silt washed into the sea also affect water color. If you catch a sal-

mon on a particular shade of pink lure, it's a good idea to record that information in your fishing log so you can try it again at the same time next year or whenever similar conditions are observed.

Baitfish grow and change throughout the year, so lure size should change to match the baitfish. Mature herring migrate to spawn in protected inlets in late winter and spring, and large lures will produce in these areas. Large lures also produce well in areas that contain resident mature herring.

- *Fishing depth*—Many anglers change lures too quickly when they are not catching fish. Their lure may be just what the fish want, but it's not being fished at the proper depth. I suggest experimenting with depth before trying a new lure. Downriggers can help trollers achieve precision depth control with their counter mechanism, which shows exactly how much line is out. Another method of measuring relative depth is by counting "pulls" as you strip line off the reel when setting out the fishing line.

My involvement with underwater T.V. camera experiments has confirmed my belief that proper lure depth is one of the most critical factors in finding fish. As we lowered our camera, we would sometimes see no fish at all for 30 or 40 feet. Then suddenly, there they were, in a band of water no more than 5 or 10 feet thick. We found the fish tended to be stratified in narrow layers, sometimes according to temperature. Using a temperature probe attached to the camera, we discovered that both coho and chinook salmon tended to seek out 53°F water in the summer months, although I'm sure the preferred temperature varies in other locations and seasons.

- *Exact location*—There are a number of special "hot spots" that produce fish at certain times of the year. I have several favorites: Two are over sandy bottoms near long spits of land. Candlefish (needlefish, sand lance) spawn and hatch out of the sand in late spring, attracting feeding chinook. Trolling or jigging over these precise spots during late April and during May is usually successful. In another area, the coonstripe shrimp come into shallow water in August and the coho are right there for their annual shrimp cocktail!

Another favorite spot is a sharp, shelving reef with a steep underwater cliff stretching down more than 200 feet. Large migrating chinook seem to "hole-up" here from June to mid-August, and my own "The Lure by Charlie White"* trolled at 100 feet tight against the reef often produces a 15- to 25-lb king salmon.

- *Time of day*—As mentioned above, many areas produce the most fish under very specific tidal conditions, but others are productive at certain times of the day. Excluding tidal considerations, the most productive time of day is very early morning. The fish have probably not been feeding all night and their stomachs are empty. They may also have been migrating at night, and have used significant amounts of stored energy, accentuating their hunger. (Did you ever notice how hungry you are after vigorous physical activity?)

Research tests in the mid-Pacific Ocean at 180° longitude, where there is practically no tidal action, showed interesting results. The objective was to catch salmon by hook and line, tag them, and release them to determine whether they migrated to North America or to Asia. The researchers found that in this area, where tidal influences were minimal, they caught 80% of their fish in the first 2 hours after daylight!

Preparation for a Fishing Trip

An important factor in successful sport fishing is thorough preparation. The angler who just grabs his tackle box and case of beer and jumps in the boat on the spur of the moment seldom comes home with a good catch.

---

*Use of brand names does not imply endorsement by the editors.

Preparing your tackle the prior evening saves precious time at the crack of dawn when hungry fish won't wait for you to find the right lure or get your leaders tied up properly. Having a check list of items to get ready can be useful. Some of the items to include on the checklist are:

- lures that are appropriate to the season and location. They should be tied up with proper leaders and bead chain swivels;
- checking reel action and examining rod guides for cracks, which can fray the line (I also cut off 6 to 10 feet from the end of the main fishing line where nicks and weak spots are most likely to occur);
- an organized tackle box with lures sorted by color and size. Be sure to have such basic tools as toenail clippers (with a straight face rather then the concave face of fingernail clippers—these are much better then a knife for cutting fishing line), longnose pliers, fishing knife and a good hook sharpener;
- polarized sun glasses (to see baitfish, euphausids, and other underwater action through reflected surface glare), sun screen for a tender nose, tide book and marine charts;
- boat equipment including landing net, downrigger, weights (plus extra wire or a spare downrigger) and rod holders.

I also plan a fishing strategy for each trip. I use data from my fishing log, tide book, weather forecast, and fishing reports from my friends and radio broadcasts to plan where I will fish and what gear I will use.

Finding the Fish

An early alarm jolts me as much as any one, but once up and moving, I love to get out on the water and watch the world come to life. Following my fishing strategy from the night before, I begin to systematically work the most productive water. Contrary to some popular beliefs, saltwater fishing, especially for salmon, is not a laid back, feet-up-and-drink-beer sport. Success demands persistence, attention to detail, and sometimes plain hard work. But the results are worth it. Solving the daily detective mystery of finding and catching fish is one of the most satisfying activities I know.

Feeding gamefish are found in a tiny proportion of the vastness of our coastal waters. We use our knowledge of the tides, weather, and fish behavior patterns to help us find where the fish are concentrated.

Tide Lines

These demarcations between currents are often highly visible, often cluttered with weeds and debris (Figure 20.3). Baitfish and salmon gather there. Migrating salmon will often follow tide lines on their way to the spawning rivers, even when the tide lines are far from their normal shore-hugging migration route.

Back Eddies

These are the quiet water areas behind a point of land where the main tidal flow sweeps by (Figure 20.4). They are good holding areas for salmon which, like other fishes, will avoid fighting the main current. Phytoplankton are swept into these back eddies where they are eaten by euphausid shrimps, which in turn are consumed by herring and other baitfish.

Sea Gulls

These are the aerial "spotter planes" that can locate salmon and baitfish far better then we can from the water's surface. When you can see sea gulls circling and squawking above the water, or actually dipping down into the water, it usually means they have located fish.

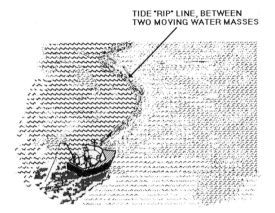

Figure 20.3. Fish can often be found near tide lines.

Figure 20.4. Back eddies provide areas of quiet water where fish may hold. The arrows indicate the direction of current flow.

Diving Birds

These sea birds (often erroneously called diving ducks or hell-divers) are also an excellent sign of salmon, sometimes even more so than sea gulls (Figure 20.5).

My own experience is that murres, murrelets and rhinoceros auklets (identified by the distinctive hump on their beak) are the most reliable indicators of baitfish and salmon concentrations. Grebes, cormorants, and others often turn out to be false alarms for me, but I know others who feel these birds are also good indicators of fish. Diving birds can also give you some tip-offs regarding the type of bait to use. Sometimes you will see them with the actual baitfish in their mouths. A good pair of binoculars can help you see if the birds are feeding on needlefish or herring and give you an idea of the approximate size of the bait for matching against your own lure. My own films and videos show dramatic underwater footage of murres attacking herring in the Strait of Juan de Fuca between Washington and the Canadian Province of British Columbia. The murres swim down under the herring, always attacking from the bottom so the herring are forced toward the surface. The murres also herd the edges of herring schools like cowboys at an annual roundup.

Sometimes, however, it is only the birds that are feeding on baitfish and there are no salmon in sight. But even if salmon are not evident, it is wise to work the surrounding area because the fish usually stay near the feed, even when they are not hungry.

Abrupt Depth Changes

Whether it is against the shore or off a shallow reef away from the main shoreline, any steeply shelved area is an excellent holding ground for salmon (Figure 20.6). Chinooks especially seem to like these areas of abrupt depth changes. Mature, migrating chinook salmon will hug steeply shelving shorelines as they move toward the spawning grounds. I have caught many large chinooks (40 lbs and up) within 10 to 15 feet of almost vertical rock shorelines (Figure 20.7).

Weed beds and kelp patches also provide an edge along which salmon will cluster. They will also hide within kelp beds, ready to rush out and attack nearby baitfish.

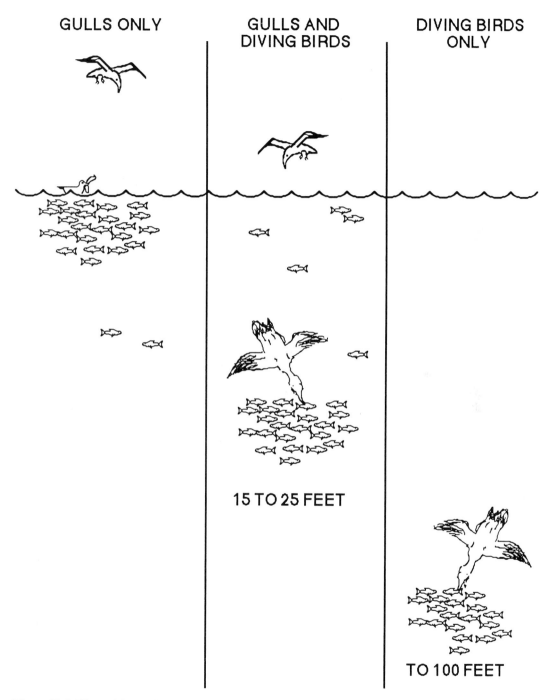

GULLS ONLY

GULLS AND
DIVING BIRDS

DIVING BIRDS
ONLY

15 TO 25 FEET

TO 100 FEET

Figure 20.5. Bird activity is often your signal that fish are present. Different types of birds can fish at different depths.

Figure 20.6. Fish are often found in areas where depth changes abruptly.

Working the Bottom

Many salmon, especially chinooks, seem to hug the bottom, except during active feeding periods in early morning or late evening. Whether it is a shelving shore or the bottom itself, chinook salmon seem to prefer being near some sort of reference point. Moreover, the bottom or steep shoreline acts as a drag on the water current, so there is less movement and the salmon does not have to work so hard to maintain position.

The bottom is also a rich source of food. On sandy bottoms, needlefish (also called sand lance and candlefish, genus *Ammodytes*) spend much of their time just above the sand or actually buried in it. Salmon hang close to these sandy areas on the lookout for a tasty meal.

Cloudy Water and Shaded Areas

Most fish seem to prefer shaded areas or somewhat cloudy water. This is one of the challenges of our underwater television research, since it is difficult to find salmon in clear, well-lit underwater environments.

When salmon were more plentiful, I used to fish Cowichan Bay on Vancouver Island for large chinooks. A steep cliff rises from the northeast side of the bay and this was often a good spot for fishing the last part of the early morning bite, when the sun covered most of the bay but the cliff shaded a narrow nearshore area. Where a muddy river pours into the sea there is often a clear dividing line between murky water and the clear salt

Figure 20.7. The author with a "respectable" chinook salmon.

water beneath. We often see this on our underwater camera as it drops beneath the turbid layer and bursts into clear water, almost like an aircraft descending from a cloud bank.

Salmon often stay in the clear water beneath a murky surface layer, protected from attack by surface predators. At Rivers Inlet in northern British Columbia, the rivers contain a colloidal suspension of glacial material which cuts visibility to only a few inches. An underwater camera lowered through this 6- to 10-foot layer, often popped into the clear water and showed both pink and chinook salmon in sizable schools just beneath the covering layer.

Rolling and Jumping Fish

This is a very positive sign of salmon! You actually see them as they roll on the surface or jump clear of the water. You can't get more direct evidence that fish are nearby and you should fish these areas intensely.

Some people claim that rollers and jumpers will not strike; perhaps they have had frustrating experiences trolling through a school time and again with no success. Although it is true that rolling fish gathered near spawning areas are often less interested in active feeding, they will strike under certain conditions. Often the biggest problem is with chum salmon, which gather in great schools in the late fall and jump on their sides, almost like a planing surfboard. These fish are often very difficult to catch, but I have used dark colored jigging lures, especially Buzz Bombs® with good success. Another excellent sign is fish "finning" on the surface. They move through the water with their fins sticking out of the water or just beneath the surface so you can see a small wave created by the moving fish. These fish are usually active feeders looking for baitfish or euphausids.

Watching Other Fishermen

When the landing net comes out on a nearby boat, you know that someone has solved the mystery of finding the fish. By all means, learn what you can from this situation. Close observation with binoculars can reveal many things to improve your success. Note the size of the sinker as the fish is brought in or look for the downrigger to give you clues about depth. Some fishermen will even count the turns as the successful angler cranks in his downrigger. Look also for the size and type of flasher or dodger and the size and color of the lure in the fish's mouth as it is brought on board.

As the boat begins fishing again, observe its trolling speed and the area on which it concentrates. You can also watch the line being let out for additional clues on lure type and depth. I also find it helpful to just be direct and ask the successful fisherman about his catch. When I made this remark at one of my fishing classes, a voice from the audience shouted, "Yeah, and they'll lie like hell!" Sometimes they will, but I find that most fishermen are willing to share information if approached tactfully. After the fish has been landed, I often troll over to the successful boat and congratulate them on their success and ask about the size of the fish. Usually they are delighted to talk about the fish and will often hold it up for me to admire. I then follow up with questions on type of lure and fishing depth. These two bits of information, along with the area of catch (which you have observed already), give you all the information you need.

Marine Radio "Chatter"

Many anglers equip their fishing boats with CB or VHF radios to exchange information on their catches. Each area seems to have its own preferred channel and sometimes the information is direct and sometimes it is "coded" so that only the "insiders" can decipher it. You can also use your radio to ask for information on fishing conditions. I remember one plaintive call from a female angler who "just had to have salmon for relatives who were coming to dinner that night." The response was instantaneous and she got enough tips to get her on the right track.

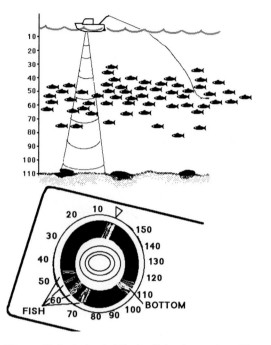

Figure 20.8. A simple "flasher" depth sounder will clearly indicate the depth of the water and will also indicate if fish are present and their depth.

Depth Sounder

To many fishermen, this electronic tool is the most valuable on board. It gives them underwater "eyes" with which to learn about bottom contours, location of salmon and location of baitfish. I used to use a simple flasher sounder for tracking the bottom when fishing deep with my downrigger (Figure 20.8). It also was helpful in spotting concentrations of fish and bait. Subsequently, I used a graph sounder, which prints out very detailed information on special paper. With a graph sounder you can often distinguish baitfish from salmon and even distinguish size and species in some instances. However, I found the graph sounders to be very fussy to work with and difficult to maintain; since I like to keep my sounder running continuously, I thought the cost of paper was excessive.

Now I use a video hydrograph and an LCD sounder, which combine the advantages of flasher sounders (instantaneous readout, no expensive graph paper) with most of the detail of graph sounders. Some of the more sophisticated models use color and have built in memories so you can recall readouts for later examination. I find that a video hydrograph gives the most detail, but it is hard to read in direct sunlight and sometimes doesn't give enough detail at depths greater than 50 to 75 feet. An LCD sounder, which is easy to read in direct sunlight, provides good detail.

Many anglers have sounders, but do not use them properly. They are sensitive instruments and will not give you accurate readings unless adjusted properly. It is wise to invest some time in reading the instruction booklet and working with your sounder before going fishing. The sensitivity controls are particularly important in that you need to adjust them so you can get enough feedback to get bottom and fish readings without all the unwanted "static" or "clutter" that results from the sensitivity being set too high.

Temperature Probe

Great Lakes anglers have used temperature probes for many years. Most large freshwater lakes tend to be stratified with warm water at the surface and quite cold water deeper. Between the warm and cold water is an area of rapid temperature change called the thermocline (discussed in Chapter 11). Salmon, trout, and baitfish are often found in this thermocline area. Locating it is perhaps the most important factor in Great Lakes salmon fishing.

My own belief was that Pacific Coast waters were cool enough and tidal mixing thorough enough that temperature was not a factor there. However, as I mentioned earlier, our recent underwater experience has shown me otherwise. During these experiments, we began to record the daily temperature when we found the salmon. Trolling at our normal depth of 30 to 35 feet in northern British Columbia waters, we found the salmon at a water temperature of approximately 53°F. One morning, we were trolling with no success and found that most of the other anglers were also having no luck at all. Our temperature probe read 55°F; we lowered the camera until the temperature dropped to 53°F and, almost magically, salmon appeared. They had moved another 20 feet deeper to find the preferred water temperature.

Fishing Methods

There are three major methods of saltwater sport fishing:

- Trolling—This is the most popular technique in most areas. In its simplest form, it means towing a bait or lure behind a slowly moving boat. The movement of the boat imparts all or part of the "action" to the lure or bait, enticing the fish to strike.
- Still fishing, jigging or "mooching"—These techniques are usually practiced from an anchored or drifting boat. The lure action is generated by moving the rod up and down in the case of an artificial lure (jigging) or by the movements of a live minnow or other bait (mooching). Some artificial and natural baits, especially plug-cut herring, get their action from the tidal current moving past an anchored or drifting boat.
- Shore fishing—Fishing from the shore is productive at jutting points of land, steeply shelving rocky beaches, breakwaters, wharves, piers, and floating docks. Most shore anglers cast and retrieve artificial spoons, spinners, or so-called "drift" lures. Others cast a weighted line to sit on the bottom and allow the passing current to "work" the lure.

  Sea perch, rockfish, flounder, sole, greenling and even skate are often taken by dropping a baited hook from a floating dock in a marina.

Trolling Techniques

I fish mostly off southern Vancouver Island and in the Strait of Juan de Fuca, where the fish, especially chinooks, tend to swim very deep most of the time. I troll with a downrigger except when surface-feeding coho arrive in late summer and fall. Then I will use 2 to 8 ounces of weight to reach depths of 20 to 50 feet. My favorite technique for catching migrating fall coho is using bucktail flies, trolled right on the surface with a small mother of pearl spinner in front. In my opinion, this is the most exciting fishing of all; you can sometimes actually see the fish rush toward the lure, strike with a vicious splash, then tail dance across the water! Listed below are some items you should consider when trolling, be it with or without downriggers, since the same general principles apply.

Choosing the Right Lure

A group of anglers working a school of fish can usually catch them on a number of different lures, but some general principles apply no matter what they use:

- Match the size and shape of natural bait in the area. Fish are more easily tempted into striking if your lure looks like the natural feed they are trying to catch. They are conditioned to seek out wounded baitfish, so the more your lure looks like that particular baitfish, the more likely the strike.
- Small spoons such as Point Defiance®, Kripple K,® and Tom Mack® match the general shape of herring (Figure 20.9). Herring grow as the season progresses, so the size of lure should also be increased.

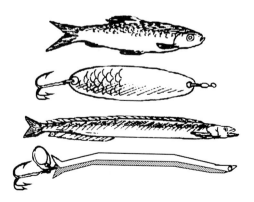

Figure 20.9. Lures are designed to imitate live bait.

Almost any type of spoon can be made more effective by adding a tiny bit of bait to the trailing hook. Cut a piece of herring or other baitfish perhaps only 1/8 of an inch wide by 1/2 inch long and impale it on the hook. If it dampens lure action, use a piece of fish skin or even a tiny piece of plastic hoochie tail to give added action.

- Plugs, both wooden and plastic, also imitate herring or anchovies (as does my own "The Lure by Charlie White").
- Long, narrow spoons, bucktail flies and thin hoochies match the needlefish or sand lance, which are usually found over sandy bottom areas. Some of the fatter hootchies and bucktail flies also imitate squid, which are found in open waters off the coast; they can also be used to simulate shrimp.

Catching Large Fish

When fishing for trophy-sized fish, the rules are different. Our underwater television tests showed that large salmon preferred a large bait, even when the natural feed was small. It seems that large predators follow a "conservation of energy" habit and don't want to use a lot of energy chasing small prey when they can gulp down one large one.

Attaching the Lure Correctly

Most quality lures are designed with great care to give the best fish-catching action when trolled properly. One of the most important details in this regard is balance. Your leader should be tied directly to the lure itself to maintain the balance designed by the manufacturer.

You should never hook a snap swivel or other device to the leading edge of the lure because this could completely alter its balance and action. It is sometimes more convenient to hook lures with snaps and swivels, but you are defeating the main purpose of the lure. You can, however, use a snap swivel between the leader and the mainline. I use a bead chain snap swivel at the end of my mainline and hook another swivel to the end of my leader. This allows me to change lures quickly when leaders are already attached to them. I prefer bead chain swivels to ball-bearing swivels because their multiple turning points keep them working even when one or two swivel points are sticking.

The action of a plug can also be changed dramatically by the position of a knot tied to the eye in front of the lure (Figure 20.10). Tying the knot snug at the top of the eye will make the lure dig deeper than tying it to the middle or the bottom of the eye. Many expert plug fishermen use knot position as a major adjustment in controlling plug action.

Tying a Proper Knot

More fish are lost because of poor knots than for almost any other reason. Some anglers just don't know how to tie a good knot, and others are just too careless to do it properly.

Several manufacturers of monofilament line have their own knots that they recommend for their particular brand. However, most knots are a variation of the jam knot, blood knot, or clinch knot. This last knot is tied by wrapping the line around itself several times then looping it back through the gap between the hook and the first loop.

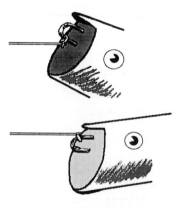

Figure 20.10. Where the line is tied affects that action of the lure.

Nylon line can score, burn, or fray when pulled tight; thus it is important to wet the line (with a bit of saliva) before pulling it tight. This lubricates the line and cools it a bit, minimizing the problem. (Figure 20.11 shows two of the knots I like best.)

Checking the Lure Action

This simple tip seems so obvious, but it's amazing how many fishermen are so eager to get their lure in the water that they just throw it over the side and strip out line without bothering to see whether it is working properly.

What is proper action? Most lure manufacturers have detailed instructions enclosed in their packages and usually give tips on proper action for each lure. By all means, read these instructions thoroughly and take advantage of the manufacturer's research and experience. After all, the manufacturer is successful only if you catch fish and, thus, should give you the best possible instructions.

Checking lure action beside the boat will also make you aware when an unusual action turns out to be productive. (Occasionally, I have been unable to get my herring strip working "normally," but I have lowered it down anyway rather than change to a fresh bait. Sometimes I am startled by a vicious strike almost immediately. Remembering the unusual lure action I saw beside the boat, I duplicate it, and have learned a whole new fishing secret for myself.)

Choosing Likely Depths

Assuming you are trolling with three lines, you can fish with one of two patterns:

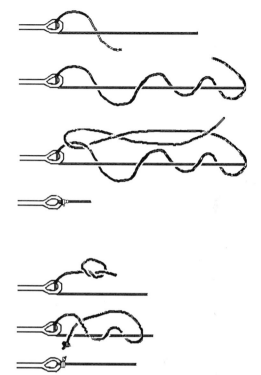

Figure 20.11. Use of the proper knot helps ensure that the line and hook do not separate at an inopportune time, like when you've hooked a fish!

- If you expect the fish to be relatively shallow, put two shallow lines on each side of the boat and a deep line down the middle.
- If you expect the fish to be deep, hang two deep lines over the side and a shallow line down the middle.

(When fishing with downriggers, I will often put a single light line over the stern to pick up any shallow feeders, especially a coho that might be investigating the propeller wash.)

Checking Your Gear Regularly

I make it a practice to inspect my lures every 15 to 30 minutes whether I'm getting any strikes or not. This allows me to be sure that my lures are not fouled with weeds or jellyfish. It also exposes the lures to fish in the complete column of water from the surface to the deepest fishing line. Often a salmon will grab the lure as it is being reeled in or let out.

If I get a strike when bringing in the line, I will often fish shallower or faster, since this could indicate a shallow fish or one which simply wanted to chase a faster moving lure.

The opposite is true if you get a strike when lowering your tackle. This may mean that the fish likes a slower moving lure, and you can adjust your trolling speed accordingly.

Following the Fish Down

At the crack of dawn, salmon are often feeding very close to the surface, usually on baitfish which, in turn, are feeding on surface plankton. As the sun moves higher in the sky, the plankton, baitfish, and salmon tend to move deeper. Although fish tend to feed more actively in the first 2 hours after daylight, they will often bite well into the day if you are willing to follow them deep enough.

I often encounter a good early morning bite on top of a 50-foot shelf near my home in Saanich Inlet on Vancouver Island. When the "bite" goes dead, I usually move offshore to the adjacent deeper water, where I find the fish are often still in a striking mood.

I remember one occasion when I had four fishing students on board and we caught four nice salmon on top of the shelf. We continued to follow the fish down the edge of the shelf, lowering our downriggers deeper and deeper. We caught our limit of 20 salmon that morning, most of them over 100 feet down, and the last one (a lovely 14-pounder) was taken just before noon with the downrigger depth counter reading 275 feet! Incidentally, that fish swam straight to the surface and put up a spectacular fight on top—think of the marvelous internal mechanism for adjusting to such tremendous changes in pressure!

Fishing a Zig-zag Pattern

Watching salmon follow lures behind our underwater camera, we discovered they will sometimes follow the lure for long periods without striking. However, changing direction will sometimes trigger a strike and, indeed, many anglers comment they often get strikes while making a turn.

Fisheries scientists conducting tests off the west coast of Vancouver Island used commercial trollers to find ways to increase catch productivity. One of their most significant findings was that catches were higher when boats fished a zig-zag pattern as opposed to those that trolled in a straight line. I will often fish along a tide line where two opposing currents mix, usually fishing back and forth across this dividing line, sometimes with considerable success.

When fishing bucktail flies, I troll straight ahead for perhaps 100 yards or more, then turn the wheel hard over to make a 90° change in direction. This causes the outside fly to whip ahead quickly and the inside fly to stop dead in the water. Often these dramatic changes will trigger an excited coho into a smashing strike.

Fishing in One Direction

Important changes in lure action take place when you change direction in relation to wind or especially the tide. Our underwater camera revealed some interesting findings regarding this action. Trolling with the tide, our lures would often speed up as we fished deeper. This is because the surface water moves faster than deeper water, pushing the boat ahead faster relative to the lure.

The opposite happened when we trolled against the tide. We often had to advance the throttle two or three times as fast as anticipated in order to keep the same trolling action on the lures. Sometimes 500 rpm on the motor worked when moving with the tide, but we would have to go 1200–1500 rpm to get the same action against the tide.

Given the above, I suggest that if you get strikes trolling in one direction (usually with the tide), it's usually best not to troll back against the tide, but just pull in the gear and run back to the starting point.

Using the Gear Shift

Mechanics tell me this technique can be hard on the engine, but it can be a very effective trolling method. A friend who was having no luck asked me to go fishing on his boat to learn what he was doing wrong. It was obvious that he was trolling too fast, but his large engine would not troll down. His heavy downrigger weight streamed out behind the boat at a sharp angle, robbing him of needed depth and forcing his lures to whirl wildly.

I suggested that we take the motor out of gear and let the downrigger and lures work deeper. Within 15 seconds we were rewarded with a strike and soon had an 8-lb chinook in the net. We used this technique for about 2 hours and ended up with our limit of nice chinook up to 14 lbs. It took about 45 seconds after putting the motor in neutral for the downrigger to reach a vertical position, then we would put the motor back in gear for 2 or 3 minutes until the 150 feet of wire was streaming behind again.

This technique allowed us to troll the lure through a wide range of depths, and it also changed lure action. In this instance, the fish were very deep and struck as the lure fluttered down to the bottom of their range, or just after we put the motor back in gear.

You can use this same technique without a downrigger, sometimes getting even more dramatic results. Trolled with 6 or 8 ounces of weight and a long line, the lures will drop down in a long arc between full trolling configuration and hanging vertically. This covers an even wider range of depths.

Varying Your Speed

Every lure has an optimum speed at which it gives the best action, but most lures can be adjusted to give proper action at several speeds. You can adjust the degree of bend in a spoon, change the knot position on a plug, and adjust the hook position on various bait rigs to give proper action at different speeds.

If you're not getting strikes at your normal speed, try going 25% to even 50% faster and adjust the lures to work correctly at that speed. You may be surprised at the results. Commercial trollers move much faster than most sport boats and have consistent success because their lures are "tuned" for those speeds. Faster speeds also allow you to cover more water, thereby exposing your lures to more fish.

Trolling much slower than normal can also be productive. This is especially true in areas where chinook salmon are "holed up" near a river mouth and are not feeding actively. In this situation, it is important to give the fish the maximum time to look over your bait as it wobbles or rolls slowly past the fish's nose.

As a general rule, I like to troll relatively slowly during periods when the fish are not feeding actively, but speed up considerably during the "bite." A hungry fish will be triggered by a fast moving bait, while a fish that is resting or digesting its food is not likely to be motivated to chase it.

Using "Gunsight" Marks to Pinpoint Location

Salmon are often found in tightly bunched schools covering an area of perhaps 100–200 feet in diameter. This is a tiny segment in the vastness of most fishing areas.

Once you locate the fish, by all means mark the spot as precisely as possible. I use range markers or "gunsight" marks to pinpoint the location. In the excitement of a strike it's often difficult to remember your position, but taking good marks can pay big dividends in getting back to the hot spot.

Most weekend fishermen believe it is good enough to look for a single shore location or try to line themselves up between two markers on opposite shores. Neither of these methods is very accurate, since a 10° error in either direction can put you as much as 900 feet off the correct spot, even if you are only 1/2 mile off the beach.

The correct method is to line up two objects viewed from the same side of the boat, one behind the other (a rock outcropping on the beach lined up with a tall tree behind, for example). Lining up 2 sets of range markers at approximately 90° to each other will give you 2 intersecting lines and a position exactly over the spot where you hooked the fish (Figure 20.12). It takes a bit of practice to pick out these marks, but it's amazing how quickly you will learn to take good marks, especially when it pays off in increased catches.

After marking the spot, I put the motor in neutral and enjoy playing and landing the fish. I then circle around, line up one set of marks, then troll until I cross the other set. Of course, I have already located the fish in the third dimension (depth) by noting the reading on my downrigger or by remembering how much line I had out.

Figure 20.12. You can return to the same position in the water if you have taken accurate sitings of stationary objects.

An alternate method of marking a location is to throw a floating object over the side immediately after getting a strike. This can be a seat cushion, bleach bottle (with a 1-lb weight attached to minimize wind drift) or any other floating object. This procedure works well when the water is relatively calm. It is also effective for marking the location of surface-feeding coho when the tide is running. The floating object will move with the school as it drifts in the tide.

Finding the Big School

When I was a fishing charter guide in the 1950s, I was out on the water almost every day, and learned a lot of special tricks to help me stay with the main body of fish. On many occasions I would have several "hot" days, catching limits for the entire party, sometimes getting limits on two different trips in the same day. With four or five persons in each party and a five-fish limit, we might take 30 or 40 fish in 1 day. Then one morning I would return to the same spot and begin catching smaller runt fish or fish with some deformity (usually a torn mouth from a previous hookup or an injury from a seal or other predator). I soon learned that this meant the main school had started to move to another location, leaving behind the stragglers.

When I began to catch these undersized fish, I would immediately begin to explore anywhere from a few hundred yards to a quarter of a mile in each direction in search of the main body of fish at the next holding spot.

Using the Sandy Bottom Technique

When fishing on a sandy bottom you can use your downrigger effectively to get the lure right on the bottom where salmon are literally picking the sand lance (candlefish, needlefish) right out of the sand. You can accomplish this by using a long narrow metal spoon such as a Koho Killer,® Koho King,® or Charlie White Needlefish Troller and attaching it 25 to 30 feet back from the downrigger release clip. Lower the downrigger until the weight touches bottom, crank up one or two turns and you're ready to fish.

The metal spoon's weight and its distance from the release clip allows it to ride just above the sand. (The release clip should be hooked on to the downrigger wire about 5 feet above the weight.)

You will need to check your lure frequently since there is a great deal of sea lettuce and debris on sandy bottoms. You should also keep "probing" for the bottom by lowering the downrigger every few minutes to check that you are still riding just above the sand. Also, your depth sounder will help determine when to lower or raise the downrigger weight.

Trolling Away from the Sun

Research tests at a trout hatchery using floating food pellets found that the trout always attacked from the sunny side, often driving all of the pellets to the far side of the pond. After studying those data, we watched our underwater television monitors more closely as the salmon approached our lures. We noticed immediately that there was a difference in their behavior when we trolled directly into the sun. They would still approach and follow the lures, but a much smaller percentage actually struck. Often they would strike soon after we made a turn away from the sun. We found this situation most pronounced in early morning and late evening when the sun was low.

Getting Deep with Light Tackle

Getting your lure to the proper depth is extremely important, but it is difficult to know just how deep the lure is actually running. As you let out your line, it goes back in a parabolic arc, with a gradually decreasing angle so the line is most horizontal where it enters the water. The degree of this arc depends on many factors, including the weight of lure and sinker, size and resistance of the lure and flasher, line diameter, trolling speed and direction, and tide stage.

In my opinion, downriggers are by far the best method to get your line down deep, get precision depth control, and still be able to play the fish on light tackle. However, there are other ways to get more depth per ounce of weight.

- *Use thin diameter line*—Thin diameter line creates less drag in the water. The difference between even a 15- and a 20-lb line is quite apparent. The 20-lb line (of larger diameter) will keep the line closer to the surface than a 15-lb line with the same weight attached.
- *Troll slowly*—This also minimizes drag, but often makes it difficult to get proper lure action. Try wobbling spoons, "flatfish" type lures, or specially rigged herring baits.
- *Use trip weights*—There are a number of gadgets on the market which allow you to drop or slip a heavy lead weight as soon as a fish strikes. However, this requires a stiff, heavy rod to hold the line while trolling. Drop weights are also expensive to lose each time you get a strike, and sometimes the net result is only a dogfish or a big chunk of seaweed.
- *Use planers*—These devices present a perpendicular surface to the line and drag it deep (Figure 20.13). They trip to a less resistant angle after the strike, but also require a stiff rod for trolling.

As I mentioned earlier, downriggers are the best method for fishing deep with light tackle. A downrigger equipped with an easy-to-read counter along with a depth sounder can be a deadly combination. Salmon are often found close to the bottom and will sometimes take only lures trolled within a few feet of the bottom itself. By watching your depth sounder, you can crank the downrigger up and down over bottom contours, keeping your lure right in the strike zone at all times.

Downriggers can be extremely helpful in locating the depth at which the fish are feeding. I often use two lines on each downrigger when I have several other people in the boat to help operate them, thus allowing me to cover 4 depths at once. For example, I will set one downrigger with two lines attached to it, one of them 5 feet above the weight and the other 20 feet above the first line. I will then drop this line to, say, 150 feet. I then rig the downrigger on the other side of the boat in the same manner and drop it to approximately 100 feet. This allows me to cover all the depths from 150 feet to 80 feet.

I will continue to change depths on both downriggers, covering the whole spectrum down to perhaps 200 feet. Once the productive depth is found, the extra lines can be removed and both downriggers targeted to the location of the fish.

I dislike using flashers and dodgers as part of my fishing tackle because, in my opinion, it takes a lot of the fun out of playing a fish. There is no doubt, however, that flashers are effective in attracting fish, so I use a technique that provides the benefits without the problems of dragging hardware on the fishing line itself. I hook my flasher on a 4- to 5-foot 80-lb test nylon leader and attach it just above the downrigger weight with a bead chain snap swivel (Figure 20.14). The flasher then rotates in the water (with no lure attached) and puts out an attracting flash and thumping vibration which brings fish to the area. The release clip is attached to the wire about 5 feet above the weight to prevent the flasher from fouling the fishing line.

BEFORE FISH STRIKES

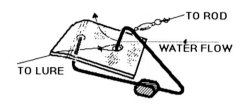

AFTER FISH STRIKES

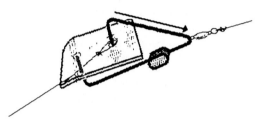

Figure 20.13. A planer can be used to help control the depth at which the lure is fished. When a strike occurs, the planer changes position to reduce resistance while you're fighting the fish.

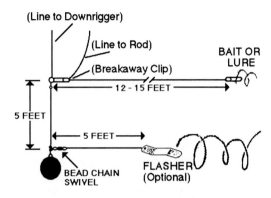

Figure 20.14. This illustration shows how I like to rig my downriggers.

Drift Fishing

Drift fishing is popular because it has several advantages:

- When you know exactly where the fish are, you can position your boat directly over the school and jig your lure right among the fish. In this instance, it can be more effective than trolling since the lure is in the fishing zone a much higher percentage of the time.
- It is economical. No fuel is required to troll and the tackle is much simpler than trolling gear. No weights, flashers, downriggers, or other expensive paraphernalia are required.
- It is quiet and peaceful—There is no trolling motor or gas fumes to distract from the enjoyment of nature. You can hear the sounds of feeding birds, the bark of a seal, the splash of a fish, and many sounds of nature that might be drowned out by the sound of the motor.

There are also a few disadvantages:

- If the fish are scattered, trolling is more effective in finding them.
- For reasons we don't completely understand, jigging is sometimes unsuccessful, especially in the winter months. There are also certain other times of the year when the fish just will not strike a jigging lure but will take a trolled lure.
- Some people find jigging to be hard work. They dislike continually jigging the rod up and down and would rather troll with the rod in a holder.

Jigging

In its simplest form, jigging involves simply casting out or lowering a weighted lure and jigging it up and down. Jigging vertically beneath the boat requires only a single action reel. If you plan to cast, you will need a heavy-duty saltwater spinning reel. Drift fishing rods are usually 7 to 10 feet in length depending on the depth you will be fishing. A longer rod with a limber tip works well when fishing shallow waters, but a shorter, stiffer rod is needed when fishing deeper than about 50 or 60 feet, since the softer pull of the long rod will be absorbed by the stretch in the long nylon line. A fiberglass spinning rod is quite suitable, but it should be light in weight—since you are holding the rod constantly, weight quickly becomes a factor.

Monofilament nylon is good for jigging even though it has more stretch than is ideal for firm hook setting. Twelve- to 15-lb test is good, with 20 lbs probably being the maximum. Anything heavier is difficult to cast, and the bulk of the line limits reel capacity.

Drift fish lures, unlike trolling lures, work best when fluttering down on a slack line. This is the most important key to proper drift fishing technique. Some lures, like the Buzz Bomb®, rotate rapidly when drifting on a slack line, sending out fish attracting vibrations. My own jigging lure ("The Lure by Charlie White"—jigging model) has a fork-shaped spinner tail which also sends out sonic vibration.

The jigging lure is lowered to the proper fishing depth, either just above the bottom or just below a school of baitfish, which you can often see on your depth sounder. Look also for schools of diving birds or sea gulls feeding on baitfish. Raise the rod tip 18 to 24 inches, then drop it quickly to produce a slack line and allow the lure to flutter down again. Keep repeating this procedure until you feel a bump or a tick, or the line feels slack when you start to pull up (this may mean a salmon has the lure in its mouth), then set the hook. Jigging lure manufacturers advise not to jerk too hard on the upstroke when jigging and recommend a slower, lazier action.

We managed to get some rare underwater TV footage of salmon attacking drift fishing lures, and our findings confirmed that short movements are better. We found salmon would attack the lures aggressively, but rapidly darting lead jigs were often too difficult to catch and the fish would give up after a few attempts. This is one of the reasons we added a spinner to our own jigging lure. It not only adds an attracting vibration, but also slows the erratic darting motion, making it easier to catch.

When casting a drift lure, you should try to place your lure near surface-feeding fish, a school of baitfish, or diving birds and sea gulls. Allow the lure to sink for perhaps 10 or 15 seconds, then retrieve 5 or 6 feet of line and let the lure flutter down again. Repeat this procedure until the lure is back at the boat.

An alternate method of retrieving is to cast out and reel in as fast as you can crank the reel handle. In certain situations, this technique can be extremely effective, even though it contradicts the basic "slack line" principle. This same fast retrieve method can be used when jigging directly beneath the boat. Simply lower the lure until it hits the bottom, then crank it as hard as you can back up to the surface.

Drift lures come in a wide variety of sizes and colors. As with trolling lures, it is a good idea to try to match the size and shape of the lure to that of the natural feed in the area. Larger and heavier lures are needed when fishing in deep water to overcome the effects of tide and current.

I have never believed that color was a major factor in fishing lures, except when surface trolling with bucktail flies. I suggest a lure in the blue, green or silver color range when fishing deeper than 50 feet and perhaps pink or red lures nearer the surface.

One last point about drift fishing: Sticky-sharp hooks are vital to success (Figure 20.15). Our underwater television camera showed us that fish were striking at our lures more often than we dreamed, but the hooks were often sliding right out of their mouths. This is especially important in jigging, when the salmon tentatively bites a heavy metal lure on a slack line. Its first instinct is to spit it out, and a sharp hook is much more likely to stick in its mouth. Use a sharpening stone, a triangular file, or an electric hook sharpener regularly.

Mooching

Another way to fish while the boat is adrift is by mooching. While jigging is done primarily with weighted artificial lures, "moochers" tend to use natural bait. Also, moochers let the current work the bait, or troll very slowly while the angler works the lure when jigging. Herring is the preferred bait for mooching, and it is either fished whole or "plug" cut (tie-ups are shown in Figure 20.16). This is an excellent method to use for large king salmon, in addition to an occasional halibut, rockfish, or lingcod. Unfortunately, mooching is also an excellent way to catch dogfish. If

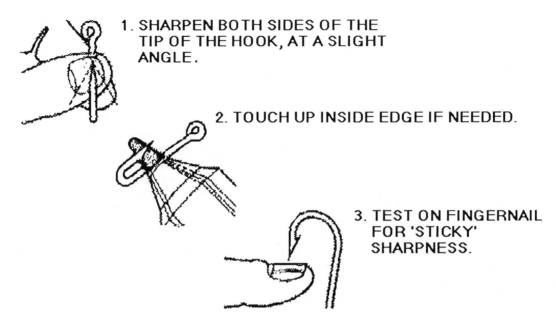

1. SHARPEN BOTH SIDES OF THE TIP OF THE HOOK, AT A SLIGHT ANGLE.

2. TOUCH UP INSIDE EDGE IF NEEDED.

3. TEST ON FINGERNAIL FOR 'STICKY' SHARPNESS.

Figure 20.15. A sharp hook is an absolute must if you want to avoid losing fish.

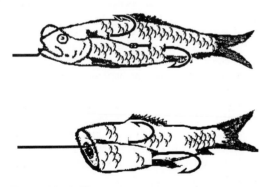

Figure 20.16. There are several ways that hooks can be threaded into herring. This illustration shows two common and effective methods.

you are mooching for kings and encounter dogfish sharks (*Squalus acanthias*), you might as well either move to another spot, switch to artificial lures, or plan to eat dogfish and lose plenty of leaders to their sharp teeth.

Live baitfish can also be used while mooching. If you are fishing with a single-hook leader, thread the hook completely through the jaw of the fish from the underside, then pull the hook back and insert it just beneath the skin behind the dorsal fin. If two hooks are used, put the leading hook through the jaw or nares and place the trailing hook just behind the dorsal fin or further back in the caudal peduncle. These hookups will allow the bait to assume a natural position while you are fishing, and the points of the hooks should be toward the boat. Remember to have the points "sticky-sharp."

Planers, downriggers, flashers and other hardware are not used when mooching. Only bait and sinkers are used. Mooching is regarded by many as a simple and "pure" technique, but it is one that may take some experience to master. Salmon often strike mooched herring in a subtle manner, "nibbling at it," rather than the solid strikes typical of trolled or jigged lures. Experienced moochers often pay out some line when they feel a nibble, allowing the fish to move away with the bait and swallow it before the angler tries to set the hook. It is best not to set the hook until the rod tip bends sharply, indicating the fish has a good grip on the bait, or until the line goes slack, which indicates that the fish is either swimming upward or towards the boat with the bait. A rod that is limber at the tip and stout at the butt is best for mooching. Either level wind or single action reels are favored.

Even though the fish can be heavy, the terminal gear should be light. Long leaders of fairly light line are used, since the fish get a better look at mooching leaders than trolled or jigged lines. Leaders of 5 to 8 feet in length are typical (as long as the rod or a foot shorter, to make landing the fish easier), and line test strengths should be 8 to 15 lbs. The leaders are tied to "banana" or crescent sinkers (3 to 8 oz.) at the beaded end. Plug cut baits in particular cause considerable rotation of the leaders, so if you have coiling problems you can put a swivel midway between the sinker and the bait. Some moochers now favor slip sinkers over the fixed sinkers in order to offer less resistance to the fish when it strikes; slip sinkers also allow moochers to feel the lighter strikes more readily.

One who chooses to fish by mooching has to remain alert. Besides having to watch for the strike, the boat also has to be attended to if it drifts into a precarious situation. Moochers try to attain a line angle of about 45° behind the boat when fishing. Some use a motor at very slow speeds (this is called motor or power mooching) to reach the right line angle and depth. When the sun is up, moochers fish deep, generally 60 to 100 feet or more. For this reason, it is best to mooch during slack tide periods or during tides that have a low magnitude of change (neap tides). Trying to mooch during a running spring tide is generally futile. If the angle of your line behind the boat is too shallow, it may be because the drift is too fast, the mainline is too heavy (has too much resistance), or the sinker is too light. Using lighter lines or motor mooching toward the bait are generally better alternatives than increasing the weight over 6 or 8 oz. If the line angle is too vertical, reduce the weight of the sinker.

Successful moochers often lower the bait all the way to the bottom, then reel in 10 turns or so to stay clear of snags. Fish often hit while the bait is falling. If no fish strike after fishing near the bottom for a while, take up more line and wait, at intervals, until you locate feeding fish.

Shore Fishing Techniques

The most popular method of shore fishing is to cast a drift lure or jigging lure from the bank with a heavy-duty saltwater spinning rod and reel. This is done best from a projecting point of land or any steeply shelving shoreline. This technique will also work on a gently sloping shore if the bottom is sandy and relatively free of weed and debris. Fishing from shore consists simply of casting as far out as possible, letting the lure sink, retrieving 5 or 10 feet, letting it sink again and repeating the process. Obviously, the amount of time the lure can sink depends on the depth of water. You can also cast out and retrieve continuously, although some drift fishing lures (like the Buzz Bomb) do not fish particularly well when retrieved in one continuous motion. Trolling spoons and various trout and bass lures can often be cast successfully from shore.

Although I have never tried it myself, it is also possible to cast from the shore and allow the current to "work" the lure. This requires a weight attached to the line perhaps 5 to 10 feet from a buoyant lure. The weight gets caught in the bottom and the buoyant lure floats up in the tide, getting its action from the passing current. (This method is used quite successfully in the Columbia and Fraser rivers. It is also a popular technique with steelhead anglers.)

Surf casting is another variation of the above technique. Surf perch, striped bass, and other species can be caught by casting a line with weight and baited hook into the surf and allowing the surge of the waves to work the bait.

Playing and Landing a Fish

There are many important factors in properly handling a fish:

- *Setting the hook*—Salmon will take a lure in a number of different ways. Our underwater studies show that more than 80% of salmon attack from the rear, sometimes just nipping at the tail and sometimes completely engulfing the entire lure. As soon as its mouth touches the hook, the fish reacts in pure panic. He will shake his head violently and try to spit it out. This is where the sticky sharp hook is most important. A dull hook will skid out of the fish's mouth and a sharp hook will dig deeper as the fish thrashes.

The resistance from the downrigger attachment will often set the hook when the fish strikes, releasing the line from the downrigger. This resistance may be insufficient, however, for some light tackle hookups or those intended for slow trolling. In my opinion, it's best to set the hook on any strike. Set the hook by a series of sharp, short, upward tugs, ending with the rod butt pointing back over your shoulder and the tip bowed straight toward the fish. This series of jerks will progressively pull the hook past the barb. (In some areas, however, barbless hooks are required for salmon fishing.)

If you are fishing live bait in which small hooks are embedded, the salmon will often grab it tentatively for a moment before turning it to swallow the bait head first. In this situation, it's important to strip out a bit of slack line so the fish feels no resistance. Then, when the line begins to move off steadily, set the hook hard.

The same is true with certain slow-trolling techniques using frozen baits or cut-plug herring. If the hooks are well buried in the bait, it is best to allow the fish to ingest it fully before driving the hook home. Nonetheless, when using artificial lures or natural baits with an exposed hook at the tail, it's better to set the hook immediately on feeling the strike.

- *Keeping the rod up*—This piece of advice has been given in almost every fishing book ever written. It is the single most important factor in successfully playing any gamefish. The fishing rod is like a giant rubber band or shock absorber. Holding the butt vertically allows the rod to bend through a full 90 degree arc, absorbing the sudden runs and lunges of the fish without breaking the line or allowing it to go slack. It gives the angler that extra split second to wind in or let out line and maintain an even pressure on the fish.

- *Playing the fish with a "light touch"*—My favorite reels are the single action type, which allow me to feel every movement of the fish, and provide complete control of the tension at all times by palming or thumbing. This is one of the great excitements of fishing, feeling the reel spinning as I palm it and apply a gentle braking action.

When using a star drag or spinning reel with a slipping clutch, the drag should be set before starting to fish, and left alone during the action. A good way to test the drag setting is to pull line off the reel and watch the rod tip. If the tip remains relatively stable, the drag is smooth, but if it jerks up and down, it is obviously sticking and is probably set too tight.

One of the biggest mistakes made with star drag reels is to tighten the drag when a stubborn fish refuses to come toward the boat. The over-tightened drag gives much more control of the fish, but a sudden lunge near the boat will often snap the line before the angler can readjust the drag setting. For this reason, it's far better to use a light drag and your thumb for additional pressure.

I fish for fun, and do everything I can to increase my enjoyment when playing a fish. To me, a single action reel provides that extra thrill and I highly recommend everyone to give it a try.

- *Letting the fish run*—When I was a fishing guide I would get just as excited as my customers when a fish struck. I would excitedly and continuously shout two pieces of advice. One was "keep your rod up" and the other was "let him run." This is the moment we've been waiting for. We want to enjoy every moment and let the fish run and play, allowing us to feel the surging rod and screaming reel. Many anglers take fishing so seriously that they can't enjoy their fish. They grind furiously, forcing the fish toward the boat as fast as possible. They somehow feel that getting maximum poundage in the boat as quickly as possible is the purpose of fishing. The actual cost of a sport-caught salmon is probably more then $50 a pound, so if meat is the object, buy it at the fish market and take up sailing or golf.

Letting the fish run also works better. The fish is easier to handle with more line out and the stretch of the nylon gives an additional shock absorbing factor.

- *Playing the fish to the stern*—Using small boats under 14 feet, it doesn't matter which direction the fish is running. You can swing the rod to clear the bow or stern as needed. However, larger boats with cabins or canvas can be a major problem if the fish runs toward the bow. You may have to put the boat in gear and turn it so the angler is always facing the fish off the stern.

If the fish runs under the boat or starts heading across the bow, plunge the rod tip down deep in the water so that the line will clear the hull and the propellers. You can then work the line around the stern (with the rod tip straight down) and move to the opposite side of the boat to play the fish as before.

- *Netting the fish properly*—Smaller salmon, under 3 lbs, can often be netted by just scooping the net under the fish's body. Larger fish, however, should always be netted head first. Moreover, the fish should be tired enough so that it is lying on its side or back and is relatively easy to control.

Many salmon, especially large ones, are lost when excited anglers try to net them too soon. Thrusting a net at a lively fish usually results in disaster when the fish panics at the sight and sound of the net in the water. With one sweep of its powerful tail, the fish can rush off with surprising force, sometimes rushing right past a portion of the net where the trailing hooks might catch in the mesh—and the fish is gone!

- *Releasing small fish gently*—If you hook an undersized fish, it will have a far better chance of surviving if it is not touched by your hands or even the landing net.

The proper procedure is to grasp the line near the hook and lift the fish out of the water. Then use a small gaff hook or even a coat hanger wire and slide it down the line to the hook and tilt the hook upward so the fish falls free. Touching the fish with your hands or the landing net will often disturb the protective slime and the scales, leaving the fish much more open to subsequent infection.

Go Get 'Em

There is no single surefire means of angling that will ensure success. Yet, for most anglers, the experience of being on the water and the scenic beauty that is so often an integral part of the fishing experience are sufficient satisfaction. Don't let anyone tell you that there isn't at least a bit of frustration involved with a fishing trip during which no fish are taken, but that kind of experience is soon forgotten and rarely dampens the enthusiasm of the angler.

In this chapter, I have tried to provide you with some useful information that should help you improve your chances of catching Pacific salmon in the marine environment. Your next task is to put the information to work. Good luck!

21

Marine Fishing:
Non-Salmonids

Gilbert B. Pauley and Frederick G. Johnson

Introduction

Now, anglers have at their disposal many tools and new technologies that were not available two or three decades ago, and these help to ensure a successful fishing trip. Space-age materials and technologies, along with modern production methods, have brought quality fishing tools within the budget of most anglers. Items that have made fishing not only easier and more fun, but more successful as well, include dacron and nylon lines, fiberglass and graphite fishing rods, modern level wind and spinning reels, depth sounders and fish locators, and a plethora of lures to fit every occasion imaginable.

Saltwater anglers catch a wide variety of fishes using several techniques: offshore trolling for big gamefishes and inshore trolling for many smaller species; bottom fishing which requires different techniques in shallow water (0 to 40 feet), medium-depth water (40 to 200 feet) and deep water (200 to 400 feet). Other approaches/techniques include pier or jetty fishing and open water fishing; surf fishing with artificial lures or bait; jig fishing along the bottom or floating plug casting at the surface. Whether your interest is in catching a 2-lb pile perch with a light spinning rod, a 10-lb bonefish on a fly rod or a 40-lb striped bass with a heavy surf rod, saltwater fishing offers something for everyone. Marine recreational fishing is enjoyed on the Atlantic, Gulf and Pacific coasts (Figures 21.1 and 21.2) by millions of people who catch millions of fish (Figures 21.3 and 21.4) of many different species.

Most of the marine recreational catch taken by U.S. anglers comes from the Atlantic and Gulf coasts. In 1987, anglers landed 337 million fish in those areas, compared with about 50 million from the Pacific Coast. This is in contrast with commercial catches, which are higher in the waters of the Pacific. As we might expect, commercial catches tend to be highest where there are large concentrations of fishes, while recreational catches are often highest where there are large concentrations of people. The five species of fishes most often caught by recreational anglers in 1987 were bluefish, summer flounder, spotted sea trout, Atlantic croaker and spot. Each of the last three species belong to the family Sciaenidae, and all five inhabit waters of the eastern United States. The marine recreational catch represents about one-fifth of the total U.S. landings used for direct human consumption. The aggregate weight of U.S. sport-caught marine fish ranges from 600 to 700 million pounds per year. While the contribution of recreational marine fishing to the food supply is significant, the contribution of sport fishing to the economy is enormous.

During the 1980s, the number of marine recreational anglers in the United States has been fairly stable at 17 million. In 1987, these anglers took over 75 million fishing trips and spent about $7.5 billion for fishing activities. Over the last three decades, the number of recreational anglers in the United States (both marine and freshwater) increased dramatically. The amount these anglers spend to pursue their sport has increased at an even greater rate and, since the mid-1980s, has

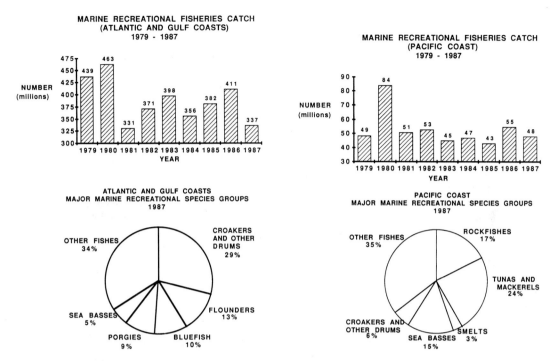

Figure 21.1. Recreational catches from the U.S. Atlantic and Gulf coasts, showing numbers of fish landed and the major groups of species involved.

Figure 21.2. Recreational catches and major groups of sport fish taken from the U.S. Pacific Coast. These data do not include salmon.

amounted to over $25 billion per year! Trends in recreational fishing in the United States from 1955 to 1985 are shown in Figure 21.5.

To enhance fish production and increase the number of fish available to anglers, many governments have sponsored the construction of various types of artificial reefs using materials such as used concrete and tires. These artificial reefs are usually well marked with colored buoys and are located in areas that can be easily accessed by boat. Piers are often built to accommodate the shore-bound angler. Sometimes piers and artificial reefs are built in conjunction with each other.

While many novice anglers have the mistaken idea that successful anglers use a "magic lure," there are similarities in the basic sport fishing techniques for various fish species in different geographic locations. You must understand the basic biology and habits of the fish, such as the types of food they eat and when they forage or when and where they spawn. Once those things are known, you will understand which lures or baits to use in various situations. This chapter contains descriptions of angling techniques used for marine fishes throughout the Atlantic, Gulf and Pacific coasts of North America.

Locating Fish

The first thing that you need to do is to locate the fish you are interested in catching. All marine fishes are found in certain areas where the habitat and associated food organisms are suitable for their particular biological requirements. For example, codfishes may be found over rocky bottoms; flounders and other flatfishes are found on flat, muddy or small gravel bottoms; greenling are found around kelp beds; lingcod are found over rocky reefs in areas of strong tidal movement;

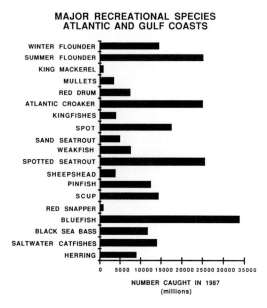

Figure 21.3. The distribution of recreational catches among the major species along the U.S. Atlantic and Gulf coasts in 1987.

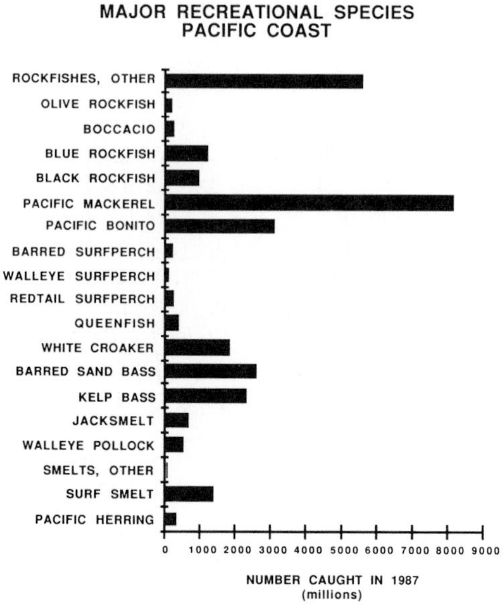

Figure 21.4. The distribution of recreational catches among the major species (not including salmon) along the U.S. Pacific Coast in 1987.

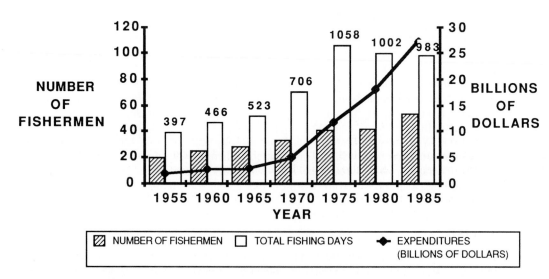

Figure 21.5. A summary of overall trends in U.S. sport fishing over the past three decades, including number of anglers, expenditures and total fishing days. This figure includes data from both freshwater and marine fishing.

spotted seatrout and weakfish are found around oyster beds and seagrass beds; tarpon, bonefish, permit and red drum are found in warm, clear, shallow water, often over sand flats; shad and tuna relate to specific water temperatures and current patterns. Even species of fish belonging to the same genus and closely similar in appearance may have evolved quite different habitat requirements and behavior. For example, the Pacific Coast rockfishes, which belong to the genus *Sebastes,* include over 60 different species. Blue and black rockfish are found near the water surface, often in association with kelp beds, and form schools; copper and quillback rockfish are found around rocky reefs at depths around 30 to 50 feet and are more solitary (Figure 21.6); and canary and yelloweye rockfish are found over rocky reefs at depths of 150 feet or more.

Many marine fishes orient to either some type of structure, such as a dropoff, an underwater reef, a point of land extending into the water; or some type of cover, such as a pier, a kelp bed, overhanging trees and grasses. Often a depth sounder will help you find these things if not readily visible, and that is where your search begins. Generally, when trying to locate fish, look for concentrations of bait and/or physical structure. Also, try fishing in places where fish go to escape strong currents.

Angling for Selected Fish Species

Many books have been written about each of the major marine gamefishes or the groups in which all of these fishes belong. The following is a summary of the types of marine game that you may expect to encounter in the waters around North America.

Bonefish (Family Albulidae)

These fish are considered the ultimate quarry by fly fishing enthusiasts. They are found in southern Florida, the Bahamas and the Hawaiian Islands. They forage over shallow, sandy flats and reefs for shrimp, crabs and other invertebrates, and they are piscivorous (fish eaters) only by chance. They may be taken by casting flies that resemble various invertebrates from boats or by wading. A 10-lb fish is a trophy, and most fish are released alive.

Tarpon (*Megalops atlanticus,* Family Elopidae)

These fish can be found in marine and brackish waters along the southern Atlantic Coast and throughout the Gulf of Mexico. Tackle is fitted to the size of the quarry, which can easily be 100 to 150 pounds in the Florida Keys. They are fished over flats and in channels with bait, artificial flies, jigs and plugs that resemble minnows. They are very strong fighters and often leap out of the water when hooked. They are usually released and are not highly regarded as a food fish.

Permit, Pompano and Jacks (Family Carangidae)

There are 36 species of these fishes along the Atlantic, Gulf and Pacific coasts, all of which are very strong swimmers and excellent sport on light tackle. They can be caught with fly fishing, spinning and bait-casting tackle. Shallow water is the place to catch permit and pompano, while amberjacks are taken from open water, reefs and around buoys. In Hawaii, anglers use special methods of getting the bait out to deep water, such as a balloon and slide rig, to get jacks or "ulua" as they are called. Many carangids are excellent food fishes.

Snooks (Family Centropomidae)

These fishes, which are very sensitive to cold water and do not survive well below 58°F, are found on the south Atlantic and Gulf coasts. Because they frequent many of the river systems, they can be found in some brackish and freshwater situations, in addition to marine waters. The largest fish will exceed 30 pounds and may be caught with a variety of terminal lures, such as plugs, spoons, jigs and flies, as well as natural bait. Wire leaders are used because snooks have sharp teeth.

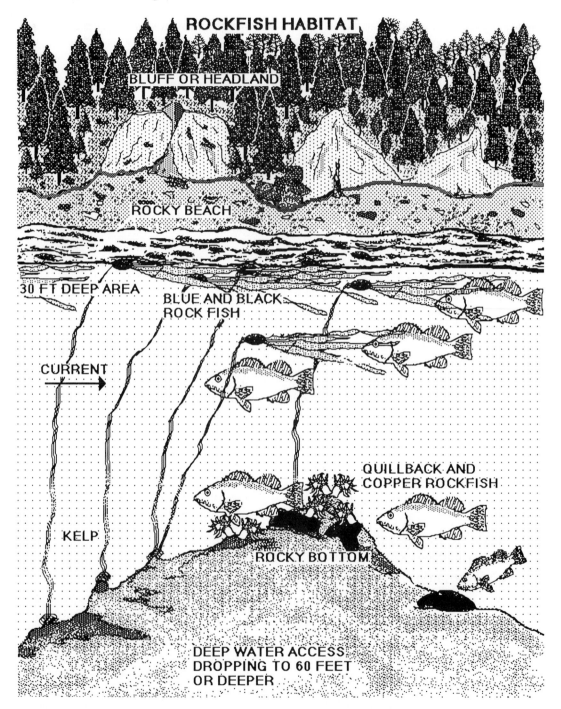

Figure 21.6. The underwater habitat of rockfishes, showing how the different species of rockfish assume different positions within the kelp bed community.

Their gill covers are also very sharp, and these fishes should be handled with care. There are four species in North America.

Barracuda (Family Sphyraenidae)

Barracudas are rapid swimmers, have sharp teeth and are found on all three coasts of North America. The Pacific barracuda and the great barracuda are the major species. The majority are taken incidentally by trolling offshore for other fish. Large fish weigh 10 lbs and are often in schools that invade shallow flats and reefs. Anglers prefer fly fishing and spinning to catch barracuda and use a variety of flies, jigs, and floating and diving plugs that are brightly colored or silvery. Chartreuse, fluorescent yellow and fluorescent red surgical tube lures are some of best for this fish. In some areas, the flesh of these fishes may cause a type of poisoning called ciguatera, but for other areas they are safe to eat.

Sharks (Class Chondrichthyes)

These primitive fishes are often hooked when bait is being used for other species of fish, but many anglers do target sharks. Several species of gamefish value—the thresher, tiger, white, hammerhead, mako and blue—are found in the Atlantic, Pacific and Gulf coastal areas. Heavy rods and reels, wire leaders and bait are needed when pursuing these fish. Since sharks retain urea in their blood to maintain osmotic balance, some find it helpful to soak the meat before it is cooked.

Croakers, Drums, Sea Basses and Snappers

These fishes all have a superficial resemblance and are often confused by anglers. Croakers and drums belong to the family Sciaenidae; sea basses and groupers are members of the Serranidae; snappers are in the family Lutjanidae. These groups of fishes are found from the mid-Atlantic southward, throughout the Gulf of Mexico and from central California on south in the Pacific. Many species in these three families are highly sought gamefish. They can be taken in shallow water down to considerable depths depending upon the species; consequently, fishing gear requirements vary.

The red drum (also known as redfish and channel bass) is a highly prized fish along the Atlantic and Gulf coasts that can reach over 40 lbs in size. These fish generally are taken in shallow water, either by casting from a boat or surf fishing with flies, floating plugs with diving lips, jigs and chrome spoons. Seatrout and weakfish (*Cynoscion* spp.) are found along the Atlantic and Gulf coasts and are caught from small boats, by wading or by surf fishing. Light spinning tackle is the most popular gear for these fishes, with anglers using jigs, spoons, plastic worms and plugs. The kelp bass and sand bass (*Paralabrax* spp.) are major sport fishes off the inshore areas in California; they are often caught with jigs, plastic worms, spoons and tin squids. Snappers are taken over open water reefs along the Gulf Coast with jigs and bait. Groupers are usually larger than snappers but are taken from the same reef areas along the Gulf and South Atlantic coasts.

Bluefish (*Pomatomus saltatrix,* Family Pomatomidae)

These fish are commonly found in migrating schools along the Atlantic Coast and the Gulf of Mexico side of Florida. Often they are taken while the angler is fishing for other major gamefish, but are targeted by many anglers who may use fly fishing gear, heavy surf fishing gear or stout rods equipped with level wind or spinning reels. Bluefish are taken in open water, over reefs and along shore with various flies, jigs, spoons and bait. They are active feeders and are known to attack the fingers of swimmers who splash about at the surface. Bluefish can weigh as much as 30 lbs.

Striped Bass (*Morone saxatilis,* Family Percichthyidae)

These anadromous fish, which may reach 40 to 50 lbs in many areas, are found in schools along the entire Atlantic Coast and to a more limited extent along the Gulf and the Pacific coasts. Optimum temperature is around 70°F and, as striped bass spawn in freshwater streams, they can commonly be found in estuaries. They are taken in the surf with a variety of lures on heavy rods equipped with spinning or level wind reels. They may also be taken in the rivers and estuaries on large plugs, bait and large flies. They are an excellent food fish and have also been fished commercially.

Tuna and Mackerel (Family Scombridae)

These are fast-moving, schooling pelagic fishes with streamlined bodies. Adults of some species commonly weigh over 100 pounds and can weigh up to 300 lbs. They are found on the Atlantic, Gulf and Pacific coasts and are taken by fast-moving lures and baits fished with stout rods and heavy-duty reels. Feathered jigs are popular lures used by tuna anglers, while plugs work well for wahoo. Popular species include the yellowfin, bluefin, blackfin, bigeye and albacore tuna; king mackerel and wahoo; and Pacific bonito.

Marlin, Swordfish and Sailfish (Families Xiphiidae and Istiophoridae)

These fishes are found offshore along the Atlantic and Gulf coasts and off California and the Hawaiian Islands. They are large fishes that reach sizes of 100 to 300 lbs. They are taken primarily from large boats by using heavy rods, reels and line with trolled baits. The angler is often strapped to a swivel fishing seat and uses a harness to hold the rod. They are strong fighters, spectacular leapers and are the fastest-swimming fishes in the world (speeds up to 70 mph). They also support commercial fisheries but, while they are fine as a food fish, some may have unacceptably high levels of mercury.

Surfperches (Family Embiotocidae)

There are 18 species of surfperches along the Pacific Coast. Although found in kelp, tide pools and around piles, most live in the surf zone. They run in schools, and individuals may reach 18 inches in length. They are excellent sport fishes and are caught on light spinning or fly fishing gear. Bait cast into the surf will also catch these fishes.

Rockfishes (Family Scorpaenidae)

Rockfishes and scorpionfishes are members of the largest family of fishes, with over 60 species along the Pacific Coast of the United States, nearly 100 species throughout the Pacific Ocean, but only five species on the Atlantic Coast. They are important sport fishes, with the rockfishes dominating the fishery. The rockfishes (*Sebastes* spp.), which have a bass-like appearance and are often incorrectly called sea bass, occupy a variety of niches; but as their name implies, they are usually found associated with rocks. Many of them are schooling fish such as the black, blue and

Figure 21.7. The results of a successful sport fishing trip in the Pacific Northwest. Curtis, Gerald and Matthew Chew proudly display their mixed catch of Pacific salmon and rockfishes. (Courtesy of Ken Chew.)

yellowtail rockfishes, while others are solitary such as the China rockfish. They are found around rocky points, dropoffs and reefs; the shallow water varieties are often found near kelp. They are found in shallow water down to several hundred feet. Most sport-caught fish probably average 2 to 4 lbs, but some species may reach 20 lbs, such as the bocaccio, cow and yelloweye rockfishes. These fishes may be taken with a variety of gear, with light spinning or casting gear usually being used in shallow water and heavy deep sea rods and reels being used in deep water. Most of the larger fish are found in deeper water. Plugs, jigs, spoons and bait are used to catch these fishes, and jigs with both a plastic skirt and plastic worm are especially effective lures.

Lingcod and Cabezon (*Ophiodon elongatus* and *Scorpaenichthys marmoratus*)

Lingcod and their cousins, the greenlings, belong to the family Hexagrammidae. They are found only on the Pacific Coast and range from shallow to deep water. Lingcod are highly prized in many areas along the Pacific Coast and are found in rocky areas and around kelp beds, especially where there are strong tidal currents. Lingcod average around 10 lbs but can reach 70 lbs. Salmon or steelhead rods and reels will normally work well for lingcod, but anglers fishing deep water should use heavy-duty deep sea gear. Lingcod may be caught using bait, live fish, heavy jigging spoons, diamond jigs, jigs with large plastic worms, and lead-head or pipe jigs. The cabezon, which is a member of the sculpin family (Cottidae), is often taken by anglers fishing for lingcod. It is a prized sport fish that does not usually get larger than 15 lbs. Both are good food fish, and they may have green or blue meat that turns white when cooked.

Cods and Sablefish (Families Gadidae and Anoplopomatidae)

Cods are found on both the Atlantic and Pacific coasts, while sablefish, also known as black cod, are found on the Pacific Coast. Fishes of both families may reach sizes up to 40 lbs, but smaller fish between 5 and 10 lbs are more common. Medium rods and reels, either spinning or level wind, are adequate, and bait, jigging spoons and jigs with plastic worms will catch both of these fishes. They often will be found in schools, and both families are important to commercial fisheries.

Flatfishes (Families Bothidae, Pleuronectidae and Soleidae)

Flatfishes, which include numerous species of sole, flounders, turbot and halibut, are found nearly world-wide. These fishes may be taken with a variety of baits and lures in conjunction with either spinning or bait casting gear. Halibut, which commonly weigh between 30 and 100 lbs, require heavy gear. They are the largest of the flatfishes and attain weights exceeding 600 lbs (*Hippoglossus*). Halibut are the most highly regarded of all Pacific Coast bottomfish and can be taken with either bait or large, weighted jigs with huge plastic worms.

Saltwater Fishing Techniques

Bait Rigging and Hookups

Live or dead bait may be used to catch saltwater fish. The size of the baitfish should be based on the size of the intended catch, with the bigger fish requiring bigger bait. Large fish are lazy and usually do not waste much time pursuing small meals. Live baitfish can be rigged in one of three ways: through the back, being careful not to injure the spine—this is best done in the midsection of the back to allow fish to swim; through both lips beginning under the mouth and coming out the top of the head in front of the eyes; and through the eye socket above the eyes. Whole dead fish may be hooked with a double hook set-up, where the hooks are passed up through both jaws, through the eye socket and over the back. The lead hook is placed just under the skin on one side and the trailing hook similarly on the other side.

Shrimps, crabs, worms, squid and octopus can all be used as live bait. Anything that can be fished live can be fished dead also, but dead bait generally does not stay on the hooks as well as live bait. With dead bait, anglers can use strips or parts of the animal, such as the tail of a shrimp or the belly of a fish. In the Pacific Northwest, herring heads are cut from the fish at a 45° angle and hooked so the "plug-cut" bait will spin in wide circles as it moves through the water. Both live and dead bait may be secured to the hook with a rubber band or wire if necessary.

Lures for the Saltwater Angler

There are a large variety of lures available to the saltwater angler that fall into a few major categories.

Plugs

Plugs, designed to be cast or trolled, are made from either wood or plastic and are supplied with two or three treble hooks. Most modern plugs are plastic and resemble some type of fish. Three types of plugs are useful in saltwater: the **topwater floating plug**; the **floater-diver** plug, which floats when resting and dives just below the surface upon retrieval; and the **sinking plug**, which sinks until you begin to retrieve it and then stays underwater during the retrieve.

Spoons

Thin-bladed spoons are made to flutter slowly through the water, because the metal is lightweight, thin and flat; **jigging spoons** and **tin squids** are made from heavy metal and are oval- or diamond-shaped to allow for quick sinking.

Jigs

Jigs are the most versatile and effective all-around saltwater lures. They have no built-in action like a plug or spoon and must be worked by the angler to achieve any type of action. They come in all sizes and a wide variety of shapes for various uses in salt water. They may have bucktail or plastic skirts and can be tipped with bait, a plastic worm or a pork rind, which may also be used with thin-bladed spoons, jigging spoons and tin squids.

Tail-spinners

Lures of this designation, which have recently been introduced to the Pacific Coast after being developed for freshwater bass fishing, are a combination of a lead head and a spinner attached at the tail-end of the lure. They are very effective on schooling fish such as black and yellowtail rockfish.

Trolling

The standard method of offshore big gamefishing throughout the world is **surface trolling**, which is done at a fast speed—5 to 10 knots—to keep the bait or lure near the surface. Baits may be either swimming or skipping on the surface, with certain fish preferring one type over the other. Fish, squid, feather jigs and hootchie jigs all work well.

Inshore trolling may target nearly any species of fish around inshore reefs, kelp beds, bays, cliffs and points of land. Trolling allows you to cover a lot of water and locate fish. Boats used for inshore trolling are smaller and are worked slower than in offshore trolling. Bait and lures can both be used with minimum weight on the line. Diving plugs are ideal for inshore trolling. **Motor mooching** is a version of this type of fishing, where a deep-drifted bait is periodically picked up by running the motor briefly and then allowing the bait to drop back down.

Deep trolling is accomplished by two methods: either with a **downrigger** or a **trolling planer** (see Chapter 20). A downrigger is a winch with a heavy lead weight (5 to 15 lbs) attached to it on durable wire. The line with the trolled bait or lure is attached by a clip or snap to the downrigger wire and the lure trails behind the wire at some desired distance. Trolling planers simply run at an angle in the water, causing water to push them downward along with their attached bait or lure; they are attached directly to the fishing line. The downrigger weight can be lowered to any depth desired for fishing, while the planer has a maximum depth to which it will dive and fish. When a fish hits on a downrigger, the lure and fishing line disconnect from the weighted wire at the attachment snap. The trolling planer, on the other hand, stays attached to the fishing line, but snaps up to run flat in the water rather than at angle after a fish is hooked.

Stillfishing

Stillfishing requires that the angler either anchor where the fish are located or hope that fish approach the angler's position. Anglers may need to attract fish to their anchored position because they are limited by the distance their lines can be cast. To solve this problem, anglers can either anchor in an area where fish concentrate, anchor over a fish "highway" or migration route, or attract fish with the use of chum or some type of fish attractor. Chumming involves tossing bits of bait into the water to cast a scent trail that attracts fish from down-current areas. Other reasons for anchoring and stillfishing include unfavorable weather conditions or strong tidal currents. This type of fishing can be done from small boats in shallow water or large boats over deep water. Bait is commonly used when stillfishing; however, artificial lures, generally some type of jig or jigging spoon, can also be fished from a stationary boat position.

Sometimes, when stillfishing for bottomfish, a perplexing problem arises when a hooked fish sounds and manages to lodge itself firmly in a crevice. You find that no matter how hard you pull on your rod, short of breaking the line itself, you cannot pull the fish free of its "hole." When this happens, there is a trick to fooling the fish into leaving its hole. It amounts to taking all slack out of the line until the line is taut and then plucking the line sharply two or three times like a guitar string. The vibration this causes translates down the line and into the fish's mouth, causing the fish to spook and bolt out of its refuge. When this happens, immediately take in line so that the fish cannot reach the bottom again. This works about half of the time, but you are not likely to fool the same fish twice.

Drift Fishing

This type of fishing can be used in almost all situations: inshore and offshore; over shallow flats and deep reefs; in open water and kelp-infested areas; from small boats and large boats. The basic object in this type of fishing is to drift and cover a large amount of water by taking advantage of the wind and tidal currents. Drift fishing is most effectively done with artificial lures such as plugs, jigs and spoons. The object is to stay over a productive area as long as possible, then motor back into a position that will allow the boat to drift back over the productive spot. This repositioning is often difficult and, while it requires practice to observe the shoreline, prominent landmarks and any navigation buoys that will lead you back to the area, this is important to successful drift fishing. Good anglers usually **triangulate** on several stationary objects that will allow them to return very close to same drift location. Often floats with weights are thrown out to mark good areas and are picked up later. Wind and tidal currents must also be factored into boat repositioning.

Kite fishing, practiced in southern states, is a variation of drift fishing in which the bait is carried out over water either unreachable by the boat or too far to cast the lure. This technique is especially good to hold live bait on the surface to attract large gamefish. (This technique is also used from the shore in the Pacific Northwest to get bait out far enough to catch sturgeon. Once the kite is over the right spot, the angler jerks hard on his line and the line falls free of the kite and sinks. A heavy weight keeps the bait on the bottom until a foraging sturgeon picks it up.)

Mooching is a type of drift fishing developed for salmon in the Pacific Northwest, where herring on a line is held in the tidal current at a 40 to 60 degree angle to the boat. The mooching technique works well for cabezon, Pacific cod and lingcod. When drift fishing in deep water (over 200 feet) a depth sounder is needed, because depth and underwater topography now come into play as

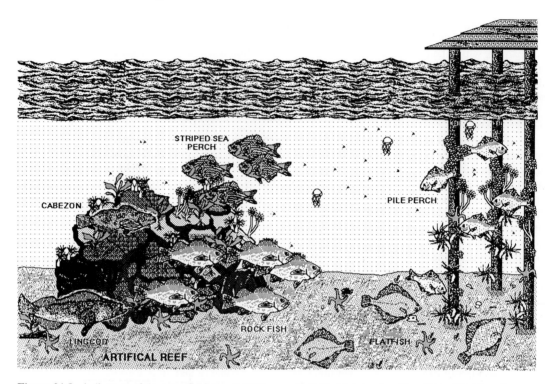

Figure 21.8. A diagram of the sport fishing possibilities presented by the construction of an artificial reef near a fishing pier. Various habitats result, and fishes shown in the figure represent those that would be expected in such habitats in the Pacific Northwest.

major factors in returning to a fish reef. Although light- and medium-weight rods and reels can be used in most drifting, heavy rods and reels, braided dacron line to eliminate stretch, wire spreaders to keep the weight and bait from tangling, and a rubber snubber to absorb the shock of a strike are required in deep water drifting. When going after trophy-sized lingcod, yelloweye rockfish and halibut, anglers will have to drift fish in deep water. Other than bait, an assortment of jigs would be the best type of lures to use when drift fishing.

Shoreline Fishing

The type of fishing that usually comes to mind here is **surf fishing**, where the angler wades out into the surf along a beach. This is practiced in many areas of the Atlantic, Gulf and Pacific coasts for fishes such as striped bass, bluefish and surf perch. Anglers can also fish from rocks that are pounded by the surf, or **jetty fish** along jetties and breakwaters, or **pier fish** from piers and docks (Figure 21.8). For this type of fishing, many innovations have been developed to help the angler catch more fish. For example, **slide-baitfishing** has been devised in Hawaii and **jetty mooching** with molded foam floats is used in Washington and Oregon (Figure 21.9). Slide-baitfishing is similar to mainland **trolley line** fishing. A weighted set line is cast out and secured to the bottom, then a baited hook on a sliding swivel is attached to the rod line and slid down the set line.

As mentioned previously, kite fishing may be used from the shore in order to place a bait farther out than is possible by casting. Two interesting variations of this strategy deserve mention. One method, which is used in Hawaii, makes use of a balloon to carry the bait to a suitable spot. The angler feeds out his line as the wind moves the balloon away from shore. When the balloon is over

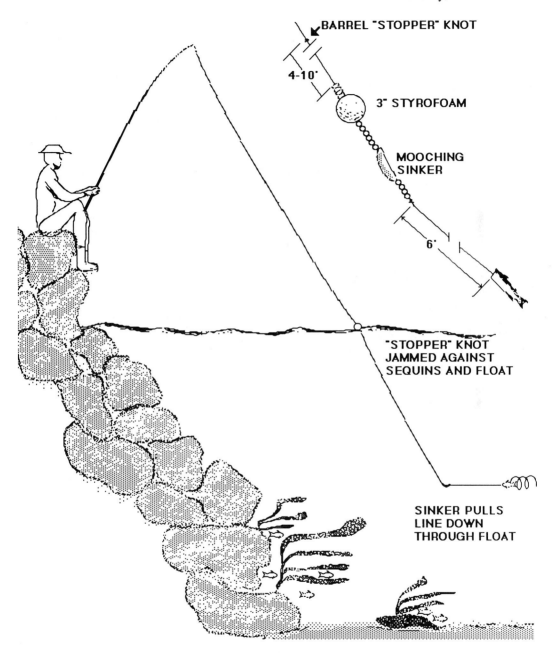

BARREL "STOPPER" KNOT

4-10'

3" STYROFOAM

MOOCHING SINKER

6'

"STOPPER" KNOT JAMMED AGAINST SEQUINS AND FLOAT

SINKER PULLS LINE DOWN THROUGH FLOAT

Figure 21.9. A jetty fishing technique, showing the tackle hookup method that involves mooching with the aid of a foam float.

the right area, the angler shoots the balloon with a small caliber rifle and the bait sinks. Another method that is more technologically refined but utilizes a similar strategy involves radio-controlled drone boats. These small boats are battery-powered and operated by a manual controller, which the angler holds on shore. The drone boat is directed over the fishing spot, the angler pulls the line free of the boat and allows the bait to sink, and then the drone boat is redirected back to the angler.

Shoreline fishing is done with both bait and artificial lures and with either bait-casting or spinning rods and reels. Sinking plugs are retrieved in an erratic or jerky motion once they have reached the desired depth. Floater-diver plugs may be retrieved at an even pace because they have a built-in side-to-side wobbling action. Surface plugs are worked to give some type of bubble trail and are most effective over either shallow and clear water areas or in areas with schools of fish actively feeding on the surface. Spoons are worked in a fast or slow erratic motion to allow them to wobble and reflect light. The most effective saltwater plugs and spoons are those that either have a metallic finish or are brightly colored. As with drift fishing, an assortment of jigs may prove effective.

Jig Fishing

Jigs are among the most versatile and effective lures used by saltwater anglers and are also the oldest recorded lures used in salt water. They are easy to use, simple to make and inexpensive to buy. Many styles of jigs are available depending upon the type of head, the type of skirt material and the desired function. Knowledge of jigging techniques can be applied to either saltwater or freshwater fishing.

Drift jigging is an open water technique that can be used either in inshore or offshore waters. In drift jigging, tin squids, "sting sildas," heavy-bodied jigging spoons and lead-head jigs with bucktail or with a plastic worm are drifted over underwater structures at a speed that is controlled by the tide and wind. The jigs are worked in an up and down motion or a lift and drop action, and most of the fish strikes occur while the jig is falling. A variation of drift jigging is to cast and retrieve a jig around visible cover such as kelp beds, still allowing the boat to move with the tide and currents. Plugs that have jigs attached to them by 12- to 18-inch-long dropper leaders may be used. The movement of the plug causes the jig to lift and drop. This is a very effective technique on surface feeding or schooled fish. Level-wind rods and reels work best when drift jigging above underwater structures, while spinning rods and reels work best when casting to visible cover or when small jigs are used.

Vertical jigging is usually done with jigs weighing 1 to 3 ounces or more, depending on the amount of wind and tidal current. This type of jigging is usually done in deep water over a reef or from a pier. The lure is worked up and down in a vertical motion with very little horizontal movement. **Shoreline jigging** is a variation of cast and retrieve jigging practiced from a boat, where the lure is cast out from the shore toward some type of structure or into the surf.

Fly Fishing

Two kinds of saltwater fly fishing exist: **Inshore fly fishing** takes place in brackish water, tidal estuaries, bays, lagoons and the nearshore ocean. **Offshore fly fishing** occurs in offshore deep ocean areas. Inshore fly fishing has the greatest number of angler participants, because it can be done by casting from the bank, by wading or from a small boat. On the other hand, offshore fly fishing requires the use of a moderate-sized personal or charter boat.

The Mecca of saltwater fly fishing is along the southern Atlantic and Gulf coasts for such species as striped bass, seatrout, tarpon, bluefish, bonefish, snook and permit. The water is clear, and anglers can easily spot individual fish or schools and then cast toward them.

As a group, saltwater flies tend to be large and very bright in color and often incorporate some type of metallic material. **Flashabou**, a tinsel-like synthetic material that comes in a wide variety of colors, has made the imitation of small fish and crustaceans an easier task for the modern fly tier.

Fly fishing (see Chapter 22) is not limited to the Atlantic and Gulf coasts. It is also practiced along the Pacific Coast in dark water with exceptional results. Blue and black rockfish are strong fishes that can reach 8 lbs. Schooling in huge numbers and feeding along the surface at certain times of the year, they are exceptional sport on a fly rod. Unlike the Atlantic and Gulf coasts, where the angler can see the fish in the shallow and clear water, Pacific Coast anglers are often limited to water having poor visibility.

In addition to flies, anglers along the Pacific Coast use fly rods to cast and drift small plug-cut herring or strips cut out of the herring's side. Innovation is endless in this fishery, as it is in others. While in freshwater, the angler can make effective casts of 25 to 30 feet that will consistently take fish; in saltwater, on the other hand, the angler must make much longer casts of 40, 60 and even 80 or 90 feet to be effective. The saltwater fly rod must be longer and must have a powerful action to throw a long line into the wind, to cast the large wind-resistant flies and to handle large, strong fish.

Spearfishing

An increasingly popular form of recreational fishing has grown along with the expanding sport of skin and SCUBA diving. (SCUBA stands for self-contained underwater breathing apparatus.) **SCUBA divers** take their air supply with them and can explore and hunt underwater for about an hour at a time. **Skin divers** take a breath of air at the surface and then submerge while holding their breath for 1 or 2 minutes at a time. Both type of divers carry either a trigger-activated spear gun or a hand-activated pole spear while spearfishing. The **pole spear** usually has a loop of surgical tubing at the butt end that the diver can wrap around his hand and use to provide thrust when spearing a fish. The **spear gun** is also cocked with the aid of rubber tubing, but the spear gun has a metal shaft that leaves the gun when the trigger is pulled. Pole spears can only be used at close range, while spear guns have an effective range of about 15 feet.

Spear guns usually have a line fastened to the detachable shaft so that if an impaled fish swims away, it can be pulled in by the diver. Barbs at the point of the shaft keep the fish from pulling free. Some divers who use spear guns target very large fishes such as California halibut or groupers as their quarry. Since these fishes can weigh hundreds of pounds, the divers have very long spools with the line attached to the shafts, so they can surface before trying to work the fish in. Usually two divers will spear a large fish and work the fish in tandem using two lines. Spearfishing, particularly for larger fish, can be dangerous, but it is also exhilarating.

A major advantage to spearfishing is that the fisherman only attempts to capture a fish that is desired, and unwanted fish need not be harmed. Spearfishing has long been a practice from boats or the shoreline by native peoples, but rapid increases in water technology have taken the spearfishermen down to the fish. The considerable effectiveness of divers in spearing fish has led to regulation of their fishing activities in many areas and has also lead to friction between anglers and divers in some instances. Regardless of the mode of capture, all anglers should share concern and care for their fishery resources.

Additional Reading

Dunaway, V. 1975. Modern Saltwater Fishing. Stoeger Publishing Co. South Hackensack, NJ. 286 p.

Eschmeyer, W.N., E.S. Herald, and H. Hammann. 1983. A Field Guide to Pacific Coast Fishes of North America. Houghton Mifflin Co. Boston, MA. 336 p.

Haw, F. and R.M. Buckley. 1973. Saltwater Fishing in Washington. Stan Jones Publishing. Seattle, WA. 198 p.

Hoese, H.D. and R.H. Moore. 1977. Fishes of the Gulf of Mexico, Texas, Louisiana and Adjacent Waters. Texas A&M University Press. College Station, Texas. 327 p.

Lamb, A. and D. Edgell. 1986. Coastal Fishes of the Pacific Northwest. Harbour Publishing. Madeira Park, B.C., Canada. 224 p.

McClane, A.J., and K. Gardner. 1984. McClane's Game Fish of North America. Time Books, New York. 376 p.

McNally, T. 1978. Fly Fishing. Outdoor Life/Harper & Row. New York. 420 p.

Oberrecht, K. 1982. Angler's Guide to Jigs and Jigging. Winchester Press. Tulsa, OK. 335 p.

Sakamoto, M.R. 1985. Pacific Shore Fishing. University of Hawaii Press, Honolulu. 255 p.

Sosin, M., and J. Clark. 1973. Through the Fish's Eye. Outdoor Life/Harper & Row. New York. 249 p.

Squire, J. L., and S.E. Smith. 1977. Anglers' Guide to the United States Pacific Coast. U.S. Department of Commerce, National Marine Fisheries Service, La Jolla, CA. 139 p.

Wilson, D., and F. Vander Werff. 1977. New Techniques for Catching Bottom Fish. The Writing Works, Inc. Mercer Island, WA. 150 p.

22

Freshwater Fishing:
Salmon, Steelhead and Trout

John Thomas

Introduction

Countless articles, books and magazines have been written over many years on various freshwater sport fishing techniques. In fact, entire libraries exist with respect to fly fishing alone. This chapter contains descriptions and accompanying figures of several universal freshwater angling techniques as they apply to the salmonid fishes of the United States and Canada.

Coldwater sport fishes included in this chapter are rainbow trout, steelhead trout (migratory rainbow trout; Figure 22.1), cutthroat trout, brown trout, golden trout, lake trout, Dolly Varden and landlocked salmon such as the king (chinook), coho (silver) and kokanee (sockeye). These salmonid species inhabit lakes, reservoirs, streams and rivers throughout the continental United States and Canada.

Specialized fishing techniques have evolved to outwit every fish listed above under virtually every imaginable fishing condition. Productive techniques have been developed for capturing these fishes from lakes as opposed to rivers, from deep water versus shallow water, for fish that feed as compared with those that must be aggravated into striking, for those that feed on large minnows in comparison with those which make tiny insects their primary prey, for season of the year (including under the ice in winter) and for various species as a function of time of day.

Many techniques will work under a variety of conditions and apply to various freshwater fish species such as those described subsequently. Before listing specifics, there are some general facts common to all gamefish and, therefore, common to all techniques, that should be explained.

Figure 22.1. Migratory rainbow (steelhead) trout often are bright silver.

Locating Fish

All gamefish relate in one way or another to structure or cover. In the simplest terms, **structure** refers to the physical bottom of a body of water and its structural makeup such as sand, mud, rock, cliffs, underwater islands, rockpiles and so forth. **Cover** refers to anything protruding from the bottom or attached to the shoreline, such as weeds, overhanging trees, logs in the water and brushpiles. Given the potential overlap, it is not surprising that structure and cover are often used interchangeably by sport anglers, who also use the term "shelter" in the same context. All of those terms are used to describe places where one will consistently find gamefish.

Deep water, docks and other man-made structures, overhanging trees, shade, underwater rocks and cliffs are all areas of shelter, structure or cover. Fish seek shelter from predators and from direct sunlight, so they typically can be found either immediately adjacent to, or within easy reach of, a sheltered area.

If the angler can locate fish food sources within a given lake, gamefish will be found nearby. Minnows, salamanders, midges, surface-dwelling insects, beetles and other such food items make up a large part of the diet of gamefishes. The angler should watch carefully for surface activity, such as that caused by schools of small baitfish jumping, or by insects hatching. On a windy day, success can be achieved by fishing the region of the lake where the surface food is being concentrated. Areas adjacent to inlet and outlet streams where food is concentrated, or next to grassy shorelines or marshy, weedy areas, may also be productive to the angler. Overhanging trees or bushes harbor various types of insect life, and fish will often be lurking below, waiting for food to drop into the water.

Fish all relate to structure, and one of the easiest types of structure to detect, owing to obvious shoreline features, is the **dropoff**. Dropoffs are indicated by the presence of steep banks. The smart angler trolls close to shore along such banks. A depth sounder will help in the location of dropoffs, underwater islands and ledges that cannot be identified from visual observations at the surface.

Most lakes stratify into three layers during late spring and stay that way until late fall (see Chapter 11). The middle layer of water, the **thermocline**, contains both a large amount of dissolved oxygen and concentrations of forage fish. Thus, anglers should troll close to or within the thermocline. The optimum temperature for peak feeding of coho and chinook salmon is 54°F, with an active range from 44°F to 58°F. For lake trout, the peak feeding and optimum temperature is 51°F, with activity from 43°F to 53°F. Fish will rarely venture out of the indicated temperature zones once thermal stratification has taken place, except to capture prey, after which they will quickly return to water of the preferred temperature. The sport angler should keep in mind that the depth of the thermocline can change from day to day as a function of wind and wave action, which tends to change the depth of the mixed layer overlying the thermocline.

Fish will also be found near dropoffs, around underwater springs and near inlet streams where highly oxygenated water is flowing into a lake. Old river channels in reservoirs are another place in which gamefish can be found. Again, such areas can best be located with a depth sounder.

Rivers and streams have their own forms of structure but do not have thermoclines. Stream fish will generally be found near cover such as deep water in pools, undercut banks, under and around logs and so forth. Since water temperature in a running stream is uniform except, perhaps, in the deepest holes or around springs, gamefish will be spread throughout areas having appropriate structure. Many river sport anglers spend the majority of their fishing time plying the waters at the downstream end of holes, where the water becomes shallower and flow rate increases just before the stream enters a riffle area or one characterized by rapids. These so-called "tailouts" provide resting and feeding areas for gamefish and can be quite productive to the angler.

Angling Techniques for Selected Fish Species

Besides the fact that gamefish relate to various types of structure, cover or shelter, there are certain species which display specific preferences when it comes to habitat and feeding habits. Entire books have been written about nearly any gamefish that one might bring to mind. Some of the important considerations associated with a few species are discussed in the following subsections.

Coho Salmon (*Oncorhynchus kisutch*)

When the temperature of the water is optimum, freshwater coho salmon orient toward the surface or in shallow water; thus, they can usually be found at depths of less than 50 feet. Coho are susceptible to fast-action spoons and the added flash and action provided by a small dodger (an attracting device that flashes and sways from side to side that is placed a foot or two above lures and/or bait). Anglers begin their search for coho salmon by starting shallow and working progressively deeper into the water column.

Chinook Salmon (*O. tshawytscha*)

Compared with coho, chinook salmon prefer deep water, often shying away from unnatural movements in the water. Longer leaders are often used as well as lighter test fishing line. Chinook salmon frequently will strike spoons that wobble slowly from side to side when trolled.

Lake Trout (*Salvelinus namaycush*)

Lake trout (mackinaw) are deepwater fish that generally lie close to the bottom or are located in the lower portion of the thermocline. They seem to prefer slowly trolled lures of the type used for chinook salmon.

Brown Trout (*Salmo trutta*)

Brown trout also react slowly to lures. They are wary of any unnatural movement or noise. In many regions, brown trout are a highly prized stream species. They have also been stocked into lakes and reservoirs. In the latter type of environment, brown trout often shy from boats, so lures should be trolled from 100 to 300 feet behind the boat. The diet of brown trout consists primarily of fish, so spoons are a good choice of lures. Spoons having a fish scale finish to them are particularly effective. This species of trout prefers to rest and feed near structures such as dropoffs and rip-rap banks. Most fishing success comes from trolling near such areas and near heavy cover such as weed beds, brush piles or sunken logs.

Rainbow Trout (*Oncorhynchus mykiss*)

Rainbow trout and their cousins, the Kamloops and Gerrard rainbow, normally are fast-moving fish found scattered at various depths in lakes and reservoirs. They are also widely distributed in streams of various sizes. Rainbow concentrate where food and/or dissolved oxygen are abundant and frequently feed on hatching insects near the surface when light intensity is low. They are often caught by flycasting.

Steelhead Trout (*O. mykiss*)

Steelhead trout are large, migratory rainbow trout. When landlocked, such as in the Great Lakes, these fish migrate up tributary streams in the fall and winter to spawn. Anadromous steelhead, which spawn in freshwater but live in salt water, also migrate into freshwater tributary streams, but they differ considerably in their habits as compared to the landlocked variety. Once they enter freshwater from salt water, anadromous steelhead feed less and less as their body chemistry changes. Although they can be caught on baits, most anadromous steelhead in the freshwater part of their cycle are caught with lures and plugs, which invade their territories and force a protective striking response. On the west coast, much debate evolves around just how much anadromous steelhead feed; however, after the fish have been in freshwater a couple of months, analysis of rubber-band-thin stomachs indicates they are practically incapable of swallowing or digesting a meal even if they wanted to. Anglers who catch anadromous steelhead in freshwater with bait do so within several weeks of the fishes entering freshwater, or they are lucky enough to trigger the latent feeding response, which remains inbred in these fish even though biologically they cannot swallow anything. A curiosity response also accounts for some bait-caught steelhead when in a non-feeding mode.

Landlocked steelhead, on the other hand, do not endure the dramatic chemical body changes of their saltwater cousins. Therefore, they aggressively take baits both in open water and in streams and rivers, as well as being caught with a variety of lures, plug and flies.

Steelhead are considered the trophies of the rainbow trout species because of their acrobatics once hooked and their large size. They also are considered by most anglers to be more difficult to catch than other trout.

Figure 22.2. A fat eastern brook trout displayed by the author. Note spot pattern characteristic of members of the char family.

Eastern Brook Trout (*Salvelinus fontinalis*)

Eastern brook trout (Figure 22.2) are widely spread throughout the continental United States and into Alaska and Canada. Although commonly called trout, they are in reality members of the char family, as are Dolly Varden and mackinaw (lake trout). Brookies are commonly caught from small streams, where they reproduce naturally, and from lakes and reservoirs, where they have been stocked. Stream brook trout are fond of small surface insects and are readily caught with small dry flies. Larger brook trout in lakes are best caught with large, trolled streamer flies, spinners, small plugs or large baits such as nightcrawlers.

Cutthroat Trout (*Oncorhynchus clarki*)

Common to the western United States, cutthroat trout are a major sport fish in both lakes and streams. This species obtains its name from the red or orange slashes just underneath the lower jaw, as if someone had "cut the throat." There are several subspecies of cutthroat, such as Yellowstone, Montana black spot, coastal, Lahontan and even an anadromous strain that lives and feeds in salt water and returns to freshwater to spawn.

Cutthroat are a popular sport fish as they are considered easier to catch than either rainbow or brook trout and readily strike small spoons and spinners with abandon. Flies imitating shrimp or small forage fish also are effective.

Arctic Grayling (*Thymallus arcticus*)

Although a rare fish to anglers in the continental United States, the Arctic grayling is a major sport fish in Alaska and Canada. However, grayling do provide some limited fisheries in the states of Montana, Wyoming, Utah, Idaho and Washington. The grayling is both a lake and river fish and is characterized by a large sail-like dorsal fin.

Grayling feed on small insects and small aquatic life forms. This, coupled with the fact that they have a smaller mouth than the other species described herein, results in a fish that is usually much more susceptible to fly fishing than other forms of sport angling. Grayling are a schooling fish, and once one is found, others generally will follow. Delicate, small dark-pattern flies and lightweight leaders are preferred in grayling fishing, as the fish are quite wary compared to other trouts.

Kokanee Salmon (*Oncorhynchus nerka*)

Kokanee salmon are the landlocked form of the anadromous sockeye salmon. They are found throughout the western United States in large reservoirs and lakes (where they have been stocked) and even in the midwest, where they have been experimentally introduced, either as a forage fish for larger species or as a sport fish.

As they are a true salmon, they exhibit a standard 4-year life cycle, with adults entering a spawning tributary in the fall during year 3 or 4 and dying after spawning is completed (whereas members of the trout family do not die after spawning). Because of this trait, huge schools of kokanee present themselves off river mouths during late summer and fall, prior to spawning, and provide good sport for anglers knowledgeable about techniques.

Kokanee spend their lives feeding upon plankton and the smallest aquatic animals and crustaceans. Because of this, they are difficult to catch with conventional tackle such as spinners, spoons or plugs. They will take small flies, however, and small lures that are either red or pink in color. A favorite technique is to troll a small thin-bladed spoon in a red-head or pink pattern slowly through a school of "kokes."

As with grayling, once a kokanee is found, there are other fish nearby, so the angler should pay close attention to the exact point and depth at which the first one is caught.

During spring and summer, kokanee can be found in the lower levels of the thermocline, where vertical jigging of a small spoon can be quite effective. Anglers jigging with light tackle for kokanee often get a big surprise in the form of a large mackinaw as these fish feed heavily on kokanee and usually can be found adjacent to their favorite food.

Freshwater Sport Fishing Techniques

Hardware Casting

Spinners and spoons are as versatile as the imagination of the angler employing them. They are easy to cast and retrieve and are effective for a wide variety of fish species. These lures are fished differently in lakes than in rivers, and there are also differences in how each type of lure is effectively used. However, they will work in virtually all water conditions: from fast currents in flowing water to the stillness of a lake or pond on a windless day. The greatest joy that comes from "hardware" casting is being able to use super light tackle, which both magnifies strikes and allows a direct confrontation between you and the fish. Plugs are another item that can be cast and retrieved.

Casting with weighted spoons and spinners provides several distinct advantages over other fishing methods. For example, a minimal amount of terminal tackle is needed (you can tie your main line directly to the lure via a snap swivel or other attachment device, eliminating the need for leaders, weights and extra knots). These types of lures are effective in shallow and/or clear water,

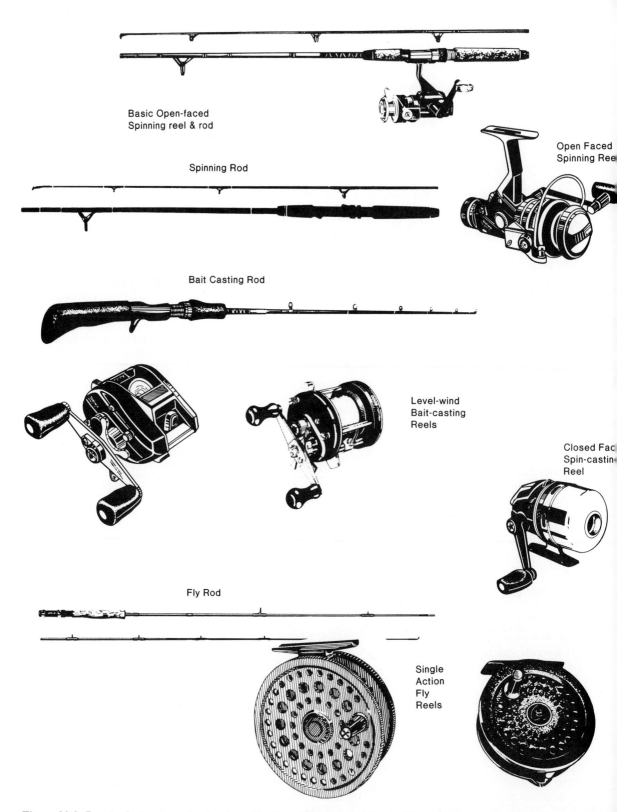

Basic Open-faced
Spinning reel & rod

Open Faced
Spinning Reel

Spinning Rod

Bait Casting Rod

Level-wind
Bait-casting
Reels

Closed Fac
Spin-castin
Reel

Fly Rod

Single
Action
Fly
Reels

Figure 22.3. Popular freshwater rod and reel combinations. (Diagrams in Chapter 22 provided courtesy of Luhr Jensen and Sons Hood River, Oregon.)

and lightweight rods, reels, lines and lures can be used. Both spoons and spinners simulate natural gamefish foods (baitfish) and, because of the vibration patterns associated with these lures, strikes often are vicious—either from actively feeding fish or from those protecting their territories from the invading lure.

In learning to master hardware casting techniques, the most important aspect is to become completely familiar with the feel and action of individual lures under a variety of water conditions. Spoons should swim and wobble from side to side, while a spinner should have a constantly revolving blade. If your spoon is spinning, you are reeling too fast and should slow the "retrieve." Conversely, if the blade on your spinner is not revolving constantly, you are reeling too slowly and should speed up the retrieve. You can go as slow or as fast as you wish within these constraints. The speed of the retrieve will often depend upon how aggressive or lazy the fish are on a given day. On some days only a fast retrieval speed will work, while on others only a very slow speed will produce strikes.

Spoons basically resemble the wide portion of a tablespoon with modifications in their shape to produce different fish-attracting actions and underwater vibration patterns (Figure 22.4). The first fishing spoon was actually fashioned from a sawed-off tablespoon, hence the name. Three basic shapes have evolved over the years: oblong, teardrop and oval. In addition, there are specialty shapes such as long and flat, flat-sided with beveled ends and U-shaped such as the Super Duper®.* Each particular spoon has its own built-in action and will swim and dart differently depending upon water current and the speed of retrieve.

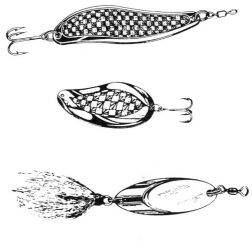

There are three basic components of a spoon: the blade, split ring and hook. A fourth component is the device for attaching fishing line to the lure, which can take the form of a swivel/split ring, welded ring, snap or snap swivel. When an attachment device comes with the spoon from the manufacturer, it should be used as it has been installed to produce that spoon's optimum action in the water.

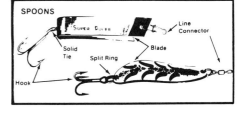

All attachment devices for spoons, including those you may add yourself, should have a rounded end that connects to the eye of the spoon so the lure can swing freely from side to side. Sharp pointed or V-shaped snaps or snap swivels destroy the action of most spoons and are not recommended.

Figure 22.4. Five varieties of spoons.

Weighted casting spinners produce a completely different pattern of underwater vibrations than those produced by spoons. The primary difference is that the blade on a spinner revolves and spins around a fixed shaft.

*Use of brand names does not imply endorsement by the editors.

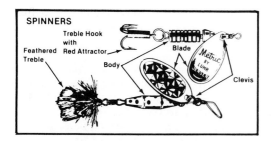

Figure 22.5. Two examples of spinners.

A spinner's basic components include the shaft on which is placed a weight, a hook (either fixed directly to the shaft or attached to the shaft with a split ring), the blade, a clevis to allow the blade to spin freely around the shaft and several beads that act as ball bearings for the clevis and blade (Figure 22.5). If a line attachment device, such as a barrel swivel, is not already on the spinner, a snap swivel should be attached to the eye of the spinner. Spinners, because of their action, must be used with an anti-line twist device. Those featuring ball bearings are the best.

Spinners can be fished effectively in a river or other location where a current is present by employing one of three distinct casting techniques: upstream, cross-stream or downstream. The spinner should be cast out and reeled slowly at first until you feel the first "tick" of the spinner blade scratching bottom. Then, speed up the retrieve for a few feet and slow it down again until the next "tick" is felt. The process should be repeated until the retrieve is completed. If you don't feel the "tick," you are reeling too rapidly and may not be close enough to the bottom to be consistently productive.

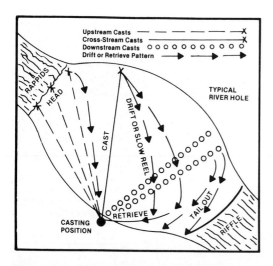

Figure 22.6. Diagram depicting various forms of river casting techniques.

Upstream Casting

Of the techniques used for fishing spinners where there is current, upstream casting appears to be the choice of experts (Figure 22.6). Upstreaming consists of casting the lure upstream at about a 45° or sharper angle and then reeling back downstream to your position. This straight upstream and straight back downstream approach is effective because fish swim into the current (face upstream). The spinner, therefore, will move straight toward the fish and can be easily seen. Also, it provides the fish with sufficient time to make necessary adjustments in the water to attack the lure, even before the spinner enters the territory being protected by the fish.

Upstream casting is only effective, however, if the retrieve is sufficiently rapid to keep the spinner's blade constantly working while obtaining those bottom "ticks" as described previously. The spinner must be worked near the bottom in order to perform well on a consistent basis.

Cross-stream Casting

A second technique employed by river and stream anglers is cross-stream casting, whereby the spinner is cast across and slightly upstream and then allowed to "swim" with the current until the first "ticks" are felt. The lure is then slowly reeled back to the rod tip in a wide arc with the current carrying the lure slowly downstream.

Cross-stream casting is an effective way to fish big water with a spinner as the lure will swim consistently just off the bottom, even in the deeper holes. Cross-stream casting is also effective for prospecting a section of river, as the technique provides full coverage of a wide swath of river from bank to bank.

Downstream Casting

The third casting technique used by spinner anglers is downstream casting. This method is more difficult to master than the previous two, since the current will constantly be pulling the spinner towards the surface and there will be few, if any, "ticks" to reveal the lure's position in relation to the bottom. Downstream casting is best employed when fishing a tailout section of a large hole or pool, where the water begins to run faster just before breaking into a riffle or section of rapids.

Tailouts of deep holes are natural resting and feeding places for gamefish and generally are from 1 to 4 feet deep. Food organisms concentrate in tailouts as they are swept therein along the bottom or at the surface; thus, the area is attractive for gamefish. Migratory fishes such as salmon and steelhead trout also use tailouts as resting places, as these areas provide the first slack water after a section of fast water.

Downstreaming is accomplished by casting the spinner across and downstream at about a 45 degree angle and then allowing the current to sweep the spinner across the tailout. The first cast should be a short one, followed by progressively longer casts, allowing the current to sweep the lure across a new path in the tailout every time until the area is thoroughly covered.

Spoons can also be fished in rivers utilizing the above techniques, although problems due to weight factors make it best to use the cross-stream or downstream methods. Upstreaming a heavy spoon will often result in a snagged and subsequently lost lure.

Lake Fishing

Areas such as inlets and outlets along with the thermocline (located at from 10 to 20 feet deep in most medium-sized lakes) are often productive fishing spots as they are highly oxygenated and may contain concentrations of baitfish and insects. Surface activity is an immediate key to where fish are feeding actively.

When lake fishing with either spoons or spinners, you must recognize that fish will be located at different depths, depending upon the type of insect hatch that is occurring, the time of day, depth to which sunlight penetrates the water and the

Figure 22.7. A large spinner outwitted this fat rainbow trout.

depth of the thermocline. River fish, on the other hand, tend to be found close to the bottom unless a major insect hatch draws them to the surface.

The depth at which a spoon or spinner is fished should be altered until the depth at which the fish are holding is found. Thereafter, that depth should be heavily fished. A standard quarter-ounce spoon or spinner on a tight line (they will sink faster on a slack line) will sink about one foot a second. You can count the number of seconds it takes to reach bottom (line goes slack at that point) and then make your first retrieve slowly and close to the bottom. On each successive cast, subtract two seconds in sinking time until you have covered the entire water depth in a particular area. Then, if no fish have been located, repeat the process in the same area with different types and colors of lures until you are satisfied that there are no striking fish. At that point, move to a different area and repeat the process. As soon as you obtain a strike, make a mental note of the number of seconds of sinking time on that cast and use it as the basis for successive casts in the same area. You will generally find several fish in an area, so it pays to work it over thoroughly as soon as you receive a strike or observe a fish following your lure.

If fish appear finicky and hard to catch (they follow the lure but won't strike), your line is probably too heavy. By switching to a smaller diameter line, which fish can't see as easily in clear water, your success rate should increase. For trout and other fish up to 5 lbs, a 4- or 6-lb test line is recommended. For fish larger than that, try 8- to 10-lb test line.

A second factor in forcing strikes from finicky gamefish is to do everything you can to keep the lure from running at a constant speed and in a straight line. A fish will not spend any more energy than necessary to catch a meal. A naturally swimming lure might be viewed as difficult to catch once the chase has begun as there is nothing apparently wrong with it. A spinner moving in an erratic fashion and giving off "panic" vibrations, however, is another matter and signals mealtime to nearby fish.

Twitch your rod tip every few seconds, speed up and then slow down the retrieve, stop the spinner or spoon dead in the water and then start the retrieve again, reeling extremely fast for a few seconds, then slowing down and so on. The more variety in speed and action you impart to the lure, the better your chances of enticing a strike.

The size and finish of the spoon or spinner selected will depend on a variety of factors including line weight, casting distance desired, fish species being sought, depth of the water and others. Some tips which should help you get started on the right track are as follows:

- A quarter-ounce lure is adequate for most lake fishing for fishes such as trout or kokanee salmon, which eat smaller organisms. Large salmon or trout, on the other hand, generally prefer a larger meal such as that imitated by a 3/8- to 1/2-ounce lure. If you have to reach bottom in deep water, a heavier lure will help. For subsurface or shallow water casting, a lightweight lure is sufficient.
- On dark days, or when there is not much light on the water, such as during early morning and late afternoon, a brass or copper finish will work well. On bright days, or in clear water, most successful anglers choose nickel finishes. Brass or copper also work well when water is brackish, murky or deep, since the light does not penetrate as well under such conditions.
- Color finishes should be matched as closely as possible to the natural food available in the lake. Minnows can be represented by metallic finishes, particularly those with red heads. Frogs can be represented by a green-and-yellow or black-and-yellow spotted finish. Rainbow trout and brown trout natural color finishes work well for predatory species such as salmon, char and large trout. It is a good idea to have several sizes and color patterns of spoons and spinners along with you to be able to match the light conditions and/or food sources on the day you fish a given lake or river.

Trolling

An angler with a boat and knowledge of proper trolling methodology can often be successful in the pursuit of gamefish.* Day in and day out, a troller who employs the correct technique, who

---

*Note: trolling is not allowed in some states or on some specific water bodies. The angler should check local regulations for specifics.

understands the habits of the fish being sought, and who is willing to experiment will outfish anglers using other fishing techniques on the same body of water.

The reason trolling is effective is basic and simple: very little of the surface area of a water body can be covered by casting or still-fishing, while the entire water body can be covered quickly and thoroughly by trolling. Trollers can also pinpoint fish concentrations with a minimum of effort. Besides being able to fish more water faster and easier, the troller has an additional distinct advantage . . . the ability to get away from the limited and frequently crowded bank access areas.

Gear for trolling includes a boat, method of propulsion (motor, oars), rod and reel, line, leader and a lure or some type of bait. Many anglers use additional revolving or flashing blades ahead of the bait or lure to attract fish. These attractors generally take the form of some blades on a cable or wire and are commonly called "lake trolls."

The lure or blade string (troll) is let out behind the moving boat until it is well astern, usually 50 to 100 feet. The forward speed of the boat will dictate the speed of the troll through the water and the depth at which the troll is located. Multiple-blade rigs (trolls) are especially effective in deep, murky waters or on days with overcast skies. The basic difference among various lake trolls is the number and shape of the blades and the length of the shaft or cable. The shape of the blade determines how fast or slow it will turn and also the particular sound vibration it will produce in the water. A round or nearly round blade swings slow and wide while narrow blades spin rapidly and close to the shaft. Narrow blade trolls are best suited for fast trolling as they have less resistance in the water.

A troll appeals to several fish feeding instincts. In addition to flash and other visual attraction, a revolving blade produces sound underwater. Particular vibrations may signal the presence of food to the fish. Blades can be used in conjunction with just about any lure or bait. Good success has often been reported when blades are paired with small spoons, plugs or worms.

A troll consists of a rudder at the front end to which the line is tied (prevents line twist), a series of free swinging blades on a wire cable or shaft, and a swivel to which the leader is tied (Figure 22.8). The leader between the lure and the troll should be at least 1 foot, preferably slightly longer, and should be composed of 6- to 10-lb test line. When trolled, the blades act as attractors. Fish following the sound vibrations and flash of the blades, spot the trailing lure and strike!

A variety of trolls may be used, depending on the type of water to be fished. The depth of the lake and water color are important factors to consider when selecting a troll. Large, deep lakes require large, multiple blades that create flash and vibrations capable of attracting fish from considerable distances, while in smaller, shallower lakes, trolls of smaller size may be desirable.

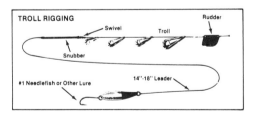

Figure 22.8. Typical troll setups showing a spoon and a spinner as terminal lures.

Trolling a Flat Line

Flat-lining is a term used to describe a technique whereby a lure or lure tipped with bait is trolled in the top 15 feet of water with little additional weight and few, if any, attractor blades. Flat-lining is particularly adaptable to ultra-light tackle where 2- to 6-lb test main line is commonly used and where rods and reels are so light that no heavy weights can be used. It is a method whereby the lure is let out behind the boat and trolled close to the surface, producing a flat angle of line to water (usually 10° to 20° angles compared with the 45° to 60° angles commonly produced when deep trolling with blades or heavy lead weights).

If desired, the angler can select from several methods that involve attaching small weights ahead of the lure. Using split shot, small rubber core sinkers, or hollow lead sinkers will provide the desired weight when coupled with appropriate trolling speed. Many small, thin-blade lures can effectively be flat-lined with no additional weight, while some require small amounts of additional lead.

Most anglers find that flat-lining a small, lightweight, thin-blade spoon with a couple of split shot is an effective technique. The trick is to troll slowly enough to give the small spoon an erratic action and still prevent it from turning over and twisting the line. It is important to attach the weight at least three feet above the lure.

Flat-lining spoons. Spoons are perhaps the easiest, most popular lure to flat-line. Weighted spoons can be flat-lined without additional weight (the line should be tied directly to the swivel on the lure). Weighted spoons will have a tendency to turn over in the water if trolled rapidly, so a ball-bearing swivel should be attached about 20 inches up the line from the lure. Small, thin-bladed spoons lend themselves well to flat-lining and will produce more vibration and erratic action in the water than will the weighted models.

Flat-lining spinners. A single-blade spinner trailed by a bait or fly can be an effective lure when used with the flat-lining technique. The most successful trollers who use such spinners add a piece of worm to the lures to increase their effectiveness or, in some cases, trail a wet fly behind the spinner. Weighted spinners can be flat-lined like the weighted spoons described above (with a ball-bearing swivel 20 inches or so above the lure to prevent the line from twisting).

Trolling with fly tackle. Fly fishing tackle can be used to troll thin-blade spoons or flies by tying 15 to 20 feet of 4-lb test leader to the leader of a fly line. Depending on the desired depth of trolling, one can use floating, sinking or sink tip line. Ten- and 15-foot sink tip lines lend themselves well to trolling in deeper lakes as they will often position the lure just above or within the thermocline. When deep trolling with a fly or thin-blade spoon and fly line, use a fast sink or 30-foot sink tip line and let the lure out 40 to 60 feet behind the boat.

Another variation for trolling flies or spoons that will get them deep in the water column without the use of heavy leads or large trolls is lead core line (Figure 22.9). Lead core is color-coded in 25-foot sections and anglers often refer to distances in terms of how many colors they have out, such as "three colors" which would be 75 feet of lead core plus whatever leader length is being used. Lead core line requires the use of a large capacity fly reel or casting reel, as it is bulky and needs a considerable amount of reel storage space. Other than the larger reel, trolling and rigging with lead core line is the same as with a sink-

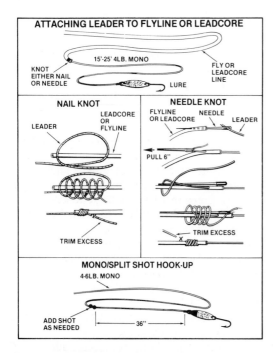

Figure 22.9. Trolling hookups for flyline.

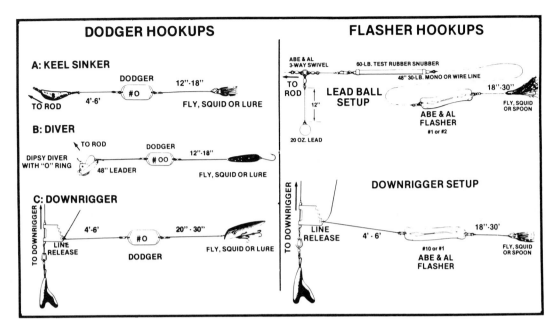

Figure 22.10. Dodger and flasher hookups.

ing fly line. Lead core line will sink rapidly and can be trolled deep, without using blades or heavy weights.

Trolling Medium to Deep

Various devices can be used to assist in getting lures to troll at greater depths. Among them are divers, downriggers and dodgers or flashers rigged with heavy lead weights.

Dodgers and flashers. With the exception of the Great Lakes, flasher-and-lure or dodger-and-lure combinations are often overlooked by freshwater anglers. While lake trolls attract fish with their flash and sound, they transfer little or no action to the trailing lure. Flashers and dodgers, on the other hand, do that job rather well.

A fully rotating flasher will act differently on a trolled lure than will a side-to-side dodger. The full rotation of the flasher transmits a wider path and a crippled baitfish action to a trailing spoon, fly, plug, spinner or bait. A short leader of 12 to 15 inches will deliver and transmit a fast, darting erratic action to a trailing lure when trolled; longer leaders, from 18 to 24 inches, will produce a more deliberate, moderate action (Figure 22.10). Lures that have built-in action of their own, such as wobbling plugs, should be used with the longer leaders. Where additional action is desired, spoons, spinners or baits should be fished with the shorter leader lengths.

For a flasher to work properly, it must be fully rotational. The rig should be lowered over the side of the moving boat and the speed increased until the flasher begins making complete revolutions (it will move from side to side like a dodger if the speed is too slow).

Dodgers impart a side to side action to trailing lures. A small dodger will have less resistance in the water than a flasher and, for light tackle, it in essence will take the place of a troll without the resulting water drag. The same rules for leader lengths that apply to flashers also work for dodgers, with a shorter leader from dodger to lure transmitting more erratic action to the lure.

Divers. One effective way to fish a spoon or spoon/attractor rig in deeper water without lead weights is with a diving sinker. Some diving sinkers feature adjustments that allow the device to be set to run straight behind the boat or to pull to port or starboard. Such adjustments prevent lines from becoming entangled during trolling. Divers also act as fish attractors because of their color and shape. Standard rigging with a diver is 4 to 6 feet of leader between spoon and diver and 20- to 30-lb test main line.

A second way of rigging a diver is to add a #0 or #00 small dodger 2 feet up the leader from the lure. A leader of 4 feet should be used between the diver and the dodger.

Trolling with Downriggers

Downrigger fishing is popular in the Great Lakes, where salmon and steelhead are sought. The downrigger provides the means of taking a lure down to fish-catching depth with light tackle and allows the angler to play the fish after a strike without additional weights and devices that might hamper the action (Figure 22.11).* Downrigger technology has developed rapidly over the past decade and anglers are finding that lake trout, walleye and other species can be caught using this aid.

Plugs are often fished behind a downrigger release with excellent results. Fast-action, darting plugs are ideally suited for trolling because they project a basic minnow/baitfish image in the water, which predatory gamefish relate to under a wide variety of conditions. Plug trollers prefer 20 to 40 feet of line between the release and the lure for optimum, erratic lure action.

Trolling Accessories

Snubbers and rudders are two accessories that may be useful to the trolling angler. Strikes that come during trolling are often vicious. If light tackle is being used along with the flat-lining technique, the rod will usually absorb the impact of the strike. If, on the other hand, a troll, dodger, or flasher is being used, or the striking fish is of a species that has a delicate mouth (e.g., kokanee and crappie), a rubber snubber is useful. A snubber is a length of surgical tubing with a swivel attached at both ends. Inside the tubing is a coiled piece of heavy line. When a fish strikes, the snubber stretches out to absorb the impact and then retracts.

Small snubbers can be used in every trolling situation, but they are particularly important when large trolls and heavier tackle are employed. Some Canadian Kamloops trout anglers even use snubbers on fly lines when trolling, as that particular strain of rainbow trout is an extremely hard striker. Many Kamloops are lost when the line is broken during the initial strike.

A rudder is an essential piece of trolling equipment when using lures or trolls that tend to spin in the water. The rudder will keep blades tracking straight and prevent line twist. Small rudders should be used in situations where there is concern about line twist. For example, an angler should use a rudder when flat-lining with a spinner. The rudder should be attached about three feet from the lure.

Special Trolling Tricks

One of the first things to remember is to *troll slowly*. A common mistake of anglers is to work the lure too fast. Large fish will not spend any more energy than necessary in capturing what they perceive to be a meal, and most lures will not perform correctly at fast trolling speeds. Many expert trollers, particularly those trolling for trout, refuse to use a motor as they feel that the speed will be too fast, no matter how far down the throttle is turned. They prefer to use oars.

Another trick is to *vary the trolling speed*. While slow trolling is critical to success, this does not mean that one should operate at dead slow speed all the time. A lure that runs through the water at the same speed and depth and that gives off the same vibrations may not produce strikes. Every few minutes, the boat should be sped up to change the lure's action and pattern of vibration and then slowed once again.

---

*See Chapter 20 for additional information on downriggers.

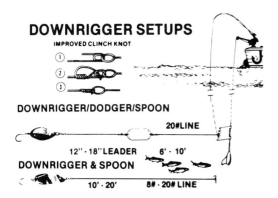

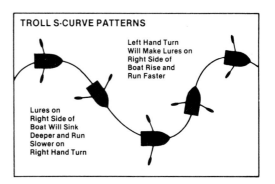

Figure 22.11. An example of downrigger setups. Choices to consider when using a downrigger include spoons, plugs, flashers and other accessories.

Figure 22.12. Diagram showing the advantage of changes in direction while trolling.

It is also a good idea to *work in S-curves*. Trolling in a straight line between two points is the least successful path that an angler can take if the object is to catch fish. Successful trolling requires frequent changes in direction—running in S-shaped curves is the best solution (Figure 22.12). Each time the lure is on the inside curve of the boat's path, it will decrease speed and drop deeper in the water column. When the lure is on the outside of a curve, it will speed up and rise in the water. Thus, the speed of the lure through the water and the vibration pattern that the lure makes can be changed without changing the boat's speed.

The angle of the line is an indicator of how deep the lure is fishing. If, for example, the angler has 50 feet of line out and the angle from the water to the rod tip is 45°, the lure should be fishing at about 25 feet. If the line angle is 20°, at least 100 feet of line would have to be let out in order to fish at the same 25-foot depth. A mental note of the line angle and amount of line out will allow the angler to return to the depth that was being fished under those conditions after each strike.

A final trick, and one which may be particularly effective at catching those fish that follow the lure but fail to strike when normal techniques are practiced, is to give the rod two short, sharp jerks, pause and then give two more sharp jerks before setting the rod back down. This routine should be repeated at intervals of about 2 minutes. A fish that is following the lure can often be enticed to make a strike when this procedure is followed.

Stillfishing

Stillfishing with live bait is a technique popular everywhere that fish and anglers can be found in proximity. Many anglers were first introduced to the sport utilizing the stillfishing technique. While many youngsters discover the joys of sport fishing by dangling a worm on a hook in the water beneath a red and white bobber, there is more to the technique than meets the eye.

All gamefish will take natural baits; this accounts for the large sales of minnows, worms, crickets and other live baits. Some states have even outlawed the use of live bait in certain waters to reduce the effectiveness of sport anglers.

Stillfishing methods can be divided into two major categories: with bait from the bank and with bait from a boat. There are two basic ways to stillfish from a boat. Most anglers anchor in a location selected for its relationship to structure (for example, near a dropoff, at the edge of a weedbed, in the vicinity of a cliff, near an inlet or outlet stream). The baited hook is dropped to the bottom and is then reeled up a few inches and maintained there until a bite is felt. A variation on that theme is to use a bobber and suspend the bait several feet (a few meters) below the surface. The baited hook, sinker and bobber are cast away from the boat and then allowed to rest motionless until a fish pulls the bobber under, or until the angler can't stand the suspense and reels the rig in to "check the bait!"

Basic rigging for trout stillfishing with live bait is simple (Figure 22.13). Monofilament line in the 4- to 6-lb test range is recommended for all but the most finicky fish, in which case a lighter weight leader might be used. In most cases, 6-lb test line may be used without additional leader and, therefore, no extra connecting knots or weak spots.

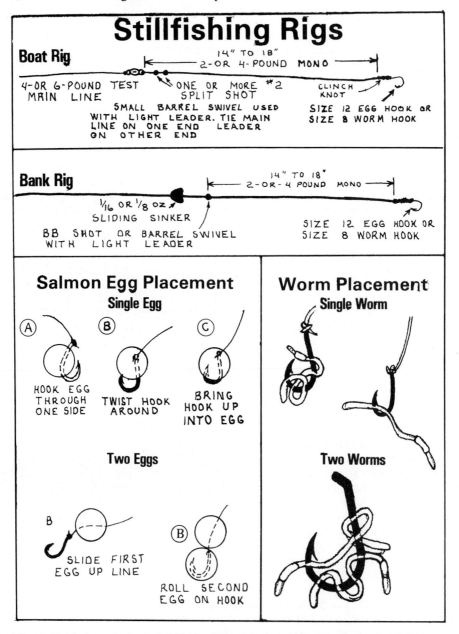

Figure 22.13. A comparison of stillfishing hookups using either eggs or worms as bait.

Anglers who like to have baits resting directly on the bottom often employ a double-anchor setup to keep their boat motionless. They will typically use a sliding sinker rig. Before the hook is tied on, a sliding sinker is threaded on the fishing line. A small split shot weight is then crimped on the line between the slip sinker and the hook about 14 inches up from the hook. The slip sinker rig is cast out away from the boat, allowed to sink to the bottom and then some of the slack line is reeled up. Some remaining slack line is left between the rod tip and the water. The basic principle is to keep the fish from feeling resistance when it picks up the bait. As the fish moves away with the bait in its mouth, the line will move through the sinker and allow the fish to move some distance before encountering resistance before the hook is set. Stillfishing from the bank is accomplished with either the bobber method or the sliding sinker method just described.

Salmon eggs and worms are the most popular baits for trout when the above methods are employed, although grasshoppers, beetles, mayflies, crickets or large grubs may also be used as bait. Where bottom vegetation is a problem, placing a small marshmallow on the hook with the bait will make the bait float above the vegetation.

Fly Fishing

Fly fishing is a way of life to many fresh-water sport anglers, complete with its own tackle, techniques and vocabulary. In fly fishing, one does not cast a weight that has been tied onto the end of the line; rather one casts the line. The line itself acts much the same as a lead weight and causes the rod to flex, load and fire out the line with a leader and virtually weightless fly attached to the end.

Spin anglers and bait casters use different weights of sinkers and lures to meet different casting and fishing conditions, and different rods to properly handle those weights. The same principle holds true with fly fishing tackle. Heavier lines are used where casting power is important for distance, and lighter lines provide for casting lighter weights and delicate flies.

Fly line weights are measured in grains contained in a certain length of line and carry numbers like No. 6, No. 7, No. 8 and so forth, with the weight of the line increasing with line number. Fly lines are also labeled as to where the weight is positioned. The weight may be called level (uniform); double taper (weight in the center tapering toward the ends); or weight forward (most of the weight toward the end), also called shooting head. Lines also carry another label that describes the type of line as floating, sinking, or sink tip (only the first several feet of the line sink while the remainder floats).

Figure 22.14. The author battling a fast-water trout fooled by a large nymph pattern fly.

Rods generally are matched by their length in feet, such as an 8-foot rod with a corresponding line weight, such as No. 8, or a size or two up or down. Fly rods, unlike those used for spinning, casting or trolling, are called upon to do much more than cast line and play fish. A primary function is to allow the angler to pick up the line and position it in various ways to provide a natural drift of the fly. On a stream, this process is called "mending."

Fly rods are built somewhat longer than other types of rods as the power to cast comes directly from the ability of the rod to absorb energy (load up) and then release that energy. The action is similar to a bow that is loaded by pulling back the string with its inserted arrow and letting go. Most fly rods are from 7 to 10 feet long, with longer ones from 8.5 to 10.5 feet becoming more popular each year.

Various flycasting techniques have their own terminology. There is everything from the simplest forward and back motion to roll casts (line is power rolled off the water and back out), single-hauls, double-hauls, false casts (used to get more line out before allowing the fly to land on the water) and so on. The specific cast used depends on the angler's level of skill, weight of the line being cast, distance desired, surrounding environment (such as nearby trees, brush and steep banks) and other factors.

Flies come in all shapes and sizes and can generally be classed into two major categories: dry flies and wet flies. Dry flies include all flies that are dressed (tied) with enough feathers, fur or other material to cause them to float. Wet flies are those that sink and can further be divided into the subcategories of streamers (minnow imitators) and nymphs (insect imitators).

Reels for fly fishing include single-action and automatic. Models range from the primitive spool to those with a host of accessories.

Fly fishing techniques, as varied and different as they are, all revolve around the principle of casting the line with leader and fly attached and presenting the fly to the fish as naturally as the insect being imitated. In rivers, this amounts to casting out into the current and then "mending" a loop of line upstream every several feet of drift. In lakes, a common dry fly technique is to cast out and allow the fly to sit naturally motionless on the surface, or to make it move every so often, thus imitating an insect dabbling at the surface.

Nymphs and streamers are used with wet lines or those with sink tips and usually are stripped in with constant, but slow movement to imitate a swimming or hatching insect or small minnow. Most fly anglers try to "match the hatch;" that is, they attempt to imitate the naturally hatching insects as closely as possible with their flies. This means attention should be paid to color, size and attitude in the water when selecting an artificial fly. Dry fly fishing is considered to require more finesse as the ultra-light leaders and tiny flies that are used must be presented very delicately to be effective.

Because of the way in which the line is cast, the fly reel exists only as a device on which the line is stored when not in use. Line is stripped off the reel and controlled with the hand not holding the rod. The reel is almost never used to play a fish, except in those circumstances when a large one is hooked and, even then, all that's needed is a sturdy reel with a simple, rim control drag system.

Much is made of getting a reel of proper weight and balance to match with the rod. It is actually more important to purchase the size reel that will hold the proper size fly line and the backing you will need for the type of fishing you want to do. Fifty yards of fly line is more than enough for trout fishing and another 25 yards with backing will do if you press the reel into service for steelhead or salmon angling.

Backing, which is nothing more than braided dacron line of 20- to 25-lb test wound onto the reel spool prior to the fly line, is used to take up space. It increases the spool diameter, promoting larger coils of fly line and less kinking. It also allows you to reel in the line faster because the reel is fuller. Backing can serve the purpose of saving a large fish that might run off all of the fly line before tiring.

Actual casting techniques can be learned from friends, guides, seminars and various books dedicated to the subject. The novice is advised to attend a sport angling show or to tune in to various angling programs that are presented on television.

One form of angling where a fly is used, but which could hardly be classified as pure fly fishing, is the use of spinning tackle in conjunction with a clear plastic bubble. This technique allows the angler to use the effectiveness of an insect-imitating lure, a fly, while employing basic spin-casting equipment. Clear, oval plastic bubbles, which can be partially filled with water for added weight, are often used for this type of fishing. Adding a small swivel at the end the main line provides a stopping place for the bobber as well as an attachment point for a piece of 1- to 4-lb test leader to which the fly is attached. Both wet and dry flies can be fished with the plastic bubble technique.

There are times when a dry fly can be fished by casting out and allowing it to rest on the surface until a fish picks it up, but the most effective technique for a fly-and-bubble rig is to retrieve it slowly, twitching the rod tip constantly. Small wet flies or dry flies with sparse hackle (feathers) are best for this twitching technique.

Jigging

A favorite sport fishing technique used for trout, salmon and steelhead is that of jigging. It is usually accomplished with metal lures or, sometimes, lead-head lures with plastic or feather tails. The primary technique is one of drift-jigging (using a boat) as opposed to the cast-and-retrieve jigging method often employed by warmwater anglers.

Drift-jigging is a technique tailor-made for open water, where fish are oriented either to bottom structure or temperature layers. With the aid of a depth sounder, drift-jigging allows pinpoint presentation of a spoon or jig within inches of a fish, providing the angler with a distinct advantage not easily obtainable with other fishing methods (Figure 22.16).

Once fish are located, either by trial and error or by using a depth sounder, the angler simply free-spools the jig or jigging spoon to the desired depth and then begins a series of varied motions with the rod that impart erratic actions to the lure (Figure 22.16). The technique consists of raising the rod tip from 6 inches to several feet, throwing some slack in the line as the tip is lowered toward the water and then raising the rod upward again. A 2-second pause is recommended after the lure has fallen. It is best to vary the distance of the upward rod motion with each sweep, so the spoon or jig produces the most erratic and varied possible series of actions. Some anglers

Figure 22.15. A 55-lb chinook salmon taken from the Columbia River, Oregon, using the jigging technique and a large spoon.

begin with short jigging motions and work increasingly and methodically into longer rod sweeps.

Strikes that come when working a jig or jigging spoon almost always occur as the lure is falling. Hesitation in the descent of a lure, a twitch of the line, a "tap," or any other unusual action as the lure is falling should be immediate reason for setting the hook. In many instances the angler is

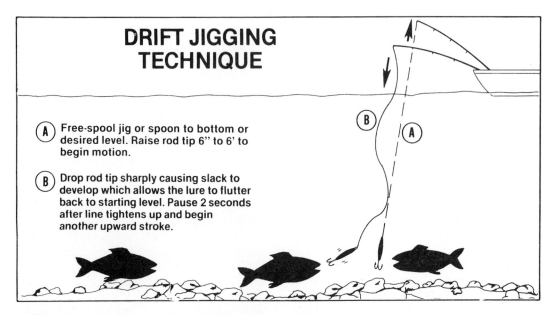

Figure 22.16. A diagram explaining the application of the drift jigging technique.

not able to detect the strike but will feel resistance when the rod is raised again. This is also a signal to set the hook.

One item that aids in detecting strikes when a lure is falling is a premium quality, high visibility line. By carefully watching the line as the jig or jigging spoon is falling the angler will be able to detect slight twitches in the line as the strike occurs.

The jigging motion described above is best used when depth at which the fish are located can be pinpointed with certainty. In cases where that information is lacking or if the area selected is associated with an underwater island or cliff face, a modified technique should be employed. After jigging several times with different motions, reel in one turn of slack before the line goes taut, make several more different motions, reel in another turn and so forth until the lure approaches the surface. Free-spool the lure back to the bottom and repeat the process. If nothing happens after a few tries, move slightly and repeat the process.

Drift Fishing

Of the river fishing techniques used for salmonid fishes, drift fishing is the favorite. The technique has gained popularity in recent years as more and more streams across the United States have been stocked with trout, steelhead and salmon. Drift fishing with buoyant drift bobbers (with or without bait) has become standard practice for many stream anglers and its popularity continues to grow.

Typical drift fishing waters consist of a series of pools and rapids with the pools (drifts) holding feeding, resting or migrating fish. Drift lures fished through the pools often produce excellent results. These buoyant small lures can be used either with or without such bait as cluster fish eggs (roe), shrimp or nightcrawlers. A lead weight is used to improve casting distance.

The basic drift fishing technique consists of casting across and slightly upstream and then allowing the gear to naturally drift downstream in the current with the lead bouncing along the bottom. When the lure has drifted back near the bank, it is reeled in and another cast and drift is made.

There are scores of ways to add weight ahead of a drift bobber. Pencil lead,* either solid or hollow core, is the most popular and is proven to be both economical and easy to use. Solid pencil lead is best fastened to the line with a lead cinch, consisting of a three-way swivel and a length of latex surgical tubing. The main line is tied to one end of the three-way swivel, leader and bobber to the other end and a section of pencil lead inserted into the tubing (Figure 22.17). If the lead becomes snagged, it will pull away from the tubing and the drift bobber and lead cinch can be retrieved.

Figure 22.17. Detail showing two ways to attach pencil lead.

Hollow pencil lead is best fastened by crimping it to a short leader dropped from a barrel swivel. If the lead becomes snagged, a sharp pull will free it from the dropper leader, allowing the drift bobber, leader and swivel to be retrieved.

Pencil lead comes in coils or long sections that can be cut to the desired lengths. Most lead available through sporting goods stores comes in diameters of 1/8, 3/16 and 1/4 inch, with 3/16 inch being the most popular for average fishing conditions and stream flows.

One of the tricks to successful drift fishing is to select the right amount of lead for the water being fished. If too heavy, lead will snag easily, while if it is too light, it will not keep the drift bobber near the bottom. The ideal weight is one that results in a "tap-tap-skip" action as the lead makes contact with the bottom and then rises a bit before hitting again. Experienced drift anglers often begin working an unknown drift with a 2 1/2- to 3-inch piece of pencil lead and then, after making a drift or two, shorten it until the drift feels right. Since snipping off pieces of pencil lead is easier than trying to add them, the initial piece should be longer than is thought necessary.

Once the lead is rigged, the drift bobber is added to a hook and leader (Figure 22.18). Depending on the river, the main line should test between 8 and 20 lbs. Leaders should test 2 lbs weaker than the main line. Leader lengths for drift bobber fishing should be from 14 to 24 inches, with shorter leaders used in average visibility and longer ones in shallow or clear water drifting.

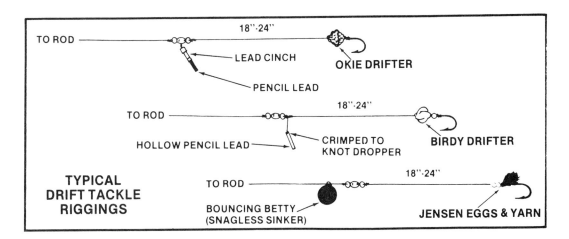

Figure 22.18. Three typical forms of drift tackle rigging.

*Unlike graphite for mechanical writing instruments, the pencil lead used for fishing is metallic lead wire.

Successful drift fishing requires that the line be close to the bottom. It is important to use heavier line than would be selected for lake fishing, as it will have to take the added bottom-scraping abrasion associated with the drift fishing technique.

Although strikes can be hard, oftentimes they are not. Some fish will only lightly mouth a bobber and this kind of strike is difficult to discern from bottom tapping. Many fish are lost simply because the angler can't easily detect the soft pickups. Any momentary slowing or stopping of the line, slack line, or tap that feels ''springy'' should be answered by setting the hook.

However, two things can be done to help hook light-biting fish. First, make sure the hooks are very sharp. Sharp hooks are critical for all types of fishing but become exceedingly important when hooking light-biting fish while drift fishing. A fish will have difficulty getting an extremely sharp hook out of its mouth. Small files have proven to be the best sharpening tools for fish hooks. For creating the ultimate hook point, hold the file parallel to the point and, with gentle, one-way strokes, remove a small amount of metal on at least two sides of the point. This will produce not only a needle-sharp point, but also a knife-like cutting edge.

Secondly, adding a tuft of colorful yarn between the hook and the drift bobber will add color contrast to the lure and make it even more difficult for a fish to spit out the hook. Upon a take, the yarn can become tangled in the fish's teeth. Attempts by the fish to eliminate the hook and yarn will send signals to the angler that mean ''set the hook.''

Backtrolling Plugs

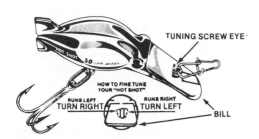

Figure 22.19. An example of an adjustable diving plug. Some of these plugs have internal rattles to produce vibrations to enhance attractiveness.

One of the most effective boating techniques developed for catching coldwater gamefish in rivers is known as backtrolling. It is used with diving plugs. These plugs, also called crankbaits, are self-planing, diving lures that incorporate built-in planing bills which cause the lures to dive when pulled through the water (Figure 22.19). The faster the movement through the water, the deeper the dive. Because of this built-in action, diving plugs are often used without additional weight. Current and retrieval speeds will control the depth to which the lure fishes.

One advantage of using a non-weighted plug is that natural river currents will guide the lure around snags and rocks while maintaining a position close to the bottom where most gamefish are located.

Line weight is important in relation to how deep a plug dives. Light line, such as 6-lb test, will create less friction in the water and allow a plug to dive deeper and closer to the bottom, while 20-lb test line will increase friction and shorten the dive depth.

The round eye snap or split ring supplied with each plug is critical for maintaining proper lure action. The line should be tied directly to the ring or snap. Tying a line to the lure eyelet or using a snap swivel will adversely affect the action of the lure.

The most important step in readying a plug for backtrolling is to ''tune'' it so that it will dive straight and true. Place the plug in the water (either from a boat or shore) and pull it through the water while observing its diving action. If the plug dives straight and true it requires no tuning and will produce the proper action. If it runs off to one side or the other it needs to be hand-tuned. Pull the plug through the water several times to determine its diving direction. If the lure runs off to the left, with the bill facing you, turn the screw eye (if it has one) slightly to the right or bend the nose fixture to the right. If the lure runs to the right, make the appropriate adjustment in the other direction. Check and retune the lure until it dives straight down.

When backtrolling, face the boat upstream above the area that is to be fished. Row or run the motor just fast enough to maintain a steady position in the current. Strip out or free spool line until the plug is 50 feet downstream of the boat. Once line stripping ceases, the lure will begin to dive. The stronger the current, the deeper the lure will dive. Constant pressure should be maintained against the plug by continued rowing or motoring.

An area is worked by allowing the boat to drop downstream while keeping the lure actively working by steady pressure against the current. The boat should move downstream more slowly than the current is moving as the speed differential will keep the plug working.

Most anglers watch the tip of their rod to determine how fast to run the boat. Some anglers even paint their rod tips orange for easy viewing. A constantly throbbing rod tip indicates the plug is working properly. The faster the tip throbs, the deeper the plug is diving. The boat should be held back against the current enough to force the plug to dive and then each stretch of water should be worked thoroughly before moving downstream.

River fish treat plugs as invaders to their territories and react to them with ferocious strikes (Figure 22.20). A lure fished too rapidly through a drift may be ignored. A slowly fished, wobbling, pulsating plug, on the other hand, may induce a strike. A slow-moving lure can be more easily caught than one that moves swiftly past a waiting fish. This technique can also be used, with modifications, from the banks of rivers. Plugs can be cast downstream across a deep hole and pulled through the hole, or pulled along the edges of the hole.

Closing Comments

Numerous studies show that 80% of sport-caught fish are taken by only about 10% of the anglers. This is due, in part, to the fact that some anglers spend a great deal more time fishing than others. There is some relationship between time spent on the riverbank or in the boat and the number of fish caught. Yet that is certainly not the entire story. The successful angler not only is devoted, but also has taken the time to learn about the habits of the fish and to present the proper lure in the proper manner to entice the fish to strike. Even the best anglers occasionally return home with an empty creel, but those who are effective most of the time continually hone their skills through practice and study.

Figure 22.20. Large salmon and steelhead often are caught by backtrolling large, wobbling plugs, which "invade" their territories and draw strong "retaliatory" strikes.

Additional Reading

Hauptman, C. Basic Freshwater Fishing. Stackpole Books.

Hughes, D. Reading the Water. Stackpole Books. (Fly fishing handbook for finding trout in all types of water.)

McClane, A.J. McClane's Standard Fishing Encyclopedia. Holt, Rinehart and Winston. (Desk reference [the "Bible"]; covers all aspects of sport fish and sport fishing.)

Slaymaker, S.R. II. Simplified Fly Fishing. Stackpole Books.

Sternberg, R. Fishing with Artificial Lures. Hunting and Fishing Library.

Stinson, W. River Salmon Fishing. Frank Amato Publications.

23

Freshwater Fishing:
Non-Salmonids

Gilbert B. Pauley and Frederick G. Johnson

Introduction

As with saltwater fishing, the average angler today has many tools and technologies to ensure a successful fishing trip. If one ponders the meaning of success in sport fishing, it would have many different meanings depending upon the goals of different anglers. To some anglers success might mean catching a whole string of panfish (Figure 23.1); to others it might mean catching and releasing a large bass (Figure 23.2); and to others it might mean having walleye fillets on the table for dinner. To many, the satisfaction of experiencing the fishing environment is reason in itself for a successful trip. In other words, success varies between anglers and therefore it is obtainable at many different levels for children and adults. Success could be viewed as beginning at the time an angler

Figure 23.1. A pile of panfish resulting from a day of successful angling. The spotted fish are crappie, while the sleeker fish with the striped pattern are yellow perch.

Figure 23.2. A proud angler showing off a fine smallmouth bass.

decides where, when and how they want to catch fish, and this emphasizes how the angler comes away from a fishing experience with education. Success could also be viewed as a place where luck, skill, knowledge, confidence and satisfaction all come together to produce a more tangible result.

Freshwater anglers have a wide variety of fish species available to them that can be caught in many ways. Many of these species can be considered trophy fish. There are sturgeon in the Pacific Northwest and catfish in the southeast that can be caught with bait; there are walleye throughout the Midwest and northern states into Canada, and trophy largemouth bass in Florida and California, as well as both largemouth and smallmouth bass in most states and provinces, that can be taken with either bait or a variety of artificial lures. Northern pike and muskellunge are found in many northern states and Canadian provinces. There is river fishing for shad and smallmouth bass and lake fishing for largemouth bass and crappie; surface fishing with flies and bottom fishing with bait; and fishing artificial jigs along the bottom or casting artificial plugs along the surface. It doesn't matter whether your interest is catching a string of yellow perch with a spinning rod, a trophy largemouth bass with a fly rod or a large muskellunge with a bait-casting rod and reel; freshwater fishing has something to offer everyone and is enjoyed by millions of people.

Freshwater fishes are found in small streams and large rivers; small ponds, large natural lakes and impounded reservoirs of various sizes; high elevation coldwater lakes and low elevation warmwater lakes. As with saltwater fishing, successful anglers do not use one particular "magic lure." Many lures will produce successfully if the angler understands basic sport fishing techniques. These basic techniques are similar for various fish species in a wide variety of environments. Anglers must understand the basic biology and habits of the fish, such as the types of food they eat and when and where they forage or where they spawn. Each freshwater sport fish species will have an optimum temperature at which it is most likely to feed, an optimum habitat type and location for spawning, and an optimum habitat where the young will grow. Once the basics are known, you will understand which lures or baits to use in various situations as well as which rods and reels are needed. This chapter contains descriptions of angling techniques used for freshwater fishes throughout North America.

Locating Fish

Before the angler can catch freshwater fish, the fish must be located. Assuming that there is an adequate supply of dissolved oxygen present in the water, the water temperature becomes the single most important factor in determining where various species of fish will be. Each fish species has an optimum temperature range within which it is biologically active, i.e., feeding and spawning. This information and the type of habitat and cover that are preferred by various fish species for feeding, spawning and rearing will contribute to successful angling in freshwater.

As in saltwater, each type of freshwater fish is associated with certain types of habitat and cover. For example, shad spend time in rivers because they are anadromous and they distribute according to specific water temperatures. Yellow bass and landlocked striped bass are best suited to large open waters of reservoirs that have a minimum of vegetation. Northern pike prefer cool clear lakes with shallow weedy areas. Channel catfish are found in deep sections of large rivers with sand, gravel or cobble bottoms. Yellow perch and walleye have adapted to a wide variety of habitats even though both prefer large lakes, with clear water preferred by yellow perch and murky water preferred by walleye. The various sunfishes have also adapted to a variety of habitats, although most prefer some type of association with aquatic weeds. Similar appearing species belonging to the same genus often have quite different habitat requirements. One example of this involves black basses belonging to the genus *Micropterus,* which includes three species of major recreational importance and three minor species. Largemouth bass are primarily found in weedy areas with soft substrate and smallmouth bass are usually found in rocky substrate areas having fewer weeds. Spotted bass have adapted to very deep water and are found down to 100 feet in man-made reservoirs.

There are other groups of fishes that show similar habitat diversifications in freshwater. Most fishes in freshwater environments show preferences toward some type of structure, such as submerged stream bed channels, and the curves and junctions in impoundments; clusters of dead and fallen trees; edges of submergent and emergent aquatic weed beds; points of land that drop off deeply into the water; drop-off areas between shallow and deep water; shallow coves with either aquatic or riparian cover; underwater islands and reefs. Knowing that the water temperature is in the preferred range and that the depth contour (revealed by use of a hydrographic map and/or depth sounder) shows some type of bottom structure has put the angler in control of understanding where 90% of the probable fish-producing water is located.

Angling for Selected Fish Species

Books have been written about most of the freshwater gamefishes or the groups to which these fishes belong. Certain species, such as the largemouth bass and walleye, have an extensive scientific and popular literature associated with them, while others such as the Sacramento perch and Rio Grande perch are species about which very little has been written. The following is a summary of the types of freshwater gamefishes you may encounter in North American streams, rivers, ponds, lakes and reservoirs.

Largemouth Bass (*Micropterus salmoides*)

This fish is pursued by more anglers and more money is spent on its pursuit than is the case with any other freshwater gamefish. It is found throughout the United States (except Alaska), the southern part of Canada, and in Mexico, Cuba and Central America. The wide distribution of this fish is based on its ability to adapt to a diverse array of habitats, such as small ponds, surface mine lakes, slow moving rivers and streams, man-made reservoirs and natural lakes of all sizes. They will thrive in both eutrophic and mesotrophic lakes, especially those that have plenty of aquatic weeds, and in reservoirs that have an abundance of flooded timber and brush (**mesotrophic** lakes are those in an intermediate state between eutrophic and oligotrophic).

Generally, largemouth are found in water less than 20 feet deep. They prefer water temperatures ranging from 65°F to 75°F but will spawn at lower temperatures. In the northern extremes of their range, they have adapted to accommodate a shorter growing season and cooler temperatures. A large strain of largemouth bass survives only in Florida, California and other extreme southern areas because it will die in water that approaches 40°F. These bass are highly efficient predators that eat other fish, crayfish, larval and adult insects such as hellgrammites, leeches, salamanders, frogs, birds and mice. They are generally found in areas of heavy cover, making them a challenge to land even after being hooked. They will strike almost any type of lure that resembles a living organism and can also be taken with live bait. They are taken with spinning, bait-casting and fly gear, and on a wide variety of lures, such as surface plugs, diving crankbait plugs, spinnerbaits, jigs, bass bugs, bass flies, plastic worms and plastic lizards. A 10-lb fish is considered a trophy. They are the object of ''Professional Bass Anglers'' in numerous tournaments held throughout North America as well as the everyday average angler. The tournaments involve top prizes in the hundreds of thousands of dollars, and this has encouraged the development of the art of sport fishing.

Smallmouth Bass (*M. dolomieui*)

Smallmouth bass are easily distinguished from their largemouth cousins (Figure 23.3) by their body coloration, maxillary (lower jaw) position and their red eyes. The smallmouth is an extremely strong swimmer and provides excellent sport on light tackle. They are considered by many to be the sportiest of freshwater fish when matched against the proper weight gear. They are caught on fly fishing gear, spinning and bait-casting tackle. They prefer water temperatures between 60°F and 70°F but will spawn below those temperatures. When they occupy the same waters as largemouth bass (generally mesotrophic lakes), the smallmouth bass will spawn first, and the two species remain segregated because of habitat differences. Smallmouth like clean, clear water and are found over boulder or cobble substrates. They are found throughout the United States and southern Canada, but they are not as widespread as largemouth bass. Smallmouth bass do well in both streams and lakes, as well as man-made reservoirs. Streams must have moderate current, and lakes and reservoirs must be over 300 acres in area and deeper than 25 feet to support abundant populations. Crayfish are their favored food by a wide margin, followed by small fishes and insects. They will take a wide assortment of live bait and lures. Lures generally are miniature versions of those used for largemouth bass, but spinners, jigs, small plastic worms and plugs that imitate minnows are the best. A fish of 5 lbs is a trophy.

Other Black Basses

Spotted bass (*M. punctulatus*) are found in the middle south and southeastern United States and have recently been successfully introduced into western U.S. reservoirs. Other black basses have a more limited distribution than the spotted bass, but all are fished using similar techniques. Spotted bass are found in small- to medium-sized clear, slow-moving streams, but they have found a real niche in deepwater reservoirs, where they spawn and live at depths below the other major gamefishes in the reservoir. Their diet is similar to that of the smallmouth, with crayfish being the most important food item, followed by small fishes and insects. Maximum size is smaller than either the largemouth bass or smallmouth bass, and a 5-lb fish is considered a trophy.

Crappies, Bluegill, Redear and Other Sunfishes

The sunfishes are related to the black basses, and both groups belong to the family Centrarchidae, which are *nest builders*. The males of each species construct the nests, guard the eggs and guard the fry until they leave the nests. Sunfishes tend to stunt severely if their populations are not cropped by either anglers or predators. Bluegill (*Lepomis macrochirus*) and redear (*L. microlophus*) are recommended for stocking in small ponds together with largemouth bass (which prey on the others) to obtain a balanced sport fish population, while most other sunfishes do poorly with largemouth in small ponds. Sunfishes, in general, tend to form schools. Crappie (*Pomoxis* spp.) schools are unique in that they often are suspended at a single depth, extending horizontally and far above the bottom.

Most of the sunfishes are distributed widely throughout the United States and southern and eastern Canada. The black crappie, white crappie, bluegill sunfish, redear sunfish and Sacramento perch are the most valuable sport fishes in this group because their maximum sizes exceed 5 lbs,

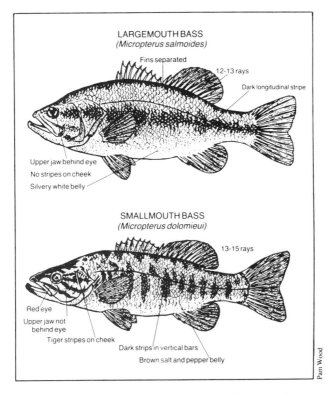

Figure 23.3. A comparison of smallmouth and largemouth basses. (Courtesy of Washington State Department of Wildlife.)

Figure 23.4. A bluegill (above) and a crappie (below) with various lures that can be used to fish for the various species of sunfishes.

and most of the others have maximum sizes less than 2 lbs (Figure 23.4). Sunfishes are a favorite of small children because they are easy to catch once a school is located. These fishes may be taken with live bait or small jigs, spinners and flies with spinning gear and fly fishing gear.

Striped Bass, White Bass, Yellow Bass and White Perch (True Basses, Family Serranidae)

White bass (*Morone chrysops*) and yellow bass (*M. mississipiensis*) are found only in freshwater. The white perch (*M. americana*) is found mainly in freshwater, but it is anadromous in some situations. The striped bass (*M. saxatilis*) is an anadromous fish that has been successfully introduced, primarily in large reservoirs, in over 30 states. The white bass and striped bass have the most extensive distribution in the United States and southeastern Canada. The other two have more limited distributions in the United States and Canada. Of these four fishes, only the striped bass, which approaches 60 lbs in freshwater, grows larger than 6 lbs.

A robust hybrid cross between the white bass and the striped bass has been stocked in reservoirs and reportedly reaches 20 lbs. These are all aggressive fishes that school, thereby giving them excellent sport fish value. The technique of finding and following schools of these fish is called **jump fishing**. True basses require flowing water to spawn successfully. The striped bass and the hybrid may be caught on live bait, diving plugs, and jigs with either bait casting or spinning gear. The smaller members of the group are most effectively taken with spinning gear.

Shad (Family Clupeidae)

The freshwater gizzard shad and threadfin shad (*Dorosoma* spp.) have no direct sporting value, but are valuable to anglers because they are prey species for larger piscivorous fishes in reservoirs and large lakes. The American shad and hickory shad (*Alosa* spp.) are anadromous fishes that are pursued by anglers when the fish make their spawning runs into the coastal rivers. The hickory shad is found only on the eastern coast of the United States and Canada, while the American shad is found on both the Atlantic and Pacific coasts. They both have strict temperature requirements for entering rivers. The American shad is the largest in size, averaging between 3 and 5 lbs. Both are exceptional sport fish when caught on light spinning gear with jigs, spinners, spoons, flies and flat-headed jigs called shad darts.

Northern Pike, Muskellunge, and Pickerel (Family Esocidae)

The northern pike (*Esox lucius*), muskellunge (*E. masquinongy*) and chain pickerel (*E. niger*) support widespread angling popularity. However, the redfin pickerel (*E. americanus*) is not of much interest to anglers because of its small size (less than 2 lbs). Northern pike inhabit cold, clear waters and are found throughout Canada and the northern United States with the exception of the Pacific Coast area south of Alaska. They can live in most freshwater environments from streams to lakes, but they require the presence of shallow aquatic weedy areas for spawning, at which time they are vulnerable to angling. Because of their predatory nature, they are easy to catch. Northern pike reach sizes exceeding 40 lbs, and can be caught on almost any large artificial lure, but large spoons, spinners, spinnerbaits, and floating plugs work best. Spinning and bait-casting gear both work well, but the equipment should be heavy duty.

Muskellunge are found in the same type of weedy water that northern pike inhabit, but the pike usually outcompete them. The two species hybridize to make tiger muskies. Muskellunge are found in upper midwestern to New England areas of the United States and in the southern part of eastern Canada. Similar large lures are used for muskellunge as for pike, and wire leaders are a must for both because of their sharp teeth. Muskellunge reach almost 70 lbs.

Chain pickerel are excellent sport fish, weighing up to 9 lbs. They inhabit clear, heavily weeded waters of states that border the Atlantic and Gulf coasts. Live minnows, streamer flies, weedless spoons, plugs and jigs are best used on appropriate spinning, casting or fly fishing gear.

Sturgeons (Family Acipenseridae)

In North America, there are seven species of sturgeons, and they are the largest fishes in freshwaters of North America. The freshwater lake sturgeon (*Acipenser fulvescens*) is found in the Mississippi River drainage northward into Canada, and the anadromous white sturgeon (*A. transmontanus*) is found along the Pacific Coast from Alaska to California and through the Columbia River and Snake River drainages (where they have become landlocked in the upper reaches). These powerful bottom feeding fishes are caught on live and dead bait using heavy duty saltwater rods, reels and 100 pound test line. They attain weights in the hundreds of pounds—up to 300 lbs for the lake sturgeon and 1,300 lbs for the white sturgeon.

Burbot (*Lota lota*)

The burbot, or freshwater ling as it is called, is the only member of the cod family (Gadidae) that is found in freshwater. They are found throughout North America to about 40° north latitude. Although they inhabit large rivers in their northern range, they are mainly found in large deep lakes with cold water. Smaller burbot feed on invertebrates, while larger ones (to 18 lbs) feed mainly on fish. They are caught at night with simple gear on various baits, jigging spoons, or jig and bait combinations.

Catfishes and Bullheads (Family Ictaluridae)

These are among the most popular of all freshwater fishes. Their popularity is based on their strong fighting ability and their high quality meat. They have poor eyesight, but are equipped with sensory barbels and other chemoreceptors that allow them to feed at night or in muddy water. Bait of various types is used to catch these fishes. Spinning gear is adequate for the yellow bullhead, brown bullhead and black bullhead (which have maximum weights of 2 to 4 lbs). Heavier rods and reels are required for blue catfish, flathead catfish, channel catfish and white catfish because the blue and flathead may weigh over 90 lbs. Large catfishes are often caught on **trotlines** or by **jug fishing**. The bullheads are capable of inhabiting areas that other fishes do not because bullheads can tolerate very low dissolved oxygen levels. Bullheads prefer water about 75° to 80°F, while catfishes prefer slightly cooler water between 70° and 75°F.

Walleye, Sauger and Yellow Perch (Family Percidae)

Three members of the perch family have sport fish value. They are the walleye (*Stizostedion vitreum*), the sauger (*S. canadense*) and the yellow perch (*Perca flavescens*), and all three require coldwater temperatures for several months of the year so the reproductive organs will mature. They spawn at water temperatures of 43° to 50°F. Therefore, they are found primarily in the northern United States and Canada as reproducing populations. They are excellent food fishes.

Yellow perch are best caught with light spinning gear using live bait or small jigs and spinners. Although known to reach 4 lbs, a large fish is about 2 lbs. Sauger and walleye are primarily fish eaters that feed at night, although they will eat other aquatic animals. The two species are most easily taken by trolling minnows, nightcrawlers or leeches, or with jigs, deep diving and minnow plugs, and spinners with heavily vibrating blades. Both spinning and bait-casting gear are effective for these two fishes.

Freshwater Fishing Techniques

Once the ability to locate fish has been acquired and fish are located, the choice and presentation of baits and lures are the major problems confronting most anglers. Locating fish is relatively straightforward and can be learned fairly easily. Choice and presentation of lures brings out all of the subtleties and nuances in fishing that make it an art. The essence of fishing is **presentation**, which is the proper placement of a bait or lure so that a fish will strike it.

Bait Rigging and Hookups

Either live or dead bait may be used to catch freshwater fish (providing it is legal to do so in a given area). Fly rods may be used to fish bait in streams, and bait-casting gear is used to fish large baits (such as chubs or suckers) for large gamefish like northern pike, walleye and largemouth bass. However, for most live or dead bait fishing, spinning gear is the best all-around choice. A wide variety of baitfish may be used to catch freshwater fishes, but in some states the use of live fish of any type for bait is illegal. Minnows may either be hooked in the back above the spine; or threaded through the mouth and out the gills and hooked in the body either anteriorly or posteriorly; or threaded through the alimentary tract starting at the mouth and allowing the hook to trail out the vent (anus). Worms and leeches are excellent baits in freshwater, as are aquatic insects, insect larvae, grasshoppers, crickets and other terrestrial insects. Insects are best fished by hooking them just under the collar or thorax. Salamanders, toads and frogs are fished by hooking either up through both lips or by hooking through the heavy muscle of the thigh of one of the posterior legs. Crayfish, grass shrimp, blue shrimp, mud shrimp, freshwater shrimp and various saltwater shrimps are excellent bait. These can be hooked under the thorax or through the tail and fished whole, or the tail may be broken off and fished by itself. When using live shrimp or crayfish, be sure to hook them so they move backwards through the water as you fish them. Many anglers break the claws off to make them more appealing to freshwater gamefish. Parts of most of these creatures may be fished effectively in many instances. Live or dead bait can often be most efficient when drifted or floated under a bobber—this technique is often the best to use around submerged brush and timber or over submerged aquatic weed beds.

Lures for the Freshwater Angler

A large variety of artificial lures is available to the freshwater angler. They fall into the same categories as those used by saltwater anglers (see Chapter 21). Generally, the lures used in freshwater are smaller than those used in saltwater, and plugs are used more extensively in freshwater than saltwater. Topwater fishing is pursued with a variety of plugs having subtle differences. Floater-diver plugs are called **crankbaits**. Sinking plugs, thin-bladed spoons, jigging spoons, jigs and tail-spinners are all used in freshwater. Plastic worms were originated by Nick Creme for use in freshwater and are used extensively in both freshwater and saltwater today. An early method of fishing plastic worms was through heavy brush and timber of newly formed reservoirs with a weedless lure made by embedding the hook point into the body of the plastic worm in a manner called the Texas Worm Rig. Spinners (see Chapters 20 and 22) are used in freshwater but are infrequently used in saltwater. Regular spinners are used for all types of gamefish, while weight-forward spinners designed to carry bait are used specifically for walleye.

Spinnerbaits, or safety-pin lures as they are sometimes called, are really a combination of a lead jig on the bottom connected to a metallic spinner blade at the top via a wire shaft bent in the shape of a V. Spinnerbaits are probably the most versatile of all freshwater lures because they may be fished with or without bait, slowly like a jig, at variable speeds from slow to fast like a spinner, or they may even be "buzzed" along the surface with a fast retrieve. **Weedless spoons** are a special type of spoon designed to be fished in dense aquatic weed beds, either over the surface of the weeds or through them. These lures usually have a plastic skirt and are fished with either a plastic worm or pork rind. **Pork rind** is often used to add weight, flavor or color to an artificial lure and consists of pre-cut and preserved sections of pig skin that are both flexible and durable.

Stillfishing

In freshwater, stillfishing is usually done by anchoring a boat or sitting on the bank of a lake or stream. This is most productive when schools of fish are in the area. Panfishes such as yellow perch, crappie or other sunfishes are generally the targets of stillfishing. If a school of panfish is located by another fishing method such as drifting or casting, then a float marker can be put out to mark the location of the fish for stillfishing. **Ice fishing** is a variation of stillfishing that is done in cold regions where the surface water freezes. An ice auger is used to drill through the ice, and all types of fish are sought by ice anglers, even large gamefishes such as walleye and northern pike. Ice fishermen have become very sophisticated, now employing depth sounders to find the optimum areas to make ice holes. Bait is most often used when stillfishing, though jigs, and jig and bait combinations can be employed.

Drift Fishing

This type of fishing is used primarily in rivers, because a current is present to drift the boat. However, drift fishing can be successfully employed on lakes and reservoirs if a slight wind is blowing (during heavy winds it is extremely difficult to fish). **Slipping** is a variation of drifting in which a motor is used to hold the boat in position to fish for a short period before idling the motor and then slipping down to the next location. **Jump fishing** is another variation of motor drifting where schools of open water gamefish are located and then either cast to or trolled through. Either bait or artificial lures can be used in drift fishing.

Trolling

Trolling is probably the most popular and the most effective way to catch freshwater gamefish, particularly large species such as walleye, northern pike and largemouth bass. There are several methods of trolling in freshwater. These methods employ the use of either a gas engine of less than 25 hp or an electric trolling motor. **Surface trolling** involves propelling the boat fast enough to keep the bait or lure near the surface, or it can be done slowly with a buoyant lure, such as a floating minnow plug. The floating minnow plug may be used with a short leader attached to a jig or fly.

Deep trolling is a method of fishing deepwater structure and is accomplished by one of three methods. Two of these, the downrigger and trolling planer methods, have been described earlier (see Chapter 20). The third method used in trolling is called **bait walking**, and employs either a slip sinker (an elongate lead weight with a hole through one end) or a bait walker (an elongate sinker having a V-shaped wire with a swivel on the end). A leader with a swivel is attached to the main line to the fishing rod, which has been passed through the sinker hole. Bait with some type of buoyancy or a floating lure is attached to the leader and the swivel doesn't allow the leader to pass through the sinker. The bait walker works in a similar manner. The leader is attached to a swivel and a buoyant bait or floating lure is tied to the other end of the leader. The main line to the rod is attached to the center of the V-shaped wire. The weight drops to the bottom and the bait or lure rides up above the bottom. Bait walkers are often called ''poor man's downriggers.'' Bait walkers were originally designed to catch walleye, but they will work with many gamefishes.

Another method of deep trolling uses the **striper trolling rig**, which was originally designed for striped bass fishing in freshwater lakes and reservoirs. This rig employs a deep diving plug to impart action and depth control to a jig or baited hook trailing behind the plug and is attached by a common leader. **Back trolling** is a technique used to obtain a high degree of control when moving at specified depths along contours and structures. This method is very effective for many gamefishes that relate to structures such as weed lines near the shoreline and drop-offs down to 30 feet.

Casting

Casting is the best angling method when fish are close to cover, concentrated in schools near shore, or in shallow water where they can be seen and are easily startled (Figure 23.5). This technique can be used from a boat or from the shoreline in either lakes, ponds, reservoirs, streams or rivers. It is effective with bait-casting gear, spinning gear or fly fishing gear. With this method, the angler casts toward visible objects and structures or known fish concentrations and methodically

Figure 23.5. Lily pad habitat provides excellent cover for largemouth bass. Accurate casts are the key to fishing for largemouth in shallow, weedy areas.

covers the area with a series of casts that vary in direction and length. The method can be used with either bait or lures, but works best with artificial lures while seeking out the fish. An electric trolling motor will allow the angler to manipulate the boat into proper position for more accurate casts.

Jigging

Jigging is one of the more difficult techniques to learn, but once mastered, it will produce consistently more types of gamefish in greater numbers and larger sizes than any other method of fishing. The basic method of jigging is the same for sunfish as it is for walleye. Jigging consists of lifting and dropping the lure along the bottom by successively raising the tip of the rod. Gamefish will usually strike the jig as it falls. For fish in heavy cover, a weed guard made of nylon or wire may be used to shield the point of the hook. Jigging is usually done with a standard lead-head jig with attached bucktail, feathers or a plastic worm; tail spinners and jigging spoons also may be used. Jigging works very well on concentrations of schooled fish suspended off the bottom over deep water, such as crappie and largemouth bass. A specialized type of jigging, called **flipping**, was developed in the reservoirs of the western United States to catch largemouth bass associated with heavy cover or off-color water. It employs the use of weedless jigs, very heavy line and short accurate casts. Jigging is done with either spinning gear or casting gear.

Fly Fishing

Fly fishing is an exciting type of fishing that can be employed for most freshwater gamefishes. Some freshwater fishes have habits that render them difficult to catch by fly fishing, however, and these include walleye, sauger and sturgeon. Fly fishing in freshwater can be done from the shoreline, small boats or **belly boats**, which are specialized, large inflated tubes that let the angler sit in the water at the level of the waterline and approach structure very quietly.

Dry flies, wet flies and nymphs are used for many panfishes and other species, and these lures have historically been used to imitate various types of aquatic and terrestrial insects. Streamers are effective minnow imitations. Recently, flies have been developed that imitate crayfish and leeches. Specialized cork-bodied flies and large deer hair flies are collectively called **bass bugs** and are used when fishing on the surface. Because of their large size and wind resistance, bass bugs require heavy equipment to properly cast them. Flies are very effective for situations involving clear water, spooky fish, shallow water areas and weedy areas.

Additional Reading

Bates, J.D. 1974. How to Find Fish—and Make Them Strike. Outdoor Life/Harper & Row. 216 p.

Fagerstrom, S. 1973. Catch More Bass. Caxton Printers. Caldwell, Idaho. 167 p.

Johnson, P.C. 1984. The Scientific Angler. Charles Scribner's Sons. New York, N.Y. 289 p.

McClane, A.J., and K. Gardner. 1984. McClane's Game Fish of North America. Time Books, New York. 376 p.

Meyers, C., and A. Lindner. 1978. Catching Fish. Dillon Press. Minneapolis, Minn. 190 p.

Oberrecht, K. 1982. Angler's Guide to Jigs and Jigging. Winchester Press. Tulsa, Okla. 335 p.

Pauley, G.B. (Editor) 1978. Northwest Bass and Panfish Guide. Western Bass Club. Seattle, Wash. 191 p.

Sosin, M., and J. Clark. 1973. Through the Fish's Eye. Outdoor Life/Harper & Row. New York, N.Y. 249 p.

Sosin, M., and B. Dance. 1974. Practical Black Bass Fishing. Crown Publishers, Inc. New York, N.Y. 216 p.

Sternberg, D. 1982. The Art of Freshwater Fishing. Publication Arts, Inc. Minnetonka, Minn. 160 p.

Sternberg, D. 1982. Fishing with Live Bait. Publication Arts, Inc. Minnetonka, Minn. 160 p.

Sternberg, D. 1983. Largemouth Bass. Cy DeCosse Inc. Minnetonka, Minn. 160 p.

Sternberg, D. 1985. Fishing With Artificial Lures. Cy DeCosse Inc. Minnetonka, Minn. 160 p.

Sternberg, D. 1986. Smallmouth Bass. Cy DeCosse Inc. Minnetonka, Minn. 160 p.

Sternberg, D. 1986. Walleye. Cy DeCosse Inc. Minnetonka, Minn. 160 p.

Sternberg, D. 1987. Freshwater Gamefish of North America. Cy DeCosse Inc. Minnetonka, Minn. 160 p.

Sternberg, D., and B. Ignizio. 1983. Panfish. Cy DeCosse Inc. Minnetonka, Minn. 160 p.

Whitlock, D. 1988. Fly Fishing for Bass Handbook. Nick Lyons Books. New York, N.Y. 157 p.

Index